W9-CGM-807

PLANT SCIENCE MONOGRAPHS

Edited by
Professor Nicholas Polunin

THE BIOLOGY OF MYCORRHIZA

PLANT SCIENCE MONOGRAPHS

Uniform with this volume are:

Grassland Improvement	A. T. Semple
*The Microbiology of the Atmosphere**	P. H. Gregory
Plant Growth Substances†	L. J. Audus
Seed Preservation and Longevity	L. V. Barton
Weeds of the World: Biology and Control	L. J. King

FURTHER TITLES ARE UNDER CONSIDERATION

* Second edition in preparation
† Third edition in preparation

WORLD CROPS BOOKS

(A Companion Series)

Volumes include:

Alfalfa	L. J. Bolton
Coffee	F. L. Wellman
Cole Crops	M. Nieuwhof
Cucurbits	T. W. Whitaker and G. N. Davis
Eucalypts	A. R. Penfold and J. L. Willis
Hops	A. H. Burgess
Mango	L. B. Singh
Mushrooms and Truffles	R. Singer
Onions and their allies	H. A. Jones and L. K. Mann
Pineapple	J. L. Collins
Rubber	L. G. Polhamus
Vegetable Fibres	R. H. Kirby
Wheat	R. F. Peterson

PLANT SCIENCE MONOGRAPHS
edited by
PROFESSOR NICHOLAS POLUNIN

THE
BIOLOGY OF MYCORRHIZA

By

J. L. HARLEY
M.A., D.Phil., F.R.S.

Professor of Botany, University of Sheffield;
Professor of Forest Science, elect, University of Oxford;
Formerly Fellow of Queen's College, and University
Reader in Plant Nutrition, Oxford

SECOND EDITION

LEONARD HILL
LONDON

Published by
Leonard Hill Books
a division of
International Textbook Company Limited
158 Buckingham Palace Road, London, SW1 W9TR

© J. L. Harley 1969

All rights reserved. No part of this publication may be reproduced, stored in a retrieval system, or transmitted, in any form or by any means, electronic, mechanical, photocopying, recording or otherwise, without the prior permission of the copyright owner.

First published 1959
Second edition 1969
This edition 1972

ISBN 0 249 44111 X

PRINTED IN GREAT BRITAIN AT
THE UNIVERSITY PRESS
ABERDEEN

To

Caroline Lindsay Smith

CONTENTS

PART III: ENDOTROPHIC MYCORRHIZAS

CONTENTS

LIST OF PLATES

PLATE

LIST OF FIGURES IN TEXT

LIST OF TABLES

PREFACE

(TO THE FIRST EDITION)

THE mycorrhizal condition is found in some members of all higher plant phyla: it appears to be at least as usual amongst the seed plants as the uninfected state. Some knowledge of the biology of mycorrhiza is therefore essential to the interpretation of the ecological and cultural behaviour of many economically and scientifically important plants.

Since mycorrhizal problems are complex it is perhaps not surprising that the amount which has been written about them is so great, but it is to be regretted that in many accounts speculative hypotheses are almost inextricably intermingled with experimentally verified conclusions. The aim of this book is to present its subject as one which is capable of being investigated by the ordinary methods of biological enquiry; it attempts to collect together the information obtained by observation and experiment, to summarize what is known, and to emphasize outstanding problems. The many hypotheses concerning the functioning of mycorrhizal organs have not been reviewed in detail, but most of those which have been based on experiment have been critically considered. As far as possible the more hypothetical pieces of writing have been relegated to separate sections which may be ignored if necessary.

It has not been my object to supplant or supersede previous accounts of mycorrhizas, but to present the subject in such a form as to interest biologists who are experimentally inclined. If they become interested, they might contribute valuable help and criticism to a subject which has often been considered vague, limited, or unscientific.

The work of writing has been made easier because I have had the opportunity of discussing problems with many of those who have contributed a great deal to our knowledge of the subject. No less helpful have been those colleagues and pupils who have corrected my errors and made me restate my views more explicitly. If I have used ideas of theirs on the assumption that they are my own, I am sure they will forgive me.

It is a pleasure to thank all those who have helped me. Dr. T. H. Nicolson, Dr. Lilian Hawker, and Dr. J. Warren Wilson, have allowed me to study their unpublished experimental results. Dr. F. A. L. Clowes read and criticized Chapter III, much of which depends greatly upon his own work. Part of Chapter VI is an account of the work of Dr. C. C. McCready, Dr. J. K. Brierley, Mr. Alastair Geddes, Dr. J. M. Wilson, Dr. D. H. Jennings, and Dr. T. ap Rees, with each of whom I worked happily for several years. Dr. S. D. Garrett read and corrected errors in Chapter II and stimulated me into finishing the book. Dr. W. O. James and Dr. V. S. Butt have given me good advice and help on plant physiology and biochemistry. My wife has made valuable suggestions concerning the order and arrangement of the chapters and has corrected the typescript.

My thanks are also due to the authors of a number of works and the editors of the journals in which their papers appeared, for permission to reproduce figures and plates and for the loan of blocks. I am grateful to Dr. F. A. L. Clowes, Dr. E. A. C. L. E. Schelpe, Dr. L. Leyton, and Dr. T. H. Nicolson, for original drawings or photographs. The source is acknowledged below each illustration.

PREFACE

(TO THE SECOND EDITION)

THIS edition is written, like the first, as a book to be read. It is not a reference book and no attempt has been made to comment on all the papers written on mycorrhiza. I have read as much as I have been able, and present only a personal appreciation of the state of knowledge of mycorrhizal associations. Many of the outstanding problems emphasized in the first edition have been examined by research workers since, and if this has been due in any way to my efforts it is sufficient justification for writing a reading book rather than a reference work where problems get immersed in a sea of detail.

Here I have revised and partly rewritten almost all the chapters; some of them have been completely recast and several are entirely new. By this means most of the important advances made in the solution of many of the outstanding problems have been incorporated, and the changes of emphasis from descriptive to experimental investigation in some parts of the subject have been given due weight. In a few problems little advance has been made since the publication of the first edition and so I include the sections and chapters on these with little change. Chapter I, because of its general historical nature, is virtually the same as before, as is Chapter XIII (XII in the first edition), but the rest have at least been extended.

One of the new chapters, that on carbohydrate physiology of ectotrophic mycorrhiza, owes almost everything to my friend and colleague, Dr. D. H. Lewis. The importance of its subject matter far exceeds its relevance to the study of mycorrhiza, for new work described will be seen to link the studies of lichen symbiosis, mycorrhiza, and fungal pathogens in the matter of their carbon metabolism.

For the last two chapters, on phycomycetous mycorrhizas, I was given advice and help by Dr. Barbara Mosse, Dr. T. H. Nicolson, Mr. G. D. Bowen, and Professor G. T. S. Baylis, to one or other of whom we owe so many of the recent advances in our knowledge of that kind of mycorrhiza.

I would like to thank the numerous friends who have helped me: Dr. D. J. Read, Dr. R. L. Lucas, Dr. D. C. Smith, Dr. B. C. Loughman, Dr. H. W. Woolhouse, Dr. B. B. Carrodus, Dr. J. Webster, Dr. J. H. Warcup, Mr. Brian Fearn, Mr. Glyn Woods, Miss Hazel Flear, and many others. My daughter, Dr. S. E. Smith, gave me advice on the physiology of orchid mycorrhiza and put unpublished results at my disposal, and my wife helped me with the final draft of the typescript.

Reviewers of the first edition of this book raised points of nomenclature concerning the meaning and use of the term *mycorrhiza*, and its plural. The word as originally used in German for a kind or organ, was given the plural *Mycorrhizen*. I have, therefore, in common with others, used the English form of the plural, *mycorrhizas*, although many writers in the U.S.A. and elsewhere use *mycorrhizae*. This is unimportant. However, it should be noted that the word has also been much used to denote the 'mycorrhizal condition', and it is so used in the title of this book and occasionally elsewhere in the text.

PART I

GENERAL CONSIDERATIONS

I

INTRODUCTION

FUNGI associate with other living organisms in a variety of ways. Many are destructive parasites causing acute or chronic disease of plants, and these have such an importance to mankind that a large part of the science of plant pathology is concerned with every aspect of their activity. There are also more permanent and benign associations of fungus and host plant, the frequency and physiological importance of which have not been appreciated until relatively recently. These are of great biological interest in themselves and also, because of the prolonged and close interaction between the component organs, the study of their structure and functioning has important contributions to make to the physiology of associated growths. Amongst these the mycorrhizal organs are the most common.

The name 'mycorrhiza' is used for a wide range of structures composed of fungal hyphae and the roots, rhizomes, or thalli of other plants, but it ought strictly to be used only where the organ concerned is a root. Associations of fungi with the thalli of liverworts or the rhizomes of rootless plants might perhaps more reasonably be termed 'mycothalli' and 'mycorrhizomata', but these names are somewhat pedantic and have never been widely accepted. One of the main distinguishing features of mycorrhizal organs is that the plants which possess them do not suffer acute disease on account of the presence of the fungi. Indeed, mycorrhizal organs are very often the main absorbing organs of many species of plants which form few uninfected roots under natural conditions.

There are several easily available and extensive accounts of the history of the study of mycorrhizal associations between higher plants and fungi. Dr. M. C. Rayner's monograph (1927) is so arranged that each part of the subject is treated historically; Hatch (1937) has given a very valuable and illuminating account of the history of the study of the mycorrhizas of forest trees, and Kelley (1950) in his book has also devoted a chapter to history. A detailed recapitulation of the story and of its many controversies can, therefore, serve no useful purpose. Nevertheless, information on certain general historical points is necessary for an understanding of the difficulties of the subject.*

Descriptions of associations between fungal hyphae and roots were published in increasing numbers during the latter half of the nineteenth century, and in 1885 A. B. Frank gave the name 'mycorrhiza' to the composite

* A very valuable recent work is 'Les Mycorrhizes' by B. Boullard, Masson et Cie, Paris, 1968.

fungus-root organs of the Cupuliferae. Closely similar organs were soon described in other arborescent angiosperms, and mycorrhizas of the same type were found in many conifers (especially in the Pinaceae) and also in a few herbaceous angiosperms. They are characterized by the presence of a complete sheath of fungal tissue which encloses the ultimate rootlets of the root system. Frank (1885) showed that these organs were always present on the root systems of the plants when these were growing in their natural woodland habitats, and he claimed that relatively few of the roots were free from a fungal sheath.

Later observations by Frank and others showed that the root systems of many other plants, previously believed to be free from fungal infection, were also colonized by hyphae. In these, no external fungal sheath was found but the mycelium penetrated the cortex of the root, entering the cells, and was connected with the soil merely by a few hyphal strands. Two types of mycorrhizas were therefore recognized: (1) ectotrophic mycorrhizas of Cupuliferae and Pinaceae, having an external fungal sheath, and (2) endotrophic mycorrhizas in other plants, lacking a sheath and constituting a commoner condition. In succeeding years it became apparent that a great many species of plants, especially among angiosperms and gymnosperms but also including some pteridophytes and bryophytes, were consistently associated with fungi which inhabited their underground organs. The endotrophic mycorrhizas proved to be more widely dispersed in the plant kingdom and more variable in form and structure than the ectotrophic mycorrhizas, which were of similar structure in all those groups of plants in which they were found.

It has proved unfortunate that ectotrophic mycorrhizas and the many kinds of endotrophic mycorrhizas have all been grouped under the single title 'mycorrhiza'. So diverse are the structures thus grouped that a succinct definition of the term is difficult to produce. And yet the fact that the name mycorrhiza has been applied to them all has led many people to assume that what is experimentally shown to be true of one type is true of all types of mycorrhiza. Indeed, the general question 'Is mycorrhizal infection beneficial to the hosts that possess it?' is frequently posed, but, as will be seen later, is not really susceptible of a satisfactory answer.

The situation is quite different with the analogous associations between fungi and algae which have been called 'lichens', and which have received separate consideration. These were classified, as functional entities or organisms, into species and genera long before their dual nature was demonstrated by S. Schwendener in 1867. In consequence, investigations of their mode of growth, their ecology, and their physiology, have not been bedevilled by considerations as to whether or not the infection of the alga is 'beneficial'.

It is perhaps surprising that, although the dual nature of the lichen thalli was well known and accepted before Frank's important papers on mycorrhiza were published, controversy should have arisen as to whether the ectotrophic mycorrhizas of the Cupuliferae were normally present under natural conditions. Hatch (1937) has given a valuable analysis of the controversy which arose following Frank's paper of 1885. The strong statements of such leading pathologists as Robert Hartig—based, it seemed, upon field observation—that mycorrhizal development was not the normal condition of the root system of beech, robbed Frank's work of much of its force. In the course of the next fifteen years or so, investigations showed that Frank was essentially correct in his description of the root system of Cupuliferae, and even Hartig and Tubeuf came around more to his view. Nevertheless, in consequence of this early period of controversy, it has never been adequately realized that many forest trees are normally dual organisms analogous with lichens, and that the organs through which they absorb materials from the soil have a layer of fungal tissue covering them. Indeed, all substances absorbed from the soil pass through the fungal sheath and thence into the host. For these plants the important question is not 'Is mycorrhizal infection beneficial?' but 'How do the mycorrhizal organs function in the absorption of substances?'

The ectotrophic mycorrhizas of all species of plants are so similar in structure that it has been found possible to generalize about their functions; they comprise a 'natural kind' of mycorrhizal organ. It is, however, impossible to generalize about endotrophic mycorrhizas in the same fashion. Within these, the first problem is, clearly, to perceive 'natural' kinds of mycorrhizas, that is, kinds having a large number of characteristics in common, and about which generalizations can accordingly be made. Only if such natural kinds of endotrophic mycorrhizas can be separated, and the frequency and distribution in nature of each kind be described, can their mode of functioning be studied with the hope of some reward.

During the period immediately preceding the awakening of interest in mycorrhizal infections, great advances had been made in the study of inorganic plant nutrition. Experiments such as those of J. B. Lawes and J. H. Gilbert had shown by 1855 that the growth of field crops was poor if there was not an adequate supply of ammonia or nitrates in the soil. Moreover, when these substances were absent the beneficial effects usually obtained by application of other nutrients were not observed. The fundamental role of nitrogenous substances in the growth of most plants, and the independence of Leguminosae in the matter of such substances, had already been explained by the presence of bacterial nodules on the roots of Leguminosae which could fix atmospheric nitrogen. It was to be expected, therefore, that the property of increasing nitrogen absorption would be ascribed to cases of symbiosis between fungi and green plants. Indeed, in 1894 Frank put forward the view that mycorrhizal plants growing in woodland soil were capable of absorbing organic nitrogen, since the soils in which they grew were deficient in nitrates. From that time onwards this 'nitrogen theory' of mycorrhizas has had its adherents, and preoccupation with nitrogen absorption has characterized much of the research. Indeed, great efforts have been made to demonstrate nitrogen fixation by mycorrhizal fungi, and claims have repeatedly been put forward, often on the slenderest grounds, of active nitrogen fixation by them.

The details of this line of enquiry into mycorrhizal phenomena will be discussed later in the present book. Actually, it is one of the commonest inaccuracies of short popular and elementary accounts of mycorrhizas to ascribe to mycotrophic plants in general the property of nitrogen fixation. The error arises, as do so many, from the tendency to generalize about all mycorrhizal phenomena without first attempting to appreciate their diversity or to classify them into natural kinds on the basis of true similarity.

In seeking to classify mycorrhizal associations into natural kinds, it is essential first to relate them to other kinds of associations between micro-organisms and higher plants which have not been given the name of mycorrhiza. We must compare them, in structure and behaviour, with the many kinds of pathological associations between roots and fungi, and we must see how they differ from the more casual communities of soil micro-organisms which exist about the absorbing organs of all plants growing in the soil. The study of mycorrhizas must begin with an investigation of soil ecology, in an attempt to perceive the range of kinds of interaction between micro-organisms and roots in their natural habitats. From such studies it becomes clear that there are not merely a few remarkable cases of close physiological relationships between roots and micro-organisms which can be easily separated as mycorrhizas, but that it is a usual and normal phenomenon for roots to be associated with micro-organisms in some way. It is only under experimental conditions that roots remain uninfluenced by micro-organisms. Indeed, those associations between roots and micro-organisms which have been given the name mycorrhiza, appear to be only the end terms of numerous less specialized tendencies for roots and micro-organisms to set up balanced relationships with one another.

The parts of this book are arranged as follows. First there is a brief survey of the ecology of soil fungi emphasizing the inter-relation between mycelia and root systems under natural conditions. In this account ecological classifications of soil fungi are considered, and the place of mycorrhizal fungi in such classifications is briefly outlined, so that a preliminary attempt is made to define mycorrhizal associations.

The second part consists of a series of chapters dealing, in detail, with the ectotrophic mycorrhizas of forest trees, considering first the structure of the fungus-root organs and their mode of growth and development, and then analysing the effect of habitat on the intensity of infection. The ecology and physiology of the fungi are described, and problems arising from this knowledge are discussed. The carbohydrate physiology and the mechanism of action of ectotrophic mycorrhizas as nutrient-absorbing organs are the subjects of the final experimental chapters of the section on ectotrophic mycorrhizas, which is concluded with a discussion of the ecological roles of mycorrhizal organs and nutritional problems concerning them.

The third part of the book deals primarily with endotrophic mycorrhizas. After an introduction, the mycotrophy of the Ericales is described because it now appears to have something in common with the ectotrophic mycorrhizas of forest trees. Then the great family of partial or complete saprophytes, the Orchidaceae, is considered. Thereafter a survey is made of less-known associations of septate fungi in higher plants, with emphasis on the problems that are raised. Lastly the so-called phycomycetous mycorrhizas, which are so widespread in the plant kingdom and so widely present in important crop plants, are considered and their physiological role and ecological importance are emphasized.

II

THE ASSOCIATION OF FUNGI WITH ROOTS

A GENERAL discussion of the problems concerning the association of micro-organisms with the roots of plants may, at first sight, appear out of place in a work upon mycorrhizas. Mycorrhizas are, however, special cases of such association, and the understanding of their peculiar properties can be achieved only if there is first some consideration of more general examples of interaction between roots and fungi. It will therefore be found that much of the information given here is truly relevant to particular problems dealt with in later chapters.*

The general proposition that the surfaces and environs of roots are colonized by micro-organisms was due to Hiltner (1904), although by that time the association of fungal hyphae with the tissues of roots had already been studied for at least half a century. Hiltner used the term 'rhizosphere' to describe the volume of soil immediately influenced by the root. The microflora of the rhizosphere has been the subject of much investigation especially since 1929, when R. L. Starkey reviewed what was then known of the influences of higher plants on micro-organisms. In recent years there has been a convergence of a number of different lines of investigation of the soil microflora and fauna, so that diverse subjects, such as rhizosphere and root-surface populations, patterns of distribution, interaction between organisms, pathogenesis, disease and disease escape, and mycorrhiza and nutrient uptake, all seem to have interdependent facets. The interdigitation of these subjects has been helped by numerous reviews and books, among which the recent symposium volume 'Ecology of Soil-borne Plant Pathogens' (Baker & Snyder, 1965) covers the widest field and S. D. Garrett's 'Biology of Root-infecting Fungi' (1956) and Krasilnikov's 'Soil Micro-organisms and Higher Plants' (1961) consider more special aspects.

If the soil in contact with roots is examined, it is found to be more densely populated with micro-organisms than the soil occurring at some distance from any roots. This can be shown quite simply by digging up a root system, shaking off the loose soil, and washing the roots in sterile water to obtain a suspension of the soil particles closely adhering to them. By using known aliquots of suitable dilutions as inocula for agar media, estimates of the numbers of viable bacterial and fungal propagules per gram of soil can be obtained. These can be compared with the numbers

* 'Sol, Microflore et Végétation' by B. Boullard and R. Moreau, Masson et Cie, Paris, 1962, is relevant to this chapter.

obtained from the soil at some distance from the root system. The number of colonies, measured in millions per gram, is consistently greater in the rhizosphere than in other soil, but varies with the kind of root and with the stage of development of the plant. There is also a qualitative difference between the rhizosphere and other soil populations, exemplified for instance in the large numbers of bacterial colonies in the former which require amino-acids for growth. This 'rhizosphere effect' is not restricted to an influence upon bacteria, and its existence and magnitude in respect of fungi and actinomycetes, algae, protozoa, and nematodes, in addition to bacteria, has been described by Katznelson (1965). It varies in nature and extent with many external factors which influence micro-organisms directly or indirectly through their effects on the host plant. The species of plant, its age, and its state of development, are especially important.

The rhizosphere populations studied in this way are relatively crude artifacts. The simple method of making a soil dilution for inoculation into an agar medium is not satisfactory. The agar chosen exerts a selective influence and largely determines which organisms are observed; moreover, the colonies obtained are derived not only from the organisms which are vegetatively active in the soil immediately around the roots, but may develop from resting spores. Harper (1950) made an extensive study of the limitations of the method and pointed out, in addition, that interaction of organisms on the agar plate may greatly influence the estimation of the proportions of the different kinds observed. It is therefore of great importance that the dilution of soil and rhizosphere inocula should be so arranged that approximately equal small numbers of organisms develop on each culture plate. In spite of all the defects in techniques, the stimulating effects of roots upon soil organisms is undoubtedly real, and has been confirmed by direct methods of microscopic observation such as the Rossi-Cholodny buried slide technique used by Starkey (1938). In recent years the use of a number of direct microscopic methods such as staining, fluorescence microscopy, and counts of micro-organisms in suspensions, together with the measurement of metabolic activity and the conscious use of selective media, has extended knowledge of the kinds of organism present and of their physiological processes.

The study of fungi in the rhizosphere presents greater difficulties than the study of bacteria, for the enumeration of fungi depends not only on the extent to which hyphae may be fragmented in making the dilutions, but also on the frequency of spore production by them. Timonin (1940) made an effort, by use of McLennan's technique (1928), to estimate the frequency of spores in the rhizosphere. He desiccated the rhizosphere soil over calcium chloride, assuming that this would kill the hyphae but not the spores. He concluded that only 10 per cent of his counts of fungi were due to spores. This is somewhat surprising, as the majority of fungi isolated by the dilution technique from the rhizosphere readily form spores in culture.

The communities of organisms described in this way are those from the soil around the roots, but there are also communities of fungi to be found upon the root surfaces themselves. These communities have the same habitat as mycorrhizal fungi and as root pathogens, and are found to have properties that are similar in many ways to those of such organisms. If roots are shaken repeatedly in successive small volumes of sterile water, they can be cleaned of detachable propagules. After about thirty such washings it can be shown that any mycelia which grow from these roots when they are planted out in agar develop mainly from vegetative hyphae. The root-surface population obtained from such cleaned roots consists of the mycelia which live in close association with the roots and is the one most likely to influence them and the soil immediately around them. Table I shows a comparison of the kinds of fungi recorded

TABLE I

PROPORTION OF STERILE MYCELIA OBTAINED IN CULTURE FROM UNWASHED ROOTS, WASHED ROOTS, AND WASH WATER. 5 mm. BEECH ROOT-TIPS. (Data of Harley & Waid, 1955.)

Experiment (date)	Unwashed roots (per cent)	Washed roots (per cent)	Washing water (per cent)
11.8.52	12·0	54·6	4·6
8.8.53	nil	45·5	nil

upon washed and unwashed roots of beech. The figures demonstrate that the fungi found in the washings and on the unwashed roots are mainly sporing species; sterile mycelia—that is, mycelia sporing only with difficulty in the media used—comprise a large proportion of the colonies observed on the washed roots.

The presence of such special populations of fungi on root surfaces caused Jahn (1934) to coin the term 'peritrophic' mycorrhiza for them. This is a term which may well be discarded as redundant, but the consistent presence of a characteristic fungal population on roots, suggested to Jahn that it should be specially named. It will be seen later that the activity of these denizens of the root surface has important implications in the study of ectotrophic mycorrhizas.

The constitution of the root-surface populations is interesting. It has been found that, besides saprophytic forms, a large number of potential parasites are stimulated to grow on and in root surfaces. For instance, Simmonds & Ledingham (1937) isolated 806 colonies from wheat roots and showed that, excluding 13 per cent of unidentifiable sterile forms, 50 per cent were potential pathogens. The occurrence of large numbers of unidentifiable sterile mycelia upon the root surface is quite usual. There always appear to be many such mycelia which find their ecological niche here, and certain of them have been shown to exert great influence on the growth of their host. The so-called pseudomycorrhizal fungi (*see* p. 85 ff.) are organisms of this kind.

The effects of the age and state of development of a root system on the rhizosphere and root-surface populations are special facets of the effects of the physiology of the host on its microbial entourage. This is illustrated by the work of Harley & Waid (1955) on the fungal populations of beech seedlings during their first year of growth. Not only was a change of population of fungi observed as the plants aged, but also the fungi differed on the surfaces of different kinds of roots. The leading apices were relatively free from fungi, and the actively-growing parts were colonized by relatively slow-growing sterile mycelia. On the older regions of the root system a large variety of rapidly growing, sporing mycelia were enumerated. Warren Wilson (1951) also examined the rhizosphere populations of beech seedlings during their first year of growth, in which mycorrhizal infection developed. As shown in Fig. 1, fungi increased in

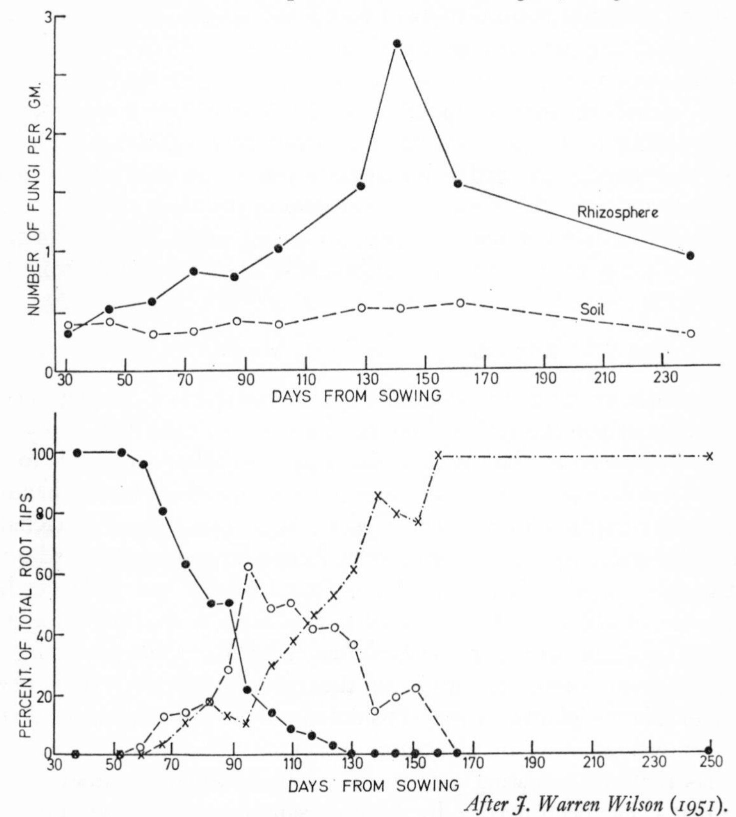

After J. Warren Wilson (1951).

FIG. 1.—Rhizosphere populations and mycorrhizal initiation in beech. *Above.* Changes in the population of fungi in the rhizophere of beech seedlings during their first season. *Below.* Changes in the structure of the ultimate rootlets during the same period. ●——● phase 1 rootlets. ○——○ phase 2 rootlets, ×——× phase 3 rootlets and mycorrhizas.

frequency at first and reached a maximum from which they then declined. This happened late in the growing-season, when mycorrhizal infection developed and bacteria increased in frequency.

It is of course by no means certain how these changes were brought about. To some extent the direct effect of the physiology of the host is undoubtedly responsible, and the relationship of mycorrhizal development with the physiological condition of the host will be discussed later. However, indirect effects must also have been involved, for Katznelson and his colleagues (1962) have shown that mycorrhizal and non-mycorrhizal roots of Yellow Birch exert different rhizosphere effects. The rhizosphere of mycorrhizal roots was characterized by the greater stimulation of bacteria than was that of non-mycorrhizal roots, and, of the fungi, species of *Penicillium* and certain sterile forms were most abundant in it. On non-mycorrhizal roots, *Pythium*, *Fusarium*, and *Cylindrocarpon*, were prominent members of the rhizosphere association. Kubikova (1965) records that somewhat similar results have been obtained with mycorrhizal and non-mycorrhizal roots of beech and spruce, and Tribunskaya (1955) makes comparable observations on pine.

Kubikova's work is of particular interest because, in it, an attempt was made to classify the fungal populations in the rhizospheres of woody plants on lines similar to the classification of higher-plant associations. Kubikova recognized synusiae, defined by the frequency of the most plentiful species, and suggested that these have some degree of reality as they constitute patterns of organisms which are repeatedly encountered. It will be of great interest to see how far the concept remains viable after further work.

Pattern Arrangement of Soil Micro-organisms

The surfaces of living roots are not the only ones bearing special populations of mycelia and of bacterial colonies, though they happen to be of great interest to plant pathologists and to be relatively simple to observe. If any definable surface is selected in the soil which can be handled in a similar manner to lengths of roots, large numbers of detachable propagules and active surface mycelia and colonies can be observed upon it (Chesters, 1948; Harley & Waid, 1955, 1955*a*), but little detailed knowledge of the constitution of such populations is available as yet. Garrett (1952), in his paper *The Soil Fungi as a Microcosm for Ecologists*, wrote: 'Recently renewed study of the root surface and rhizosphere floras of higher plants, when considered in conjunction with Watt's (1947) novel analysis of structure of communities of higher plants, suggests that the repeating pattern of soil fungus-flora is on so small a scale that it has been masked by the wide sampling techniques employed in general studies. This repeating pattern is made up of individual patches around unit substrata.'

The very elegant soil-plate method of studying soil fungi (Warcup,

1950), using as it does small soil granules or particles for inocula, lends itself to a demonstration and study of the pattern of soil populations. Indeed the combination of this method with direct methods of isolation of rhizomorphs, sclerotia, and hyphae, as well as with dilution-plate methods, has introduced a new dimension into the study of soil micro-organisms. Amongst Warcup's interesting observations have been those on the distribution of basidiomycetous mycelia, which have been mapped by his method. Within the zone colonized by these mycleia, local populations of microfungi have been demonstrated (Warcup, 1951*a*) which exhibit some quantitative and qualitative differences from those of the surrounding soil. Similarly, his surveys of the fungal populations of horizons in natural soils (Warcup, 1951) have served to illustrate the existence of communities of micro-organisms that conformed in distribution to visible and analytically determined soil habitats.

The soil consists of a gross differentiation into horizons within which exist finer specialized biological niches. In a natural (as opposed to an arable) soil there is a cyclic, often annual, addition of organic matter to the surface in the form of leaf-litter and other plant debris. There is also a further cyclic formation of new roots, and an ageing and death of old roots within the soil itself. The horizon differentiation is a resultant of the local addition of organic matter, its breakdown, the movement of its products downwards, and their fixation in the deeper layers. Movement is brought about by the leaching of soluble or suspensible materials and by mixing due to the activity of the soil fauna. In a mature soil, the differentiation of horizons has become relatively stable and there is a cyclic set of changes of rate which is the resultant of seasonal changes of accretion and breakdown modulated by seasonal climatic factors such as temperature and rainfall. In podsolic soils, which show clear morphological differentiation of horizons, it was early shown, even with dilution-plate methods, that the distribution of micro-organisms conformed in a general way to horizon differentiation. Gray and his colleagues (1933–35), and Timonin (1935), showed that the organic horizons of podsols were more biologically active than the leached horizons or the C horizons.

Burges and his colleagues have more recently made a detailed study of this matter (*see* Burges 1963 and 1965 for references) and a list of so-called dominant fungi is given by them. Of very great interest are the estimates, made by Parkinson & Coups (1963), of oxygen uptake in the various horizons. Oxygen uptake was greatest in the surface litter and F layers and far smaller in the lower layers. Moreover, the surface layer of the youngest litter showed a great rise in microbiological activity in autumn and winter when its oxygen uptake rose some sevenfold, presumably due to the accretion of new plant debris. The F layers showed a somewhat delayed increase, so that there were high rates in the spring. The magnitude of the changes of rate in the lower layers was small.

The breakdown of plant debris is accompanied by the liberation of ammonia from amino-acids and of metallic ions from organic compounds and complexes. These tend to be leached downwards, but there is a very efficient utilization by the micro-organisms and plant roots, so that they are not lost in the drainage waters. In the biologically active layers of woodland soil there is an intense competition between the micro-organisms and between these and the organs of higher plants, and the phases of rise and fall of activity are related to the availability of essential elements or substances in limited supply. The localization or pattern arrangement of micro-organisms within any horizon is related to the availability of substances which are required in different amounts by different organisms or whose requirement is a peculiarity of certain species.

The Ecological Classification of Soil Fungi

Apart from these community studies, autecological investigations have been pursued especially in connection with pathological problems. Attempts have been made to classify the species of soil fungi into ecological classes. Some of these place the fungi in groups according to the character of the substrates that they affect, e.g. sugar fungi, cellulose- and lignin-destroying fungi, keratinophylic fungi; others are reminiscent of the life-form classification of angiosperms and have similar limitations. They are, indeed, of value mainly in respect of the particular problems on which their originators are engaged. Two groups can usually be distinguished: unspecialized, widely-distributed soil fungi, and specialized fungi restricted to peculiar substrates. The first are usually ephemeral, developing rapidly upon newly decaying substances with a high content of readily soluble matter, and give place to less rapidly spreading forms that are capable of causing the breakdown of more resistant substrates. The second group includes fungi of specialized sites, such as the surfaces of living organisms, of lignified material, and so forth.

From the point of view of a plant pathologist, the extension of such attempts at ecological classification is well exemplified by the scheme proposed by Garrett (1950):

<table>
<tr><td rowspan="3">Root-infecting Fungi</td><td colspan="2">Mycorrhizal Fungi</td><td rowspan="2">Root-inhabiting Fungi</td></tr>
<tr><td rowspan="2">Root Disease Fungi</td><td>Specialized Parasites</td></tr>
<tr><td>Unspecialized Parasites</td><td rowspan="2">Soil-inhabiting Fungi</td></tr>
<tr><td colspan="3">Obligate Soil Saprophytes</td></tr>
</table>

Garrett lists several important criteria for distinguishing the 'root-inhabiting fungi' from 'soil-inhabiting fungi', the most important of which criteria relate to their behaviour as saprophytes. He remarks that 'the root-inhabiting fungi are characterized by an expanding parasitic phase in the living host plant and by a declining saprophytic phase after its death.'

Inability to colonize dead substances in the face of competition is thus made the decisive criterion distinguishing root-infecting fungi. Some such fungi, e.g. *Ophiobolus graminis*, having very limited powers of saprophytic competition, can be eliminated by crop rotations or by suitable soil treatment. Others, spreading by rhizomorphs, and dependent on a food-base for spreading, also pose peculiar problems to pathologists. Such specialized fungi contrast, from a pathological point of view, with *Rosellinia* and *Rhizoctonia* (*Corticium*) *solani* which are capable of existing as aggressive saprophytes and also as parasites.

In Garrett's group of unspecialized parasites there are, however, many indeterminate fungi of which the exact position in this ecological classification is often in doubt. Garrett devotes much of his paper on the ecology of the root-inhabiting fungi to a discussion of some of these. Such species as *Fusarium culmorum* and *Rhizoctonia bataticola* may indeed be soil-inhabiting fungi by virtue of their ability to colonize dead tissue in unsterile soil, but they also seem to have a fairly restricted ecological distribution and are part of the population of certain plant residues and of the living roots of host plants. Garrett writes (1950): 'A great variety of soil-inhabiting fungi have been isolated from root systems of apparently "healthy" plants, as in studies of wheat root systems by Broadfoot (1954), Samuel and Greavy (1937), Simmonds and Ledingham (1937), Glynne (1939) and White (1945). A diversity of casual parasites has been observed in roots of tobacco and strawberry plants suffering from root rot. Millikan (1942) isolated a number of weak pathogens such as *Corticium solani*, *Fusarium culmorum* and *Curvularia ramosa* from lesions on the roots of wheat plants suffering from severe zinc deficiency.'

Burges (1939), in his attempt at producing a list of biological groups of fungi present in the soil, disposed of the organisms included in Garrett's 'unspecialized parasites' in two classes, viz. 'casual parasites' and 'facultative parasites'. His scheme was as follows:

(*a*) Root parasites.
(*b*) Casual parasites and mycorrhizal fungi.
(*c*) Facultative parasites and primary saprophytes.
(*d*) True soil fungi.
  (i) Sugar fungi.
  (ii) Humus fungi.

The definitions of groups (*b*) and (*c*) are 'numerous other fungi which can penetrate the roots and give rise to infections which do not seem to cause the plant much damage' (casual parasites), and 'fungi involved in the micro-succession . . . after initial attack by a characteristic root parasite' (facultative parasites).

We are not here concerned with further attempts to improve these classifications. All that is necessary is to realize that, amongst the plexus of important forms, are (i) obligate saprophytes showing variable substrate preference, and (ii) obligate root-inhabiting forms, dependent on living tissue, and showing specific substrate demands. Between these is a multiplicity of fungi which occur with great regularity on root surfaces and in the tissues of living roots, but are not obligately dependent on them; these comprise the matrix of the root-surface communities. The influence of these populations as such, and of certain component species of them upon their hosts, is ecologically most important.

It is noteworthy that in each of these ecological classifications, mycorrhizal fungi figure as a group without being clearly defined. Garrett regards them as 'symbionts rather than harmful parasites', or as the end term of parasitic specialization. It is the object of the remainder of this chapter to try to obtain a working definition of this group of fungi.

The Stability of Soil Populations

Although the whole community of micro-organisms in the soil is made up of a large number of unit patterns, each changing with time and with the condition of the particular unit substrate on which it develops, the whole is stable in the same sense as is a community of higher plants. A considerable distorting influence is necessary to change its form or activities in a permanent and significant manner. Of course, changes may be brought about by manipulations which fundamentally change the chemical constitution or physical properties or aeration of the soil. The addition of a considerable amount of plant debris, of carbohydrates, or of nitrogenous compounds, will in particular have a lasting effect, as also will treatment with disinfectants or partial sterilization. The introduction, even in great numbers, of viable propagules of new microbial or fungal species to a soil will, however, usually have only a short-lived effect.

An example of manipulation through changing the chemical and physical properties is afforded by the work of Martin *et al.* (1942), who added a considerable amount of plant residues to the soil and traced the rise in activity of certain fungi and the disappearance of constituents of the residues. During the first period Phycomycetes, including *Mucor* and *Rhizopus*, increased in numbers, and sugar and starch were rapidly decomposed. There followed a rise of *Penicillium* and *Aspergillus*, coupled with a loss of cellulose and hemicellulose, and lastly a rise of other fungi coupled with decomposition of lignified material.

Others have added selected species of micro-organisms to soil in an attempt to establish in it useful forms such as *Azotobacter* for nitrogen fixation, or species known to inhibit the growth of disease organisms. Katznelson (1940, 1940*a*) added various kinds of fungi and bacteria, and showed that they all disappeared rapidly from all the five soils which he used, but that the rate of disappearance depended upon the nature of the soil. As might be expected, he succeeded in establishing some species by dint of changes in the nature of the soil, e.g. through the addition of plant debris or dried blood, or by other manurial treatment. Similar results attended his efforts to establish *Azotobacter*. Waksman & Woodruff (1940) carried out similar experiments in the course of work on antibiotic-producing organisms. The addition of non-indigenous bacteria to soil resulted in a temporary stimulation of certain of the indigenous species and the disappearance of those added. Some of the stimulated species of the indigenous flora were found to be antagonists, producing end-products causing death of introduced species. This kind of result formed the basis on which the soil enrichment method of isolating antagonistic micro-organisms was improved.

Attempts have been made to introduce selected bacteria into the root region of various crop plants in order to stimulate desired biologically dependent processes in the immediate environs of the roots. This subject was briefly surveyed by Katznelson (1965), and in his paper the important references to experimental work will be found. In particular, the possibility of the stimulation of nitrogen fixation by *Azotobacter*, and also of phosphate solubilization, have been investigated. The results obtained in soil culture have often been somewhat disappointing, and at best only small but variable stimulations of the growth of the host plant have usually been achieved by many observers. In the case of *Azotobacter* inoculation, it is moreover far from certain that such growth stimulations as have been observed can be ascribed to nitrogen fixation in the rhizosphere. As Katznelson points out, the important problem lies in the ability of the desired organisms to establish themselves in the rhizosphere. This seems to have been achieved most often by special means such as the use of plant-adapted cultures or by incorporating specific substances in the rooting environment. Discussion of the Russian work carried out on this subject is given in great detail by Rubenchik (1960), who discusses the possible use of bacterial inoculation of agricultural crops. This process seems to be used in Russia but not to any extent elsewhere.

The absence of unqualified success in these cases does not imply that organisms carried on the seed-coat do not become established in the rhizosphere; indeed, seed-borne diseases may be extremely important. In such cases, however, organisms natural to the rhizosphere and root surface are allowed to become firmly established in the early stages, before the root-surface community is built up, and they consequently may become dominant and cause abnormally high and early disease incidence. In contrast to this, Simmonds (1947) showed how the natural flora of wheat-grain surfaces might have an effect of preventing attack by *Helminthosporium*. As a general rule, however, it is as difficult to bring about a high incidence of a disease by inoculation of the soil with pathogens as it is to establish biological control of a disease organism by heavy inoculation with antagonistic micro-organisms.

The subject of 'biological antagonism' in disease studies was familiar to plant pathologists long before the importance of antibiotics for the control of human disease was realized. As early as 1926, G. B. Sanford suggested that the effect of green manure in suppressing *Streptomyces* (*Actinomyces*) *scabies* attack on potato might be due to the stimulation of competitive species: this was confirmed by Millard & Taylor (1927). The difficulties experienced in the experimental production of disease in field soils has been demonstrated frequently to be due to the antagonistic effect of the normal soil flora upon the pathogen. Some of A. W. Henry's results with *Helminthosporium sativum* are shown in Table II. Even the addition of a trace of unsterile soil to a pot of sterile soil greatly reduces the incidence of disease.

TABLE II

THE EFFECT, ON THE PATHOGENICITY OF *Helminthosporium sativum* ON WHEAT, OF ADDITION OF UNSTERILE SOIL TO POTS OF STERILE SOIL. (Data of Henry, 1931.)

Amount of unsterile soil added per pot	Recovery of pathogen per 100 attempts	Emergence per cent	Seedling height in cm.	Infection per cent
0	100	74·4	11·5	47·6
trace	30	94·8	17·9	7.8
1 gm.	0	92·4	18·2	7·0
5 gm.	0	86·4	17·5	3·1
50 gm.	0	88·4	16·4	5·5

In spite of such results and some early hopes of success, biological control of disease by inoculation alone has not yet proved to be a practical measure. *Trichoderma viride* has, for instance, been shown to cause the death of other fungi in culture. A carefully controlled series of experiments by Wiendling & Fawcett (1936) failed to demonstrate that inoculation with *Trichoderma* could alone be used effectively to control *Rhizoctonia solani* as a cause of damping-off disease of seedlings of *Citrus*. On the other hand, successful biological control of a number of diseases has been obtained by a combination of soil treatments, with or without inoculation, which have significantly altered soil populations and increased their biological antagonistic effects (*see* Baker & Snyder, 1965).

One obtains the impression, then, that pathogens, like other soil fungi, exist in their own peculiar ecological niche under natural conditions. They cause occasional acute diseases or chronic minor lesions of the root. Disease in epidemic proportions is only associated with, or only follows, considerable soil changes due to human interference or continued growth of one crop. These treatments presumably bring into existence a soil population in which widespread development of the pathogen occurs on extensive suitable sites.

The Effect of Root-surface Communities on the Host Plant

It has been assumed that the presence of a peculiar flora around the surfaces of living roots must indicate that conditions for the growth of micro-organisms are different in this region from elsewhere, and that the rate of soil processes must be different here on account of the nature of the community. The following reactions have been reported to be increased in rate in the rhizosphere: nitrogen fixation, cellulose decomposition, the solution of phosphorus compounds, nitrification, and the release of carbon dioxide (Starkey, 1929, 1929*a*, 1929*b*, 1931, 1931*a*, 1938; Thom & Humfeld, 1932; Krasilnikov *et al.*, 1936; Katznelson, 1946).

Warren Wilson (1951) examined the rhizosphere population, the production of carbon dioxide, and the presence of soluble nitrogen in the rhizosphere soil, of developing beech seedlings. The numbers of fungi in the rhizosphere rose to a peak and then declined during the same summer season, whilst those in the surrounding soil remained fairly steady. There was considerable fluctuation in the number of bacteria in the rhizosphere soil, but it was always four to eight times as great as that in the control soil, and it reached its peak later in the season. The production of carbon dioxide by rhizosphere soil (measured by micro-diffusion technique) was much greater than that of the control soil (by a factor of 3 to 4). The rate of production of carbon dioxide by the rhizosphere soil was, moreover, doubled by the addition of root extracts, but that of the control soil was affected to a smaller degree. Warren Wilson also observed that, as compared with the control soil, very much higher concentrations of soluble nitrogen were extractable from the rhizosphere soil as ammonia, amides, and amino-acids. It is, however, possible that his method of extraction of nitrogen caused the death of the organisms of the soil population and, as these were present in greater numbers in the rhizosphere, greater quantities of soluble nitrogen were found there.

Changes such as these, in the availability of nutrients and in the magnitude of other factors which affect plant growth, have been discussed by Katznelson (1965). He pointed out that organisms capable of bringing about the solution of organic and inorganic phosphates, the transformation of nitrogen compounds, and the chelation or fixation of other nutrients, can be shown to be present in large numbers in some rhizospheres. Their presence does not, however, allow assumptions to be made concerning the chemical effects of the rhizosphere population on their host plants, without detailed experiment. The availability of nutrients to the host depends on the simultaneous presence of other organisms utilizing the materials rapidly and on the resultant net availability in the root region. Effective experimental studies are relatively few. Certain of these have achieved fame because of their striking nature.

Gerretsen (1948), in an interesting paper on the influence of micro-organisms on the phosphate intake of plants, grew a variety of crop plants under sterile and unsterile conditions in pots of soil to which various insoluble phosphates were added. Much greater growth and higher phosphate absorption were observed in the unsterile soils. He isolated bacteria in pure culture from the root surfaces of the plants, showing that if some of these were used to inoculate sterile soil they stimulated phosphate absorption. He was also able to show that roots which retained their normal rhizosphere flora brought phosphates into solution in the root region far more rapidly than did uninfected roots (Table III). In iron-deficient soils there was a secondary effect caused by the precipitation of iron by excess of phosphates. In such cases the non-sterile cultures produced stunted plants suffering from iron chlorosis.

TABLE III

THE EFFECT OF BACTERIA ISOLATED FROM THE ROOT SURFACE ON GROWTH AND P UPTAKE OF MUSTARD. P SUPPLIED AS $Ca_3(PO_4)_2$ AT pH 6·8. (From Gerretsen, 1948.)

Treatment	Height in cm.	Mean dry-weight in gm.	P_2O_5 absorbed in mg.
Sterile	40·3	11·3 ∓ 1.4	87·9 ± 13·6
Inoculated	98·4	18·6 ∓ 1·8	134·9 ± 12·0
Mean per cent increase	+ 144 per cent	+ 64 per cent	+ 55·6 per cent

By contrast, Katznelson (1965) has pointed out that work by himself and his colleagues and others has given negative results on phosphate availability, while with other nutrients confused or variable results have been obtained. Two interesting and well-known cases of negative effects are the deficiency diseases 'grey-speck of oats' and 'little-leaf of citrus'.

Grey-speck disease has been ascribed to manganese deficiency and certainly the symptoms appear on manganese-poor soils. The symptoms are much aggravated in unsterile conditions, as Gerretsen pointed out (1937). Timonin (1948) has shown that manganese-oxidizing bacteria are stimulated in the rhizosphere of oats, and appear to account for the onset of 'grey-speck' symptoms in susceptible varieties on unsterile soil at levels of manganese availability at which no disease develops in sterile soil.

The complexities of 'little-leaf' of *Citrus*, a zinc-deficiency disease, were similarly explained by Hoagland *et al.* (1937) and Ark (1937) as due to a biological factor—bacteria in the root region—which increased zinc deficiency.

The effect of the presence of micro-organisms on absorption of phosphate has lately been the subject of physiological and biological study by Bowen & Rovira (1966), Rovira & Bowen (1966), Barber (1966), and Barber & Loughman (1967). The interest has arisen not only because the micro-organisms on the root surface may affect uptake quantitatively, but they may also affect the proportion of phosphate translocated to the shoot and the apparent incorporation of phosphates into esters and other compounds within the tissues. The quantities trapped by the micro-organisms and the metabolic turnover of phosphate within the micro-organisms, if ignored, may materially affect interpretations of uptake mechanisms.

Both pairs of observers showed that non-sterile root systems absorb phosphate more actively than sterile roots. However, the pattern of incorporation into phosphate fractions within the living system shows that much of the excess uptake is into the bodies of the micro-organisms. There is indeed a higher rate of incorporation into ribonucleic acids, phosphoproteins, and phospholipids, in the non-sterile root systems, and a smaller incorporation into the easily soluble phosphate fractions—especially inorganic phosphate.

The effect of micro-organisms on the measured translocation of phosphates to the shoots was to decrease it in the experiments of Barber & Loughman with barley, but to increase it in those of Bowen & Rovira with tomato. The influence of the external concentration of phosphate on these effects was studied by Barber & Loughman (1967). The effects of micro-organisms were more marked at low concentrations of phosphorus but tended to be relatively small, though perceptible, above 0·5 p.p.m. KH_2PO_4 (*c.* 0·015 mM.) (*see* Table IV), and reinforce the contention that

TABLE IV

THE PERCENTAGE OF ABSORBED PHOSPHATE TRANSLOCATED TO THE SHOOT OF STERILE AND NON-STERILE BARLEY SEEDLINGS GROWN IN VARIOUS CONCENTRATIONS OF KH_2PO_4. DURATION OF EXPERIMENT 24 hrs. (Results of Barber & Loughman, 1967.)

Phosphate concentration in p.p.m.	Percentage translocated to shoot: Sterile	Non-sterile
0·001	20·4	2·3
0·01	24·0	9·7
0·05	40·0	17·8
0·1	47·5	34·1
0·2	66·0	50·1
0·3	65·6	56·7
0·4	69·9	66·0
0·5	70·0	63·6
1·0	73·3	73·6
10·0	74·6	78·8

the population of the root region is likely to exert its greatest effect, whether inhibitory or stimulatory, in conditions of low or limiting nutrient availability.

Besides their effects on nutrients, the rhizosphere organisms may have other chemical effects. Many are known to produce, in common with other micro-organisms, a variety of organic substances having auxin, antibiotic, vitamin, toxic, and enzymatic activities. The importance of these activities to the growth of the host plant is the subject of much research. It is of interest for instance to determine the extent to which the production of, and resistance to, antibiotics helps to determine the composition of rhizosphere populations. It is also of prime importance to know whether the host can develop resistance to disease by the absorption of antibiotics. Most higher plants are essentially vitamin autotrophic, at any rate in adult life; however, certain groups, e.g. Orchidaceae, are vitamin heterotrophic at seed germination, and the production of vitamins by their attendant mycorrhizal fungi will be considered in a later chapter. The presence of vitamins in the root region may also be due to their release by the host plant, and their importance will be discussed in a later section.

One of the effects of auxins or morphogenic factors in the root region is the curling of root-hairs in legumes by *Rhizobium*. Bowen & Rovira (1961) made a study of the general effect of micro-organisms in the root region on the growth of roots and root-hairs. In all species studied, the growth of the primary root was reduced and micro-organisms capable of causing the stunting were found in each of four soils. Bacteria inhibiting root growth in clover were also found to be abundant in the same range of soil types. These morphogenic changes in the host-plant were such as might materially affect its establishment and growth under natural conditions, but the actual mechanism of their causation is not clear; for, as Bowen & Rovira pointed out, a number of kinds of substances—auxins, antibiotics, and toxins—have all been shown to affect root growth and root-hair development. As will be seen later, the effects of morphogenic factors of various kinds produced by mycorrhizal fungi have been found to be, or have been believed to be, important in the development of mycorrhizal structures.

Explanations of the Association of Fungi with Roots

Various explanations have been put forward to explain the association of fungi with roots and the occurrence of root-surface and rhizosphere populations. Those in the following list are by no means mutually exclusive:

(1) Roots provide a continuous surface for the growth of forms requiring a fixed surface on which to spread.

(2) Dead root-cells from the root-cap, root-hairs, etc., provide food materials and essential micronutrients or growth factors.

(3) The *p*H of the root region is relatively constant.

(4) The roots absorb oxygen, water and salts from the soil, and release carbon dioxide. The uptake of salts is selective and tends to cause local changes in the soil solution.

(5) The roots excrete various organic substances.

These suggestions all seem reasonable and some have, indeed, been shown to be true in certain cases. Little need be said concerning the first. The selection of particular species by surfaces is one of the criticisms of the buried-slide technique for the observation of soil fungi. The direct observation of R. L. Starkey and others has verified the congregation of bacteria and fungi around the sloughed dead tissues of roots. Whether the pH is here more constant, and whether, if this is so, it constitutes an important factor, is doubtful. It seems unlikely that the mere existence of a zone having a fairly constant pH in the middle of the range of common soil pH values, would lead to the selection of a population so different from the general soil flora as is that of the root region. Indeed a constancy of pH—if it really exists—is likely to be the resultant of many activities such as carbon dioxide production and ion uptake and release by the root and its micro-organisms, rather than a factor affecting them. These exchanges of substance which occur rapidly in the root region may perhaps give rise to unique conditions there, but nothing is yet known in detail of the effect of any one of the changed factors.

It is doubtful whether oxygen supply is an important factor affecting directly the rhizosphere population and its activity. The oxygen availability in the surface layers of soil is not usually far different from that in air. This has been shown by analysis of soil air in arable soil (*see* E. W. Russell, 1950) and in woodland soil (Penningsfeld, 1950; Brierley, 1955). Indeed in most soils, except heavy clays, the oxygen in the soil atmosphere rarely falls below about 10 per cent by volume even at considerable depths. It is only in the lower regions of clay, or after heavy rain in the lower regions of soils of high porosity, that oxygen concentrations fall much below this limit. It seems unlikely, therefore, that even in the rhizosphere where gas exchange is very active, the oxygen supply can as a rule be an important factor. It is of interest that Bisby *et al.* (1933) found that the fungi which they isolated from the deepest layers of soil were more frequently facultative anaerobes than those isolated from the surface horizons. This was not confirmed by Burges & Fenton (1953), who worked on a light sandy soil. On the other hand, lack of oxygen may limit fungal penetration into substrates. Waid (1962) tested the ability of fungi isolated from decomposing *Lolium* roots to grow and respire in atmospheres of low oxygen concentration. Those from the deeper tissues were more active at low oxygen concentrations than those isolated from the outer cortex and root surface.

By contrast, carbon dioxide concentration varies very greatly in different parts of the soil, at different seasons, and in different conditions. It may rise to as high as 5–10 per cent by volume of soil air at the sites of rapid production, so that in the rhizosphere or on rapidly decomposing substrates it may very well be much higher still. Burges & Fenton (1953) have shown that fungi may be differentially affected by carbon dioxide concentrations, which may therefore be an important factor in the distribution of soil fungi. The manner of the effect of carbon dioxide concentration may vary; it is inhibitory to some, but may stimulate growth and perhaps have a morphogenic effect on other species of fungi. Carbon dioxide concentration may also affect root growth. The roots of different species are not equally affected by carbon dioxide concentration which may therefore possibly be an important ecological factor for the growth of green plants.

The release of organic and inorganic substances from roots is certainly important and has been the subject of much investigation in recent years. The results have been summarized by Rovira (1965), who has contributed much to our knowledge of root-exudates. The term 'root-exudate' must be taken to include substances originating both by release across the cell-membranes of living cells and those originating from senescent cells of the secondarily thickened regions, root-hairs, and root-caps, when they are sloughed off. Rovira has emphasized the need to study exudates under sterile conditions, for only in this way can a true appreciation be obtained of the substances which arise directly from the host plant. The organic substances exuded from various plants include a number of carbohydrates (monosaccharides, disaccharides, and oligosaccharides), some twenty-one different amino-acids, organic acids (including citric, succinic, fumaric, and malic acids), vitamins, growth factors, hydrolytic enzymes, nucleotides (such as adenine, guanine, uridine, and cytidine), auxins, glycosides, and a number of other substances of which some have specific effects on fungi and nematodes. Inorganic substances including phosphates, and metallic ions, such as potassium, are also released.

Very little is known about the factors which affect the process of exudation from roots, and still less about the mechanism. Rovira (1959) examined the effects of light, temperature, and the supply of calcium, on exudation by tomato and subterranean clover. In both species the exudation of amino-acids altered quantitatively when the light-intensity was increased from 10 per cent to 100 per cent of daylight, and when there was a temperature rise of about 10°C. In Rovira's experiments calcium had little effect on exudation, but in those of Stenlid (1947), who used pea roots, the rate of exudation was sensitive to low concentrations of metals. Calcium chloride at 0·0625 M. reduced exudation by 20 per cent as compared with exudation in water. In the presence of potassium chloride and other alkali-metal chlorides, exudation was temporarily depressed but later stimulated. In the presence of very low calcium chloride concentrations, these effects of potassium were very greatly reduced. The effects of calcium no doubt depend in part upon the well-known but unexplained effect of calcium upon the permeability of membranes.

Stenlid (1947) also showed that exudation was increased under conditions of low oxygen supply or in the presence of the inhibitor KCN. Indeed both these observations and the effects of temperature, also observed by Stenlid, seem to indicate a connexion between exudation rate and diminished oxygen utilization in respiration. Katznelson and his colleagues (1955) observed that plants of pea, soybean, wheat, barley, and tomato, exuded greatly increased quantities of amino compounds during recovery from incipient wilting. These relatively short-term experiments on rates and quantities of exudation, show that the release of substances from living cells is a real component of the process. The method of Pearson & Parkinson (1961), using chromatographic paper to collect exudates, has shown that different parts of the root have different rates and proportions of exudation. The growing parts are especially active in young sterile seedlings, but the older parts of the root, producing senescent cells, become progressively more important, and their exudates become qualitatively different later on.*

There is no doubt that the substances released from roots may directly affect their root-surface populations. Katznelson *et al.* (1955) showed by direct testing that the exudates of their experimental plants stimulated the growth of selected amino-acid-requiring bacteria. In other work, the effects of root exudates have been investigated by using model systems such as the celloidin cylinders of Timonin (1941) or the repeated watering of soil with exudates as used by Rovira (1956). These methods have resulted in the creation of artificial populations resembling the rhizosphere populations in respect of the relative frequencies of nutritionally specialized groups of bacteria.

The final constitution of the flora and fauna of the root region is not only the resultant of the direct effect of the root and its processes. The activities of the organisms themselves, including their utilization and release of substances, complicate the physical and chemical conditions, altering in kind and in quantity the conditions imposed by the root itself. The very kinds of substance—nutrients, vitamins, enzymes, antibiotics, indeed a multiplicity of compounds of organic and inorganic nature—can be produced by micro-organisms themselves. The general explanation of the existence of peculiar rhizosphere and root-surface populations is clear, whereas the explanations of particular examples or of special details are usually still obscure. It will be found that the explanation of the association of almost every kind of mycorrhizal fungus with its host has been sought in the study of root exudates. What is needed there, as in the general case, is much more information on the biochemical antecedents of the exudates, and on the external conditions that affect their formation and release.

The Penetration of Host Tissues and the Production of Disease

So far we have only considered the associations of micro-organisms with the root region and the surfaces of roots. Some of the fungi of these zones are potential pathogens which form restricted lesions even on

* *See* Appendix §2.

healthy roots. The research upon strawberry root-rots, especially in Canada, has shown the possibility that, in 'sick soils', disease may be occasioned by abnormal groupings of fungi in the root region, some of which cause penetration of hyphae and restricted areas of disease. It is not inevitable, however, that penetration of the tissues must cause acute disease symptoms, nor is it inevitable that a significant decrease of health will result from it. Keyworth (1947) has shown that resistant hop plants may harbour in their vascular tract the fungus *Verticillium albo-atrum* without showing any disease symptoms, whereas susceptible varieties suffer from 'wilt' by similar infection. Garrett (1956) has stated that even *Ophiobolus graminis* may survive indefinitely in the tissues of perennial grasses which are little affected by it and may even, whilst infected, remain dominant in grass communities. Susceptible hosts, too, may be little affected, provided they are maintained in a state of vigorous growth by adequate supplies of plant nutrients.

TABLE V

DIFFERENCES BETWEEN THE BACTERIAL POPULATION OF THE RHIZOSPHERE OF *Leptosyne* AND THE CONTROL SOIL AS EXEMPLIFIED BY CRYSTAL-VIOLET SENSITIVITY. (Data of E. M. Osborn & Harper, 1951.)

Samples		Bacteria & Actinomycetes in millions per gm.	Per cent resistant to crystal violet
Rhizosphere	1	260	57
	2	110	30
	3	56	27
Control soil	1	15	13
	2	15	11
	3	45	12

Waid (1957 and 1962) has studied the association of fungi with the roots of grasses obtained from grass swards. He has shown not only that a community of surface-inhabiting fungi develops upon those roots, but also that internal infection of the cortical and stelar tissues occurs. The deep-seated tissues are colonized by sterile hyphae especially as the root ages, and the population extracted from them differs greatly from the population of the surface of the cortex. The physiological behaviour of the internal fungi seems to be somewhat different from that of those from the surface, and they are less inhibited in growth and respiration by low oxygen supplies.

These are by no means unique examples of internal infections which produce no disease symptoms. Tervet & Hollis (1948) made a survey of the presence of bacteria in healthy storage organs of plants and showed that—with few exceptions—the organs which they studied contained bacteria internally. Similarly, Allington & Chamberlain (1949) studied the internal populations of bean leaves and showed that they might contain populations of potential parasites which were pathologically inactive in resistant plants. The presence of such organisms is well known to those who study the respiration of starving leaves, and Charles (1954) has shown how their activities may greatly increase the difficulties of interpretation of experimental results obtained under conditions where they reproduce actively in the tissue.

Fungi have also been recorded as denizens of stem, root and leaf tissues in a variety of plants. F. J. Lewis (1924) found endophytic infections by unknown fungi in the buds, stems, and roots, of various conifers. The hyphae penetrated the parenchymatous tissues of the cortex, medullary rays, pith, and phelloderm, without causing any apparent damage. The parasitic genera *Phoma*, *Phomopsis*, and *Colletotrichum*, have been reported as causing systemic infections in *Coffea arabica* (R. W. Rayner, 1948) and in *Casuarina* (Bose, 1947).

W. Brown (1936) contrasted the stages of parasitism of facultative and obligate parasites. The obligate parasites usually pass through a phase of symbiosis with their host—that is, a phase where the host tissue is not killed and perhaps may not appear to be greatly influenced by the presence of the parasite. Gäumann (1950) says of the loose smuts (*Ustilago tritici* and *U. nuda*): 'In the plants attacked it produces no obvious injury and correspondingly no special disease symptoms, but only a greater development of inter-cellular spaces, vascular bundles, stomata and internodes, together with an increased transpiration and respiration rate.' *Botrytis anthophila* on Red Clover (Silow, 1933), and *Microascus trigonosporus* (Whitehead *et al.*, 1948) as well as *Epichloë typhina* (Sampson, 1933), all show similar prolonged symbiotic periods during which systemic infection, almost indiscernible by external examination, gradually develops. In some of these cases the symbiotic phase is greatly prolonged. For instance, on some hosts, *Epichloë* rarely gives rise to disease. These examples merge with examples of symptomless infection such as that in *Lolium*. A variety of fungi have been isolated from the seed-coats of species of *Lolium*, races of which are regularly systemically infected by fungal hyphae (Günnewig, 1933; Sampson, 1935, 1938, 1939; Neill, 1940, 1942); but it is often not possible readily to differentiate between infected and uninfected plants through external inspection. The transmission of the infection is upon the fruit-coat, where hyphae develop in the superficial tissue, and there is no phase of sporing.

The range of examples of this kind could be greatly widened by mentioning the many symbiotic bacterial infections such as those of Rubiaceae, and of *Dioscorea macrura* which houses bacteria in its leaf tips. Enough has been said to indicate that the external and internal surfaces, both outside and within the living cells, are potential sites for the establishment of vegetatively active micro-organisms. The impact of these upon the metabolism of their hosts is very diverse: it appears to range from completely symptomless association to acute disease on one hand, and to stimulation of host development, directly or indirectly, on the other.

AN ATTEMPT TO DEFINE THE MYCORRHIZAL CONDITION

In view of the considerable diversity of kinds of association between various organs of higher plants and fungi, it may be wondered why or how any particular kind can be delimited and named 'mycorrhiza'. In some measure this is an historical accident. As was pointed out in the introductory chapter, the name was originally given to an easily defined kind of composite organ found upon the root system of Cupuliferae and Pinaceae. In these cases a great many structural peculiarities were found to be possessed by the infected roots; and the infected condition was observed to be the normal one of absorbing roots. A full account of these structural peculiarities will be given in a later chapter, but it must be stated here that their morphology and histology, and every aspect of their growth and development, is characteristic enough to separate them fairly sharply from most other kinds of association between fungi and roots. The contrast is as sharp, in these cases, as that between the association of bacteria with the rhizosphere and the organized dual system of bacteria and root tissue in the nodules of Leguminosae.

The extension of the term 'mycorrhiza' to other kinds of association—first begun by A. B. Frank—depended once again primarily upon the observed presence of a particular kind of mycelium as a normal associate in natural conditions with the root of a particular species, so that some anatomical change resulted from the association. As more and more cases were described, the contrast between the forms called mycorrhizas and the more casual association between fungi and roots became less and less clear.

In so far as associations of absorbing organs and fungi, constant in structure and development and consistently present and functional in natural conditions, are recognizable, the name mycorrhiza can be legitimately applied to them. But the use of this name should not be taken to signify a similar mode of functioning of all the organs to which it may be applied.

Each of the schemes of classification of soil organisms which we have considered earlier in this chapter, places mycorrhizal fungi as a separate group in the system. Each accepts the mycorrhizal condition as a naturally-occurring, non-pathological state in which there exists a prolonged association of a symbiotic kind, perhaps similar to that observed as a stage in the life of obligate parasites. Garrett (1950) writes: 'Evolution of the root-inhabiting relationship has culminated in the mycorrhizal association in which a state of true symbiosis has been achieved.' The implication here, as also in N. A. Burges's work, is that mycorrhizal fungi have evolved from aggressive parasites by progressive selection of nonlethal varieties so that the symbiotic stage becomes indefinitely prolonged. A very good case can be made for this assumption. Indeed, examples of apparently intermediate types have been mentioned in the previous paragraphs.

The evolutionary origin of the mycorrhizal conditions is not, however, a subject of so great an interest as the study of the physiology and biology of mycorrhizas. Undoubtedly much can be gained by comparing highly specialized mycorrhizal associations with obligate associations of all sorts, both parasitic and non-parasitic. Indeed it is a matter of great importance that recent developments in carbohydrate physiology and metabolism have shown a parallelism of behaviour (*see* Chapter V). Work with ectotrophic mycorrhizas, lichens, orchid mycorrhizas, and obligate parasitic infections by fungi, has been mutually stimulating and synergistic. In fact there are clear indications of a much greater similarity of functioning in this and other respects than had ever been previously imagined. Much can also be gained by comparing what is known of the root surface and rhizosphere associations with ectotrophic mycorrhizas, where the bulk of the fungal tissue is on the root surface and around it. This point was what I had in mind when I wrote, 'We are dealing, then, in ectotrophic mycorrhiza with a special case where, on a given root, one species of fungus in particular is so affected by local conditions as to become dominant in the root-surface zone' (1948). This is, I believe, a statement of fact and does not mean to suggest any evolutionary origin.

PART II

ECTOTROPHIC MYCORRHIZAS

III

THE ECTOTROPHIC MYCORRHIZAS OF FOREST TREES—STRUCTURE AND DEVELOPMENT

THE structure, development, and distribution of the mycorrhizas of many of the forest trees of temperate regions are so similar that they can be grouped together as a natural kind, and it is reasonable to assume that their modes of functioning as absorbing organs and their ecological roles are also similar. Until recently the species of trees known to form ectotrophic mycorrhizas belonged mainly to the Pinaceae among the gymnosperms, and to the Betulaceae, Fagaceae, and a few other angiosperm families. Most of the species were temperate and northern in distribution. It is now certain that ectotrophic mycorrhizal infection is more widespread than was hitherto thought, in both a taxonomic and a geographical sense. A few examples will illustrate this. Peyronel & Fassi (1956–1957) and others have described ectotrophic mycorrhizas in tropical Caesalpinoideae in the Congo. K. B. Singh (1966) has recorded the occurrence of similar infections in Dipterocarpaceae in Malaysia. Chilvers & Pryor (1965) have amplified an earlier account (Pryor, 1956) of the occurrence of ectotrophic mycorrhiza in the genus *Eucalyptus* where all the species that they have examined were mycorrhizic and the structure of their mycorrhizas resembled those of *Fagus.** It is therefore likely that, although it may be reasonably assumed that the physiological role of all of them is similar, there will be variation in detail of genetical and environmental origin.

As was explained in Chapter I, A. B. Frank originally gave the name 'mycorrhiza' to the composite organs in the Cupuliferae, so it seems logical to describe the structure of some of these in full and to comment on the similarities and differences found in other groups.

The structure of the root system of the European Beech (*Fagus sylvatica*) provides a good example which has been widely studied on account of the economic importance and ecological interest of this tree. The structure and mode of growth of the root apices of adult trees have been described in detail by Clowes (1949, 1950, 1951, 1954). In the paragraphs that follow, the observations of Clowes and myself which were made for a set purpose have been amplified by observations made upon beech mycorrhizas during their collection and dissection for physiological investigation.

The root system of the beech colonizes the soil within the immediate

* *See also* B. K. Bakshi, *Ind. Forester*, 92, 19-20, 1966.

area of the tree canopy and produces a large number of fine roots in the surface layers of the soil. The accumulation of rootlets near the soil surface is especially evident under woodland conditions. There appears to be a differentiation of the ultimate laterals into 'long' and 'short' roots;

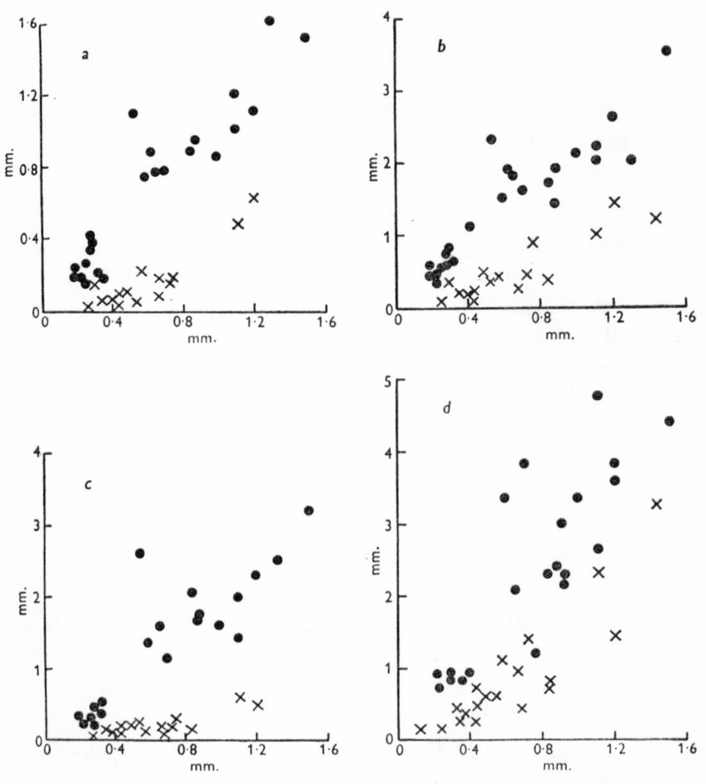

After Clowes (1951). By permission of New Phytol.

FIG. 2.—Relation of root size to level of differentiation in *Fagus* roots. Dots—uninfected roots; Crosses—mycorrhizal roots. The ordinate = the distance from the plerome apex; abscissa = diameter of root tissue; *a* = cortical vacuolation; *b* = vacuolation of the outer cortex; *c* = tannin impregnation of endodermis; *d* = lignification of protoxylem.

that is, into roots of potentially indefinite growth in length, which are the main roots of the system, and their branches which are of restricted growth in length and are relatively short-lived. The short roots are often termed the 'feeding roots', and to them is ascribed the main function of absorption. Indeed, as their surfaces constitute the greater part of the surface area of the root system, this must be true. It should not be assumed, however, that the long roots have solely a pioneer and anchoring function, for their apical regions at least are capable of absorbing salts and water and are often equipped with root-hairs.

In *Fagus* there is no distinction in quality between the two types of root, so that if a comparison is made of long and short roots that are uninfected by fungi, the types clearly grade into one another. Clowes (1951) has shown that if such characters as the position along the longitudinal axis at which the stelar and cortical differentiation occurs is related to the diameter of the root, all non-mycorrhizal laterals of adult trees fall into one series. Examples of this gradation of long into short roots is shown in Fig. 2. The ease with which the two types are recognized depends on there being many more very short and very long rootlets than roots of intermediate lengths, for reasons which appear below.

In natural soils it is mainly the shortest roots which become modified, after fungal infection, into mycorrhizal organs. When this has happened, the whole of the rootlet is enclosed by a fungal sheath and the apex is surrounded and covered with fungus. It is important at this point to refrain from assuming that the differences in histological structure between host tissues of mycorrhizal and non-mycorrhizal roots are solely caused by fungal activity and the following paragraphs are only descriptive.

Mycorrhizal short roots usually develop many branches, each of which is also of restricted growth in length, so that short lateral systems are formed which are completely enclosed in fungus. The appearance of these systems is variable. They are found most abundantly in the humus layers of the soil where their size in length, diameter, and branchiness, reaches its maximum. Even in rich mineral soils, mycorrhizal infection of short roots occurs, but here they are smaller and usually poorly branched. It is possible to classify types of mycorrhizal infection of beech by their external morphology. Types may be separated by mode of branching, diameter, colour, etc., but generally the kinds so grade into one another that it will be more convenient to describe their common points of structure and anatomy before a classification is presented.

In transverse section the commonest types of mature mycorrhizal axis have the structure shown in Fig. 3 and Plate 3 (*b*). The whole organ is bounded by a layer of fungal pseudoparenchyma which varies in thickness between 20μ* and 40μ in axes 300 to 500μ thick. This fungal sheath comprises about 20 to 30 per cent of the total root volume and 34 to 45 per cent of the total dry-weight. The rootlets have a high water-content (about 85 per cent of the fresh weight) and a high content of potassium, nitrogen, and phosphorus. The sheath is usually differentiated, often fairly abruptly, into two regions. These are (1) an external layer of large cells with somewhat thickened walls between which inter-hyphal spaces are small or absent, and (2) an inner layer which is more hyphal in structure, and loosely packed, with less thickened cell-walls and wider inter-hyphal spaces. In the inner layer, the cytoplasmic contents of the hyphae appear to be denser and the nuclei stain more intensely. The outer

* $\mu = 1$ micron $= \frac{1}{1000}$ mm.

layer is externally fairly smooth, but it has some hyphal connexions with the soil. Hence if it is viewed under a dissecting microscope, groups of hyphae can often be observed which frequently form slightly flattened strands connecting the sheath with the soil. The inner layer of the sheath is connected with a network of hyphae which run between the cells of the outer cortex. In most roots this network of hyphae, called the 'Hartig

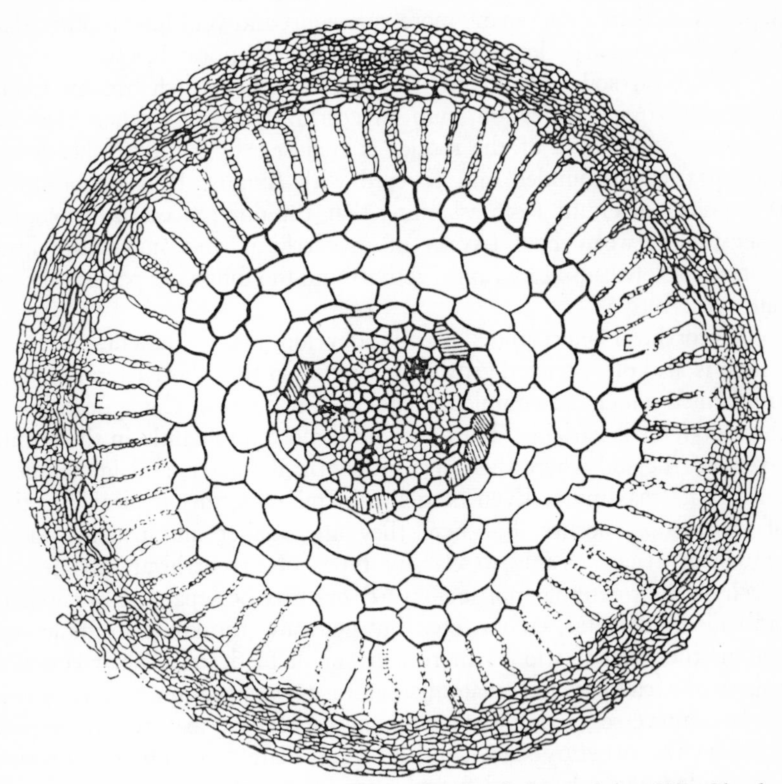

After Clowes (1951). By permission of New Phytol.

FIG. 3.—Transverse section of mycorrhiza of *Fagus*. Note: Hartig net in radially elongated epidermis and tannin impregnation of inner tangential walls of the epidermis. The shading shows endodermal cells filled with tannin.

net' (after Robert Hartig), extends only between the epidermal cells and the first few layers of the cortex, but in roots of small diameter it extends into deeper layers, even up to the endodermis. This last is, however, never penetrated so long as the organ appears healthy.

The cell layers of the root which are surrounded by the Hartig net are often composed of radially elongated cells. Such layers are most conspicuous in roots of large diameter when fungal penetration does not go far into the cortex, but are absent from slender mycorrhizas. The fungal penetration of the cortex often appears to be limited by a tannin barrier which may consist of impregnation of the cell-walls, or of droplets in the cytoplasm of the radially elongated cells.

Hyphae penetrate into the cortical cells to some extent in all types of beech mycorrhiza. The degree of penetration varies; it may be almost absent, but in some forms it is a regular feature of infection. As a general rule, the mycorrhizas with extensive penetration are large in diameter and irregularly branched: their axes seem to become secondarily thickened, or woody, closer behind the apex than in those forms with little penetration.

The longitudinal sections shown in Plates 2 and 3 amplify this description. The apex is completely overarched by the fungal sheath. Neither a Hartig net nor intercellular penetration is present in the apical region where the epidermis-cap complex is limited to a few loose layers of cells. The radial elongation of the epidermal layer associated with inter-cellular penetration is oblique, and the extent of the meristem is so small that stelar and cortical differentiation is apparent quite close to the apex.

Clowes (1949, 1951) has contrasted the structure of these mycorrhizal short roots with that of non-mycorrhizal roots borne in equivalent positions on a mother root. He first confirmed the earlier observation (Harley, 1937) that a proportion only—a proportion depending upon habitat conditions—of the short roots develop into mycorrhizas, so that infected and uninfected roots of similar origin exist side by side. Hence a trustworthy comparison of them could be made, provided a suitable basis could be found. As uninfected roots differ quantitatively, in the spatial relationship of developing tissues, according to their diameter (*see* Fig. 2), a similar basis of comparison is necessary for mycorrhizas. In Fig. 2 the distances from the stelar (plerome) apex at which various stages of maturation are achieved in mycorrhizal and non-mycorrhizal roots, is plotted against root diameter. In every case—for vacuolation of the cortex, tannin impregnation of the endodermis, and lignification of the protoxylem—differentiation was observed nearer the plerome apex in the mycorrhizal axis than in uninfected roots. The differentiation of the xylem also differs somewhat in the two kinds of organ. The earliest xylem elements to become lignified in the uninfected roots are usually annular, annular-spiral, or scalariform. In mycorrhizal roots the first-formed xylem elements are usually pitted, rarely scalariform or annular-spiral.

This spatial precocity must not too readily be attributed to more rapid differentiation in mycorrhizal roots. Certain observations suggest that it is associated with a reduced rate of growth in length. For instance, Clowes (1951) treated uninfected beech roots of a variety of sizes with colchicine, and showed that differentiated cells were found closer to the apex of the treated roots, that the xylem vessels nearest the apex were pitted, and that many of the cells of the cortex became elongated in a transverse plane, as in mycorrhizas. These changes were associated with a reduced rate of growth in length in the colchicine-treated roots. The effects of colchicine on *Fagus* roots are reminiscent of the observations of G. Smith & Kersten (1942), who obtained similar changes in *Vicia faba* roots by reduction of their growth-rates with soft X-ray, and of other comparative studies of differentiation in fast- and slow-growing organs. From them we cannot immediately conclude that infection *causes* the spatial precocity observed in mycorrhizas, for it may be that infection is only developed upon those root axes which already show differentiation close to the apex. This problem will be considered in detail later.

Mycorrhizal short roots are, in contrast to uninfected short roots, usually abundantly branched, and branches develop on them even when they are totally covered with a fungal sheath. The branch initials develop from the pericycle immediately opposite the xylem poles and each young branch, as it passes through the cortex, forms a bulge in the sheath which is visible on the surface of the mother-root. The fungal sheath above the bulge does not break under tension but extends so that throughout its growth the young branch remains totally encased in fungal tissue. As the cells behind the apex of the young branch mature, intercellular penetration by hyphae occurs to form the Hartig net, and the branch soon has the same structure as the parent axis.

Although the long roots of beech have been reported to be free from mycorrhizal infection, this is not invariably true. Especially at times of maximum root growth, which occur in spring and autumn, a large number of long uninfected roots bearing root-hairs are always present in the root system of adult trees. They are derived from the apices of long roots, from short roots which have resumed active growth, and from mycorrhizal roots which break through the sheath and form an uninfected axis. In spite of discussion concerning the renewed growth of mycorrhizal roots, there is no doubt whatever of its occurrence in beech. Not only are mycorrhizas found in the soil from which the host axis protrudes in an unsheathed condition, but also excised mycorrhizas kept in distilled water or in sugar solutions in spring will form an extension of the host axis up to 1 cm. in length. Uninfected laterals formed in any of these ways may bear root-hairs which arise from the epidermal cells during vacuolation. During periods of less active growth, the long roots are colonized by a sheath of fungus which may eventually close over the apex, but no Hartig net is formed. The bulk of the fungus lies on the surface only, and apart from occasional hyphae which penetrate the cells and form swollen structures within them, there is little penetration of the host tissue. Clowes has named this type of organ 'the superficial ectotrophic mycorrhiza', and has shown that its development does not inevitably inhibit the formation of root-hairs. He has described and figured short root-hairs projecting into the layer of crushed root-cap tissue which lies between fungus and epidermis (Fig. 4).

Branches formed on superficially infected mother-roots pass through the cortex and bulge out the sheath as in other types, and may often remain totally ensheathed. The closing of the fungus over the main apex, which occurs during periods of slow growth is, however, usually temporary. At the onset of active root-growth the host axis again escapes and grows forward; in consequence the apex is not usually rounded, as it is in short mycorrhizas, and the root-cap tissue is relatively well developed at all times.

Clowes (1949) made a very detailed analysis of the activity of the apical meristem of *Fagus* roots, both infected and uninfected. In essentials, the architecture of the apex and the sequence of division of the cup-shaped group of initials is similar in the two types. The striking differences in their appearance in section are explicable in terms of differences in the differentiation of cells. The spatial precocity of differentiation in mycorrhizal roots, as described above, gives the impression of great dissimilarity although no fundamental difference is present. Another striking feature of mycorrhizal roots is the relative absence of root-cap material although, as Clowes pointed out, a distinct tissue developing distally to the initials is present in them. This cap tissue is restricted in thickness because it becomes decomposed and incorporated in the fungal sheath. The central core of the cap tissue of uninfected roots consists of a columella, which originates from a series of initials lying horizontally below the plerome pole. The epidermis and lateral region of the root-cap originate from the lateral cells lying in a ring around the columella initials. A similar epidermis-cap complex is recognizable in mycorrhizal roots, between fungal sheath and cortex, as an organized series of cells lying 20–110μ distal to its initials and as a thin layer extending 30–380μ proximal to the initials.* The epidermis itself is recognizable within the sheath throughout the full length of the mycorrhizal axis, but the cap tissues soon become disorganized (Plate 2). There seems, therefore, to be a continual dissolution of tissue by the

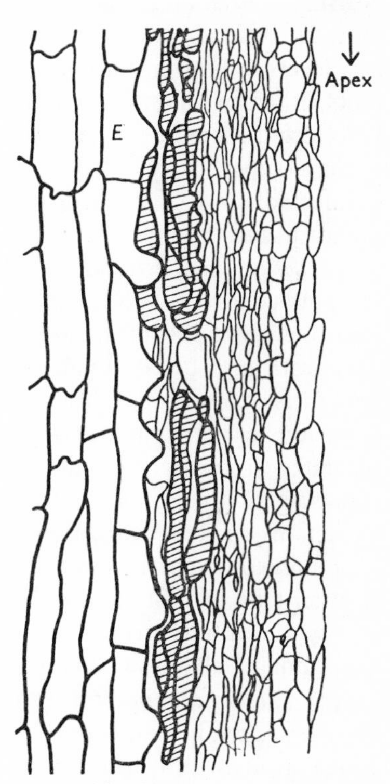

After Clowes (1951). By permission of New Phytol.

FIG. 4.—Longitudinal section of superficial ectotrophic mycorrhiza of *Fagus* in the region of root-hairs. E—epidermal cells. The shaded cells are sloughed root-cap cells in the fungal sheath.

* *See* Appendix §4.

fungus as growth and differentiation proceed, and this process may be significant in the nutrition of the fungal sheath (*see* Chapter V).

Beech Mycorrhiza in the Soil

The kinds of primary root, lacking secondary thickening, which are found in the adult root system of beech in natural conditions, will be considered in order of increasing intensity of infection.

(1) *Uninfected roots* retain the potential for unlimited growth and are present at all times, but are especially abundant in spring and autumn. Under this heading may be grouped totally uninfected axes, together with the extensions of infected axes which escape from the sheath. In soils bearing a litter of considerable depth, many of these uninfected roots grow horizontally through the compacted leaves and also upwards into new layers of litter in spring. Some of them are the main apices of pioneer roots, others are homologous with short roots which become mycorrhizal. Uninfected roots may bear mycorrhizal laterals and their main axis may become superficially infected or, occasionally, the apices of otherwise uninfected roots of small diameter may become fully infected, forming a sheath and Hartig net.

(2) *Superficial ectotrophic mycorrhizas* are roots of the same kind as uninfected roots, except that in their case the external layer of fungal sheath encloses the apex. Roots of all diameters may be so infected, and the sheath may be continuous to just behind the apex even when active growth is proceeding. The leading pioneer roots, several millimetres in diameter, are often infected in this way. Their sheath tissue is easily separated from the host, and large flaps of fungal pseudoparenchyma, about 0·5 × 3 cm. in dimensions, can readily be obtained from them by dissection. In some cases the sheath near the apex is an extremely thin weft a few microns thick with perhaps a few locally thicker patches. This is often true of roots of small diameter which can, then, be recognized as superficial mycorrhizas by the relative paucity of root-hairs. The behaviour of all these superficial types is like that of uninfected roots.

(3) *The diffuse infected system* consists of lateral systems of root branches usually composed of both fully infected mycorrhizal roots which are light-brown in colour, and, interspersed with them, uninfected and superficially infected roots. It is as though the fungus controlled the growth of some of the laterals which become completely infected, whilst other homologous branches retain their powers of growth. These systems are much-branched, as if the fungus either reduced the elongation between points of origin of branches or else stimulated the frequency of branching. This diffuse type of arrangement of mycorrhizal and other laterals is common in the surface layers of most woodland soils, but is most abundant on relatively moist brown-earth sites in the litter and humus layers.

(4) *The pyramidal infected system* is similar to the diffuse system except that uninfected apices are rarer in the former. The laterals with their branches take the form of a Christmas tree, every branch being totally enclosed in the sheath and bearing daughter branches of similar structure which increase in length backwards from each apex. This type grades almost imperceptibly into the diffuse type, and might be considered as a state of greater fungal control of form. It is found most abundantly in the F and H layers of acid woodland soils where there is a mor humus, and in other soils where deep humus and litter accumulate in hollows. The diameter of the axes is somewhat variable but is greatest in the upper H layers where rootlets may be 0·5 to 0·75 mm. thick. Here the upward-pointing apices of the branches tend to grow into the lower leaf-litter. In soils showing the greatest mycorrhizal development, large flat aggregations of totally infected roots of this type, up to 15 cm. or even 20 cm. in diameter, may be excavated from the H and F layers. In deep layers of leaf litter, the systems are of smaller diameter and are very regularly branched.

(5) *The coralloid infection* is only distinguishable from the last as an extreme case. The diameters of the branches, which are less regularly arranged, are usually greater (up to 1 mm.), and close behind the apices secondary thickening of the main axes occurs, so that the complete system has a springy feel when handled. The internal structure shows a great development of intra-cellular hyphae penetrating the cortical cells, as if the fungus were extremely virulent and active. This type of system is found on raw humus, but it is relatively rare.

(6) *The nodular infection* is another kind in which the fungal partner appears to be extremely active. The racemose branched mycorrhizal axes are bound together in a nodule by mycelium. Nodules up to 1·5 cm. in diameter have been found in waterlogged soils, but are so rare that their importance is negligible. Their existence in beech is worth noting, however, on account of the occurrence of similar infection in Pinaceae.

(7) *Apically infected roots.* The types so far described are encountered in the surface layers of podsols or brown earths, where a mass of the feeding roots of the beech develop in the surface horizons. Deeper in the soil itself, the laterals of the root systems of adult trees bear mycorrhizal systems which are smaller in size and especially in diameter. These are often infected at their apices only, so that the root axis shows the normal structure of an uninfected root with root-hairs in its proximal region, but bears at the apex a swollen region that is completely ensheathed and has a Hartig net. This kind of root is found abundantly on seedlings during their early growth at the beginning of infection, and is also common in deep soil layers—especially where these are calcareous or base-rich.

(8) *Loose weft infection.* Associated with roots of the last type may be found small lateral systems that branch racemosely and have an abundant weft of hyphae around them. The sheath is thin and a Hartig net is

present. This type is very likely to be of the same general nature as the last, and perhaps constitutes a stage in infection from the soil.

(9) *Other types of infection on adult trees.* Two further types of mycorrhizas have been described on adult beech trees. The first is a black mycorrhizal rootlet, usually little branched, and bearing stiff projecting hyphae from the sheath. The sheath of dark pseudoparenchyma has a remarkable structure that is frequently met in the make-up of ascomycete fruiting-bodies. When pressed flat under a coverslip, it appears to consist of centres from which long cells or hyphae radiate in all directions. These mycorrhizas are found especially on poor podsols in very dense stands of beech. This type of mycorrhiza has been described (*see* p. 4) as occurring upon many other angiosperm and gymnosperm trees, and is caused by the fungus *Cenococcum graniforme* (*Mycelium radicis nigrostrigosum*).

The other type is a secondary infection in which hyphae colonize the surface of existing mycorrhizas. Hyphae that are recognizable by their colour, which is often yellow, brown, black, or pink, form complete or partial sheaths outside the primary sheaths. In 1937 I confused the black type of secondary infection with that due to *Cenococcum*. The mistake was pointed out by J. Warren Wilson (1951). The black secondary infections are due to sterile black or greenish mycelia which are commonly found associated with beech roots of all kinds, and especially with those occurring in poor and in garden soils. Many strains, varying somewhat from one another in cultural behaviour, have been isolated from uninfected roots as well as from the surfaces of mycorrhizas, where they can be seen under the microscope as sterile black hyphae lying on the surface or amongst the surface cells of the root or fungal sheath. The fungi conform to the type named *Mycelium radicis atrovirens* by E. Melin, who first isolated them from roots. Robertson (1954) has shown them to be an important constituent of the root-surface population of the long roots of pine, and Levisohn (1954) has described their pathogenic potential which will be discussed later.

This description of the feeding roots of the beech may give an impression of great variability of structure. The variability is much less great than appears from the foregoing account, because certain of the types of root are present in great preponderance in most soils where beech is growing vigorously. These are the various long roots and the pyramidal and diffuse types of mycorrhiza on which secondary infections develop in their old age.

A description of the range of types of mycorrhiza occurring in *Fagus sylvatica* has also been given by Boullard & Dominik (1960), using the general classification of ectotrophic mycorrhizas proposed by Dominik (1955). They studied the root systems of beech in a number of different kinds of beech forest. The same impression was gained of variation, and also of the preponderance of a few types (in group F of Dominik) associated with vigorous stands.

Mycorrhizas of *Pinus* and other Genera

Similar types of ectotrophic mycorrhiza are met in other plants besides beech. The recent work on the structure of the mycorrhizas of *Eucalyptus*, by Chilvers & Pryor (1965), illustrates this point. All the species of this large genus which have been suitably studied form mycorrhizas. According to these observers, the 154 species in which infection is known to occur represent about one-quarter of the total number of species and belong to 40 out of the 47 taxonomic series of W. F. Blakely. As with beech, the root system comprises both long roots and short mycorrhizal systems. The long roots may be invested with a sheath in a similar manner to the superficial ectotrophic mycorrhizas described in *Fagus* by Clowes. Amongst the lateral mycorrhizal systems diffuse, pyramidal, nodular, and apically infected, systems are formed. In addition, unbranched infections due to *Cenococcum*, and primary and secondary infections, occur.

In a study of the differences between the structures of uninfected roots and pyramidal mycorrhizas, Chilvers & Pryor showed how closely similar *Eucalyptus* roots are to those of *Fagus*. Fig. 5, which is derived from their paper (1965), could equally represent the mycorrhizal roots of *Fagus*—as may be seen by comparing it with Plate 3. Chilvers & Pryor also concluded that in many respects the architecture of the apices of roots and mycorrhizas are similar and differ only in quantitative terms. The spatial precocity of mycorrhizal development could be attributed, like that of *Fagus*, to slower growth: 'Ageing roots, roots growing slowly through unfavourable media, and roots artificially restrained by overdoses of colchicine and naphthalene acetic acid, all exhibit the same phenomenon.'

The many differences between the ectotrophic mycorrhizas of different species have often been over-emphasized because the full range of structural variability had not been described. In the early work, too, faulty technique and inadequate description led to some confusion and discussion. In particular, the degree of intra-cellular penetration by the fungus and the possible digestion of the hyphae by the host cells, were much discussed. It is now clear that digestion within the host cells is, at most, very rare in occurrence in ectotrophic mycorrhiza. On the other hand, penetration of the hyphae into the cell lumina occurs to variable degrees in almost all species. Certain types of mycorrhiza, such as the coralloid form of the beech, have been called *ectendotrophic** on that account and may be associated with fungi of somewhat dissimilar physiology from the purely ectotrophic forms (*see* p. 71).

The external forms of the mycorrhizas are similar in most species and composed of racemose branch-systems; but in the genus *Pinus*, alone amongst ectotrophic mycorrhizas, the mycorrhizas are dichotomously

* It should be noted that Lobanow (1960) uses the term 'ectendotrophic' to describe mycorrhizas called 'ectotrophic' by most authors.

branched. In some other respects, too, the pines are peculiar. They exhibit a much more marked heterorhizy, for the differentiation into long and short roots is more sharp. The plan of the root systems of seedlings, and the arrangement of branches on a single main lateral of an adult

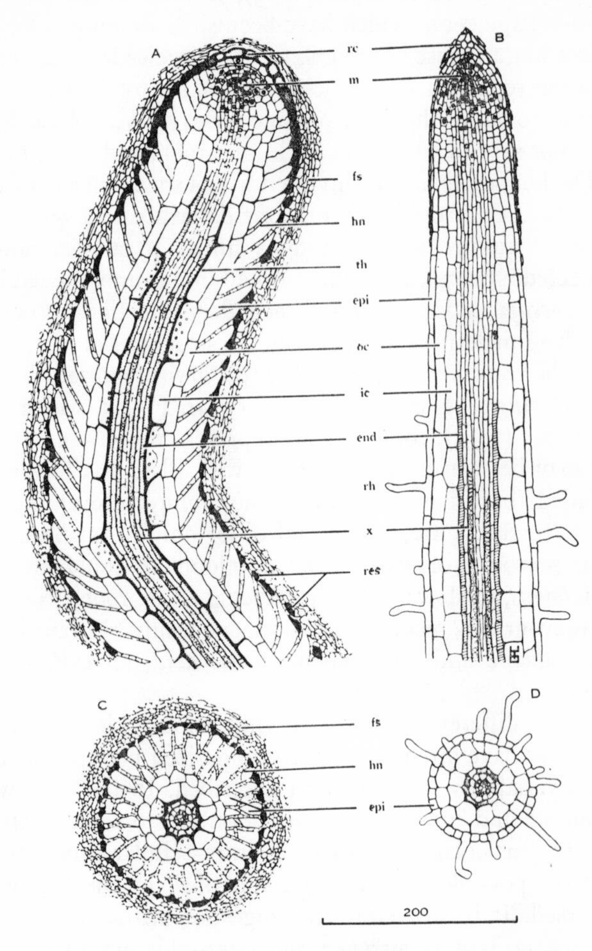

After Chilvers & Pryor (1965). By permission of Aust. J. Bot.

FIG. 5.—Schematic comparison between mycorrhizal and non-mycorrhizal roots of *Eucalyptus*. rc—root-cap, m—meristem, fs—fungal sheath, hn—Hartig net, th—thickened layer, epi—epidermis, oc—outer cortex, ic—inner cortex, end—endodermis, rh—root-hair, x—xylem, res—cell of root-cap complex in the fungal sheath.

pine, give the impression of great dominance of the apical region, so that the branches for some distance behind it are short and of limited growth. The short roots, from the time of their initiation, appear to show inherent morphological differences from the long roots (Aldrich-Blake,

1930; Hatch & Doak, 1933; R. J. Preston, 1943). However, as Aldrich-Blake pointed out, there are no marked qualitative differences in anatomy between long and short roots in *P. laricio* var. *corsicana*, although sometimes the resin canals are absent in short roots. Nevertheless he described clear quantitative differences. Thus the mean diameter of the stele between protoxylem poles of long roots was $117 \pm 6\mu$, while of short roots it was $75 \pm 3\mu$—a large, highly significant difference. Usually only the short roots develop a sheath and Hartig net and become fully-formed ectotrophic mycorrhizas. The long roots, although often infected, do not develop a sheath.

Hatch & Doak (1933) provided a clear analysis of the root system of pines which also emphasizes the difference in structure between the two types of root. These authors used the term 'long root' to include (i) the tap-root of seedlings, (ii) laterals called 'mother-roots', and (iii) the prolongations or direct continuations of mother-roots called 'pioneer roots'. All these types are normally equipped with root-caps and grow and increase the spread of the root system. The mother-roots are frequently branched and give rise to long as well as short and abortive laterals. The pioneer roots are less branched because the majority of their branch initials abort. There is present, always, a great excess of short roots over long roots, and they are mostly monarch in stelar structure and may lack a well developed root-cap. They take three forms: (i) uninfected short roots which bear root-hairs, continue slow growth, and dichotomize 'frequently but not profusely'; (ii) mycorrhizal roots which are enveloped in a fungal mantle, develop a Hartig net and occasional intra-cellular hyphae, exhibit change of shape of the cortical cells, and grow with repeated dichotomy; and (iii) pseudomycorrhizas which are attacked by fungi such as *Mycelium radicis atrovirens* and soon cease growing.

Melin (1927) made a classification of the types of mycorrhizal short roots found on pine which has since been used in descriptions and evaluation of intensity of mycorrhizal development. It has been modified so that it can be used for other trees. The following account, which is adapted from Björkman (1941), includes one addition to the original scheme, together with a comparison with *Fagus*.

Type A mycorrhiza is unbranched or forked in pine, and simple or monopodially branched in fir and birch, having a dark brown to greyish-white colour. The colour depends upon the type of fungus. A sheath 10 to 25μ thick encloses the root, in which a Hartig net develops. Hyphal strands connecting the sheath with the soil are formed only in a few special cases. The diffuse and pyramidal types occurring on beech may be included here.

Type B mycorrhiza consists of branch systems of which only the apices develop a complete type *A* mycorrhizal structure. The proximal part of the axes may be of 'pseudomycorrhizal' structure. Usually this type is not usefully distinguished from the type *A* in experimental records. The apical infections of beech fit well into this category.

Type C mycorrhiza was originally described as 'nodule mycorrhiza' (Knollen). Large nodules, formed of concrescent short root branches, are enclosed in one complete sheath system, yellow-grey in colour. From the sheath a few trailing hyphae run into the soil. This type of mycorrhiza, even when relatively little developed, is characterized by the thick yellowish mantle. The type was synthesized by Melin (1923) with *Boletus* sp. The nodule mycorrhiza of beech, rare in occurrence, is similar except in colour.

Type D mycorrhiza comprises the black mycorrhizal roots and is separable into two kinds:

Type DN, which is a black or black-brown organ with a well developed sheath and Hartig net. Thick dark hyphae run stiffly out from the sheath into the soil. This type is formed by *Cenococcum graniforme* (*Mycelium radicis nigrostrigosum* Hatch) and is common on beech and a great variety of other trees.

Type DA, which is a secondary infection, like those described above for beech, where dark hyphae, *Mycelium radicis atrovirens*, colonize the external layers of type *A* mycorrhiza.

This scheme emphasizes the similarity in this connexion of pine to other trees, even though it appears to differ in detail. Too much emphasis must not be placed upon the apparent difference of degree of differentiation between long and short roots, because much of the descriptive work on *Pinus* has been done upon seedlings which, whether in *Pinus* or in other genera, show more marked heterorhizy than do adult plants.

Dominik (1955) has proposed a more detailed system of classification of ectotrophic mycorrhizas which is based primarily on the structure of the fungal sheath or mantle. The characters used are: its differentiation into layers, its density, whether it is plectenchymatous, prosenchymatous, ornamented with setae, or giving rise to strands, etc.

The main divisions of this classification are further subdivided, especially according to the colour of the fungal tissue. The aim of the system is to provide groups which are more likely to be caused by a single fungal species than are those of other classifications.*

Robertson (1954) made a detailed study of the root system of *Pinus sylvestris* trees thirty years of age. He observed the acropetal development of short roots upon the long roots. Those in the distal region were relatively simple or once-forked, whilst those in the proximal regions were repeatedly dichotomized, exactly as figured by Hatch (1937). This difference of development of short roots along the length of the long roots was explained by Robertson in terms of the distribution of fungal infection in the long roots. A Hartig net was present within the cortex although no external sheath occurred. This condition of infection had previously been described in *Pinus* by Laing (1923) and Möller (1947),

* *See* Appendix §3.

in *Abies firma* by Masui (1926), and has since been seen in other conifers, e.g. *Pseudotsuga taxifolia* (*P. douglasii*), by Linnemann (1955). Robertson also observed (as did Möller) that only the tips of the long roots were usually free from fungus. On those long roots where most of the short roots developed into dichotomous mycorrhizas, a complete Hartig net was found in the outer cortex and the branch-initials became infected as they passed through the cortex. Robertson therefore concluded that the development of mycorrhizas on a long root depended on the presence of a Hartig net in the cortex. Short branches formed in the tip region, usually free from infection, remained uninfected and simple in form; short branches formed in the proximal regions developed into mycorrhizas and grew into dichotomous systems.

The differential distribution of infection in the long roots of *Pinus* explains certain features of Hatch & Doak's description (1933) of the root system—especially the three kinds of short roots (simple, abortive, and mycorrhizal), and the fact that pioneer roots are unbranched because of abortion of lateral root initials. It can now be seen that the structure and development of pine mycorrhizas is not dissimilar in any fundamental respect from that described for beech. The infection of an adult root system is permanent. The branches of long roots are infected from the mother-root—from the sheath in the case of beech, or from the Hartig net in the case of pine. The range of types of infected short roots is similar in initial structure and in the development of secondary infections. The presence of the superficially sheathed long roots and of the racemose mycorrhizas in beech, and of the long roots with the Hartig net and the forked mycorrhizas in pine, are the most striking differences. Less striking, but probably more important, is the presence of aborting laterals on pine which do not seem to occur in quantity on beech root systems in the adult state.

So far as can be ascertained, dichotomy of mycorrhizas is restricted to pines; racemose mycorrhizas are the commonest forms in both angiosperm and gymnosperm root systems. In general the structure of all types conforms with the ones described, but there are a few exceptions. For instance, a kind of birch mycorrhiza was described by Melin (1923*a*) which contained a differentiated cell layer in the cortex in which structures similar to *haustoria* were observed. Some degree of hyphal digestion seemed also to occur in this cortical layer. How far such mycorrhizas are widespread is not known. Certainly the birch mycorrhizas usually encountered are very like those of other Betulaceae and Fagaceae. One of the most striking facts which emphasizes these ectotrophic mycorrhizas as a natural kind is provided by the *DN* mycorrhiza in which *Cenococcum* is the fungal partner. These were observed by Hatch on eleven tree species in the U.S.A., and they have since been described upon the roots of many more species of trees forming ectotrophic mycorrhizas (*see* Trappe, 1962).

Mycorrhizal Development in Seedlings

The factors affecting the development of mycorrhizal infection in seedlings will be described in a later section. Here the differences of behaviour of the short roots of seedlings and adult trees must be described, since much of the experimental work with ectotrophic mycorrhizas has made use of seedlings. Those who have worked solely with seedlings and young trees have emphasized two features, namely (i) the sharp differentiation of long from short roots, and (ii) the frequency with which non-mycorrhizal short roots abort, become parasitized, or are converted into pseudomycorrhizas by fungi such as *Mycelium radicis atrovirens*.

The differentiation of long from short roots on fully grown trees is sharper in pine than in beech; but beech, in the seedling state, does exhibit a more obvious heterorhizy than in later life. In the early stages of growth, when the tap-root and long roots are uninfected, the short roots become infected from the soil so that a sporadic, patchy development of mycorrhizas occurs on the system. Those rootlets which become infected remain active and branch repeatedly, growing slowly in length, the mature tissue differentiating close behind the apex in the manner described for mycorrhizas of adult trees. These infected systems are not so short-lived as those short roots which remain uninfected, so that mycorrhizal infection often appears to increase the short-root production. This may only be due to prevention of abortion of the rootlets by mycorrhizal infection, with the result that, at any particular time, the root systems will bear a greater number of short roots than will an uninfected seedling of equal age. As Aldrich-Blake (1930) showed, short roots in pine which would, if they remained uninfected, hardly escape from the cortex of the mother-roots, develop as mycorrhizas, grow, dichotomize, and persist for a longer period than uninfected ones.

Another contributory factor to the apparent stimulatory effect of infection on short-root production in seedlings, depends upon the influence of soil nutrients on root growth and branching, as well as upon the intensity of infection. In very rich mineral soils, branch roots are produced less abundantly and mycorrhizal infection of the branches is also less intense, than in poorer soils. In this case again the correlated decrease in the development of short roots with decrease in the intensity of mycorrhizal infection, is not due to the latter being the cause of the former.

The sequence of development of mycorrhizas in pine and spruce seedlings during the first two years of growth was described by Laiho & Mikola (1964). They took samples of seedlings from nursery beds at short intervals of time. The first short roots formed about 16–20 days after germination, and thereafter increased steadily in number during the first year. In the second year they again increased in number, a very rapid midsummer increase in pine being due to their branching (which did not occur in spruce). Mycorrhizal development started in the first year from three to four weeks after the formation of the first short roots, and was correlated with the setting of the first two leaves. It continued steadily during the growing-season, so that about 60 per cent of the short roots were infected at the end of the first season, and about 90 per cent at the end of the second season.

Laiho & Mikola emphasized that the onset of infection of short roots was somewhat different in the two seasons. In the first year the infection in spruce and pine is mainly from the surrounding soil; hyphae penetrate the tissues, first forming a Hartig net and only later developing a sheath. In the second year the infection is mainly from the parent root, which has hyphae either on its surface or in its tissues, and a fungal sheath is often formed immediately in pine, but not in spruce (where the period before sheath formation is much extended). In each case growth in length occurs (together with dichotomous branching in the case of pine), and as the base of the root ages its infected cortex becomes suberized and forms a protection over the stelar tissue. The whole short root may live and grow for at least three or four years, whereas the cortical cells survive for perhaps two seasons only.

An extremely detailed examination of the development of infection on the roots of beech seedlings was made in Oxford by J. Warren Wilson.* He confirmed the earlier descriptions of growth of seedlings in their first year, showing that three periods of growth could be observed. These were (i) rapid development of roots, hypocotyl and stem, depending mainly on consumption of cotyledonary stores and culminating in the expansion of the first leaves; (ii) a period of dry-weight increase and leaf expansion without further morphological development; (iii) a final period of loss of leaves and the onset of the winter resting phase. Between the periods of weight increase and incipient dormancy there was an additional phase if light intensity was good, in which 'lammas' shoots were expanded. Mycorrhizal development begins after the end of the first period and continues until dormancy commences. Careful examination of the root-tips of developing seedlings showed that they changed with time during these periods. On first escape from the cortex they possessed a well developed root-cap and apical meristem. As they aged, differentiation occurred nearer and nearer to the meristem—until a condition not unlike that seen in the host tissue of mycorrhizal apices was developed. Wilson divided this sequence of stages of root development as follows:

Tissue system	Distance in μ, from root apex, of first differentiated cells		
	Stage 1	Stage 2	Stage 3
Cortex	93– 112μ	50– 82μ	0– 47μ
Endodermis	100– 117μ	57– 85μ	27– 63
Root-hairs	120– 345μ	45–135μ	0– 75μ
Lignified xylem	750–1340μ	250–525μ	87–250μ

* *See* Appendix §2.

Estimates of the abundance of rootlets on the three stages at different times during seedling growth, suggested that they are indeed developmental stages passed through by each root (Fig. 1). This was confirmed by direct observation and by studying the sequence in potted seedlings grown in soil from a beech woodland. The frequency of the three phases in seedlings growing in natural woodlands is shown in Table VI.

TABLE VI

THE PROPORTION OF ROOT-TIPS IN STAGES 1–3 ON SEEDLINGS OF *Fagus* IN FOUR WOODLANDS AT TWO STAGES OF SEEDLING GROWTH

Stage of seedling development	Woodland	Percentage of root-tips in Stages 1–3 on seedlings in four beech woodlands, 1951		
		Stage 1	Stage 2	Stage 3
Leaves half expanded	1	7·3	23·3	69·4
	2	6·8	20·8	72·4
	3	8·7	21·3	70·0
	4	8·2	21·3	70·5
Leaves fully expanded	1	0·0	4·0	96·0
	2	0·4	4·2	95·4
	3	0·6	3·7	95·7
	4	0·3	7·5	92·2

Results of J. Warren Wilson.

The development of Stage 3 roots into mycorrhizas is their common fate, but those in the earlier stages are rarely modified directly into mycorrhizas by fungal infection. In one instance, Warren Wilson examined surfaces of root-tips for the presence of hyphae in sufficient quantity to give estimates of the presence of loose wefts of hyphae, fungal sheaths, and

TABLE VII

THE PROPORTION OF ROOT-TIPS OF *Fagus* IN STAGES 1–3 COLONIZED BY FUNGAL HYPHAE

Appearance of fungal infection	Percentage of root-tips infected		
	Stage 1	Stage 2	Stage 3
With investing hyphae	0·7	1·4	5·6
With sheaths	0·4	2·4	32·6
Mycorrhiza	0·0	0·0	27·4

Results of J. Warren Wilson.

full mycorrhizal development, on rootlets in each stage. The seedlings used were in the phase of rapid dry-weight increase when mycorrhizal development was occurring (Table VII).

On the young roots of Stages 1 and 2, no particular aggregation of hyphae was noted, but in Stage 3, no less than 27·4 per cent of all the roots examined were mycorrhizal.

Warren Wilson correlated with their diameter the rate of growth in length of roots and their passage through the three morphological stages. He found positive correlation. Roots of the largest diameter continued to grow fast for a long period and remained in Stage 1 for a long time. These observations confirmed the view of Clowes that the rootlets of beech are not rigidly divided into two kinds, long and short, but that there is a continuous sequence of types. Hence it seems likely that even where the two types of root appear to be very sharply different, the short root is simply one which soon ceases elongation, so that differentiation of the tissues occurs up to close behind the tip. The long roots are those which continue to grow for one or more growing-seasons. The apically infected long roots are those which cease growth after a rather prolonged growth period. The apparent differentiation is simply due to a preponderance of extreme types.

The existence of fungi upon Stage 2 and Stage 3 roots, poses the question whether the fungi bring about the cessation of growth and in some way promote the differentiation of the apex. Warren Wilson compared the roots of seedlings grown in sterile soil with others grown in unsterilized soil, and observed that the rootlets in both treatments behaved in a similar manner. No mycorrhizal roots were found in his sterilized soils, although Stage 3 roots were converted into mycorrhizas in unsterile soil.

These phasic changes, according to Warren Wilson, are not necessarily brought about by fungal stimulus. There are good grounds for taking this view. Changes in the physiology and biochemistry of meristems as they age have been recorded, as for instance in the work on oxidase systems of barley by James & Ward (1957). Again, Street & Roberts (1952) and Hannay & Butcher (1961) have observed the ageing of the meristems of *Senecio* in tissue culture. Both attached and excised apices may pass through stages of modification of structure and rate of growth which may be influenced by carbohydrate availability and supply of growth factors. The difference in the structure of the host in respect of the distance between meristem initials and fully differentiated cells in beech mycorrhizas, as first described by Clowes, is not necessarily a direct effect of the mycorrhizal fungus; mycorrhizal development may take place after the root-tip has adopted this structure.

The first visible phase of mycorrhizal infection is the formation of a weft of mycelium over the surface. This weft thickens to form a sheath, especially over the apex; later, the typical Hartig net develops between the cells. Warren Wilson has suggested that there is a great difference in his account of mycorrhizal development in seedlings from that of Clowes in adult trees and in seedlings. Clowes (1951) expressed the view that fungal infection might cause the change of host structure:

The earliest recognizable stages in the formation of ectotrophic mycorrhiza show the presence of very simple interrupted sheets of pseudoparenchyma, one cell thick, lying on the surface of the root. The fungal 'cells' are relatively large and appear like those of the cells of the outer layer of well-developed sheaths. The fungus occurs in patches behind the apex and seems never to extend over the apex. The hyphae do not penetrate between any of the cells. . . . At this stage in *Fagus* the apical meristem is of normal extent for uninfected roots of its diameter and the cortical and other cells are elongated as in uninfected roots. With the formation of a more continuous mantle, one or two cells thick, the hyphae penetrate between the epidermal cells and the meristem is reduced in longitudinal extent as in a typical mycorrhiza. . . . The fungus never seems tc form a Hartig net between cells that have elongated normally, but it does so sometimes where cells are intermediate in length between the elongated state and the extreme mycorrhizal state.

It is interesting to note that Lobanow (1960) quotes O. M. Trubezkowa & W. W. Michalewskaja as having described the early stages of infection of oak in a very similar way. There the fungus first forms a loose but not complete weft around the root and this gradually thickens over the entire root. Then the hyphae grow into the intercellular spaces of the root-tissue but leave the apical meristem uninfected. At first the cells of the outer cortical layers have an unchanged shape; but they rapidly become radially distorted as the hyphae penetrate the tissues, and they behave as though they were struggling against the mycelium as it enters.

TABLE VIII

COMPARISON OF MYCORRHIZAL AND UNINFECTED EPIDERMAL CELLS FROM DIFFERENT PARTS OF A SINGLE AXIS OF ROOT OF *Eucalyptus*. Data of Chilvers & Pryor (1965).

	Diameters of epidermal cells in μ			Volume
	Radial	Axial	Tangential	(μ^3)
Uninfected axis	8·5	41·8	15·4	5472
Mycorrhizal axis	23·6	18·6	15·3	4390
Significance	$P > \cdot 001$	$P > \cdot 001$	—	—

Chilvers & Pryor (1965), in their study of eucalypt mycorrhizas, point out that there are both quantitative and qualitative differences between uninfected and mycorrhizal roots. They agree that the quantitative changes may well occur irrespective of fungal infection, and that infection may well be initiated after quantitative modification of the apex. However, certain qualitative changes are, they argue, the resultant of fungal infection. Amongst these they include the thickening of the radial and inner tangential walls of the inner cortex of mycorrhizas. These thickenings form a layer which is not penetrated by the fungus and appears to constitute a defence reaction on the part of the host. A second qualitative change is the radial elongation of the epidermal cells. In Table VIII, the measurements of the sizes of epidermal cells of apically infected axes are given. It is clear that radial expansion is nearly thrice as great in the mycorrhizal zone. This does not in itself confirm that it is wholly a resultant of fungal activity, and indeed in Warren Wilson's preparations a considerable degree of expansion occurred in the absence of infection.

Hence in an adult root system characteristically infected by mycorrhizal fungi, the surfaces of all roots may be colonized by the symbiont which forms wefts and sheaths of greater or lesser thickness and organization, but the full development of mycorrhizas only occurs when the apex tends to the Stage 3 condition of Warren Wilson. A study of the preparations made by both Clowes and Warren Wilson leads to the view that both sequences of the development of mycorrhizas may occur in beech. If we accept, as we must, Warren Wilson's conclusion that Stage 3 short roots become converted into mycorrhizal organs, we are driven also to conclude that the fungus in this case prolongs the life of Stage 3 apices and promotes growth and branching in them. Such a conclusion would resolve many of the difficulties in the interpretation of the behaviour of root apices and agrees fully with the conclusions of other observers (Aldrich-Blake, 1930; Hatch & Doak, 1933; Hatch, 1937; Robertson, 1954) who have been of the opinion that fungal infection prolongs the life of short roots by delaying the ageing of the tissues and promoting further development and branching. Hatch (1937) assembled a great deal of information on this point for *Pinus strobus* and, upon it, built his well-known hypothesis concerning the ecological and physiological action of mycorrhizal roots. It is most elegantly and convincingly demonstrated in his Plates XIII and XVI which are shown in our Plate 4.

These important conclusions will be re-examined later, when the effects of fungal metabolites on root structure are discussed.

Hypertrophy in the Tissues of Mycorrhizas

The results of all these observations on mycorrhizal roots agree with the conclusion first reached by Hatch in 1937, that the surface area of the root system of infected seedlings exceeds that of uninfected seedlings. This increase is occasioned by many changes: first by the prolongation of the life of infected roots and their greater degree of branching; secondly by the addition of the sheath as an external layer; and thirdly by the expansion of diameter due to the radial extension of the epidermal or cortical cells.

There has been some discussion of this third feature, for the cortical cells are sometimes described as hypertrophied. Strictly, the term hypertrophy should only be used where cells or organs are increased in size, i.e. in volume. The change in shape of the epidermal cells in Clowes's preparations involved an increase of radial diameter and a decrease of longitudinal diameter; they were, in adult beech, not hypertrophied. In beech seedlings the equivalent cells described by Warren Wilson were often of increased volume, i.e. the radial expansion was not totally offset by decreased expansion in other diameters.

In Table VIII, the figures obtained by Chilvers & Pryor agree with those of Clowes, for there is a net decrease in volume of the epidermal cells in the *Eucalyptus* mycorrhizas. There is little to be gained by arguing about the use of the term hypertrophy, provided the criteria on which any description is based are fully stated. Increase of rootlet diameter, volume, and surface area, may be occasioned by radial expansion of cells in the epidermis or cortex (*see* Figs. 3 and 5), and this is an important consideration in respect of surface areas for absorption.

The Intensity of Mycorrhizal Infection in Natural Soils

The trees forming ectotrophic mycorrhizas are typically trees of temperate forests which develop on brown earth and podsolized soils. In these forest soils, an aggregation of absorbing roots, mycorrhizas, fungal mycelia, and other living bodies, is found in the upper horizons—especially in the humus layers. The roots of the trees are usually mycorrhizal, so that non-mycorrhizal laterals are few. The proportions of the two types of roots do, however, vary somewhat with soil conditions; mycorrhizas are far more abundant in raw humus and less so in base-rich agricultural soil, as A. B. Frank first pointed out. Quantitative numerical estimates of the change of intensity of mycorrhizal infection with soil type are few—partly because of the difficulty of obtaining suitably representative samples, and partly because of the difficulties of finding a suitable basis on which to score the differences.

An indirect approach was used in the study of beech roots when it had been found that the nitrogen content of uninfected laterals fell far below that of infected laterals. By estimating the nitrogen content of the fine roots from various soils and from various horizons, it was shown that the degree of mycorrhizal development was greatest on the raw-humus sites and least on mull soils (Harley, 1940). In other cases actual counts of subjective groupings have been made for comparison of different sites. Sarauw (1893) observed that the roots of forest trees were heavily infected in forest thickets but were less so in agricultural and garden soil. E. Stahl (1900) decided that the relative poverty of available mineral salts determined the prevalence of mycorrhizas in woodlands. In the epoch-making studies of tree mycorrhizas by Elias Melin (1917–27), it was early (1917) shown that in raw-humus soils the roots of forest trees were entirely mycorrhizal, but in mull soils infection was less abundant. Hatch & Doak (1933) stated that in good forest soils of pH less than 5·5 they rarely observed uninfected short roots, and that long roots were wholly mycorrhizal in these situations. These findings have since been confirmed by almost all observers, and indeed Björkman has made detailed numerical estimates of the numbers and kinds of mycorrhizas in various Swedish soils and has come to identical conclusions.

Most of the experimental work, especially that performed more recently, has made use of seedling roots, and substantially the same results have been obtained. Further studies—also with seedlings—have attempted to determine the factors affecting the intensity of mycorrhizal development. Some have successfully correlated mycorrhizal development with nitrogen availability and nitrogen mobilization in the soil (Melin, 1936; Richards, 1961, 1965; Richards & Wilson 1963). The factor of soil pH stressed by Melin (1923, 1925) may very well work as Richards (1961) indicated through nitrogen availability and perhaps nitrification rate (*see* Lindquist, 1932). Worley & Hacskaylo (1959) have given results showing that mycorrhizal formation in *Pinus virginiana* seedlings is affected by water supply. The frequency was greatest in soil kept moist by frequent watering, and least in dry conditions. In addition, the type of mycorrhizal development was different in moist and dry soils. These results and many others are of importance in particular cases but do not lead to a general hypothesis of mycorrhizal formation. We owe to A. B. Hatch and E. Björkman broad generalizations, based on their experimental work, which have given a great impetus to research.

Hatch (1937) worked with *Pinus* seedlings grown in natural soils and under experimental conditions. Using material of *Pinus sylvestris* grown by Dr. P. R. Gast on a range of humic soils in Sweden, Hatch obtained a positive correlation of intensity of infection with increasing mobilization of soil nitrogen similar to that previously obtained by Melin (1927). In a later experiment, Hatch added a complete nutrient solution in various amounts to a natural forest soil and obtained greatly diminished mycorrhizal infection. The apparent contradiction between these two sets of experiments was resolved by manurial treatments on a range of soils, the results of which are shown in the histogram (Fig. 6). The ridge soil was initially nutrient-deficient and manurial treatment had little influence on mycorrhizal development in it. On the other two soils, great variations in mycorrhizal infection was occasioned by variation of treatment. Hatch considered that balanced or high mineral nutrition of seedlings diminished mycorrhizal intensity and that a low or unbalanced nutrient supply increased it. In the starvation range, however, a condition present in some of his own earlier experiments and in those of Melin and H. Hesselman, a reverse correlation occurred and the small, starved seedlings responded to manurial treatment by greater growth and greater mycorrhizal development.

On the basis of these and other experimental results, Hatch suggested that internal nutrient status, especially with respect to nitrogen, phosphorus, and potassium, was the prime factor in determining the intensity of infection. When the internal concentrations of these elements were high and balanced, the seedlings were resistant to infection by mycorrhizal fungi. As an example of this effect he showed that the roots of seedlings which were allowed to penetrate from a highly manured cartridge of soil into very nutrient-deficient mineral soil outside, were little infected, like those inside the cartridge, and not intensely infected as were the roots of seedlings grown in nutrient-deficient soil. These results of Hatch's explained in a satisfactory manner most of the previous observations by himself and others.

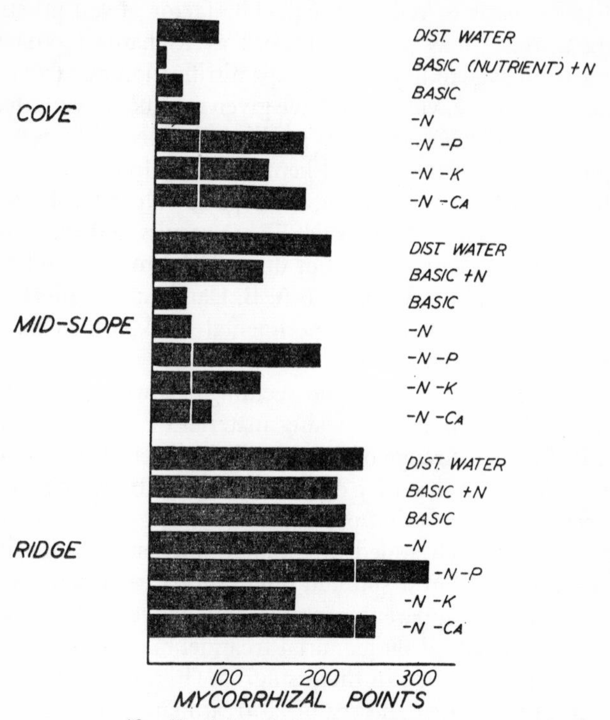

After Hatch (1937). By permission Black Rock For. Bull.

FIG. 6.—Diagram indicating the effects on mycorrhizal development of manurial additions to the soil in the Black Rock Forest. Basic = N, P, K, Ca, Mg, S, Fe; + = Extra N added; – indicates omission of element.

In a few experiments Hatch tested the effect of light intensity upon mycorrhizal development. His results showed that there was an increase in the numbers of infected roots on seedlings in high light intensities which did not represent a general increase of 'mycorrhizal intensity'. The percentage of short roots becoming infected did not increase, but the increase in the number of mycorrhizas was proportional to the increase in the number of short roots.

It should be understood that there are two methods of scoring the degree of infection upon seedling roots. The first method is simply to estimate the percentage of short roots which are mycorrhizal. The second is to determine the percentage of mycorrhizal root-tips present, each branch of each mycorrhiza being counted as a mycorrhizal tip. The total score of short root-tips used for calculation is the number of mycorrhizal tips plus non-mycorrhizal tips. These two methods give results of different value. In both cases the nutrient status affects very significantly the mycorrhizal development, but Hatch only used the first method in estimating the effects of light intensity, and obtained the negative results described.

E. Björkman in Sweden followed Hatch's work with a series of experiments conducted during 1937–42, using in particular seedlings of *Picea abies* and *Pinus sylvestris*. His method of scoring was the second method of Hatch, counting all mycorrhizal tips as separate units (the mycorrhizal points method). Björkman (1942) found that the light intensity in which the plants were grown exerted considerable effects on mycorrhizal development. On all the soils used, increase of light intensity from 6 per cent of daylight to 49 per cent of daylight caused greatly increased infection. In no case were mycorrhizas found at 6 per cent of daylight, but the degree of infection rose sharply up to 23 per cent and more slowly thereafter with increasing light.

Using natural soils, ranging in type from bog soils to mull soils, Björkman also made an extensive study of the effects of added nutrients. Soils, and especially forest soils, which were deficient in available nitrogen, produced seedlings having a high mycorrhizal score. Addition of ammonium salts and nitrates diminished the intensity of infection and increased the growth of the seedlings. On soils rich in available nitrogen but poor in available phosphate, extensive mycorrhizal development also occurred; here the addition of ammonium nitrate did not greatly alter the intensity of infection nor the rate of growth of seedlings. In this case mycorrhizas were decreased by the addition of phosphate, especially at high light intensities. Björkman summed up this group of experiments as follows: 'a severe lack of available nitrogen or phosphorus hampers the formation of mycorrhiza as well as growth, but moderate scarcity of one or other of these nutrients is a condition for mycorrhizal infection.'

In other experiments, the results with phosphate were not quite so clear-cut, for addition of phosphate increased the formation of mycorrhiza in nitrogen-poor soils (e.g. in spruce mor and in a mixture of sand and pine humus) when it was added alone, but decreased mycorrhizas when added with ammonium nitrate. These results are, however, similar to those of Hatch, in that they show that unbalanced mineral nutrition may favour the formation of mycorrhizas.

In contrast to Hatch, Björkman came to the conclusion that the internal nutrient status of the seedling was not of itself the prime factor. He pointed out that, if it is assumed that susceptibility to infection is dependent on the carbohydrate status of the root tissues, his experiments on light intensity and those on nutrient supply can all be explained. Increase of photosynthetic rate, resulting in increased carbohydrate levels, promoted infection. High manurial treatment, especially with nitrogenous fertilizers, by increasing synthesis of proteins, diminished the carbohydrate status; the greater the fertilizer supply, the higher was the light intensity needed to bring about equivalent mycorrhizal development (Fig. 7). If this view were correct it would be expected, as Björkman found in his experiments, that the supply of potassium and other metallic nutrient elements might not greatly affect mycorrhizal development and that phosphate supply would have a lesser—and

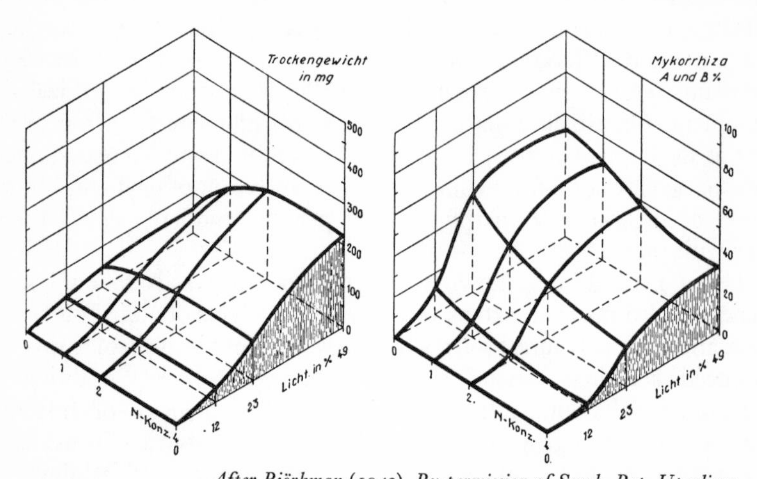

After Björkman (1942). By permission of Symb. Bot. Upsaliens.

FIG. 7.—Diagrams indicating the growth (left) and mycorrhizal development (right) of one-year pine seedlings in various light intensities with various NH_4NO_3 additions to a raw-humus soil.

perhaps more complicated—effect than would nitrogen supply. Björkman showed by analysis that high light intensities resulted in increased reducing power and starch content in the roots of pine, and that reducing sugars were decreased at any given light intensity by addition of nitrogen and phosphate to the soil. Similarly, the partial ringing (strangulation) of seedlings with wire resulted in a reduction in the reducing power of extracts from their roots. All these effects had the expected correlations with mycorrhizal development (Björkman, 1944).

LIGHT AND MYCORRHIZAL DEVELOPMENT

Many lines of experimentation have given results compatible with those of Hatch and Björkman. Hence their views have had an important influence in encouraging further work. But it is not yet fully agreed that the hypothesis of Björkman, which states that infection is primarily correlated with the internal concentration of soluble carbohydrates in the roots, fully explains the facts.

There seems to be little disagreement that, as a general thesis, mycorrhizal intensity is greatest in soils of moderate deficiency of nutrients, especially of nitrogen and phosphorus. Nor is there any reasonable doubt that there must be adequate conditions to promote a suitable rate of photosynthesis. The primary difficulty is in the interpretation of the interaction of these factors of nutrient supply and light intensity, and whether it is their effect on the internal carbohydrates in the roots that is important. It is therefore essential not to interpret those experiments that have shown an undoubted correlation of intensity of infection with light supply as fully upholding Björkman's hypothesis without careful examination of the problem of sugar analysis.

Huberman (1940) gave time-tables of normal growth of spruce and pine seedlings. In these, mycorrhizal development followed the setting of the primary needles. Similarly Robertson (1954) showed that *Pinus sylvestris* formed mycorrhizas only when the primary needles were as long as the cotyledons. The first mycorrhizas of *Fagus sylvatica* were recorded by Harley (1948) and Warren Wilson (1951) as being formed after the first foliage leaves were set, and more recently Boullard (1960, 1961) and Laiho & Mikola (1964) observed the same features in *P. sylvestris*, *P. montana*, and *Picea*. In all these cases there was a correlation of the initiation of infection with the probable onset of active photosynthesis.

Using *Pinus virginiana*, *P. taeda*, and *P. strobus*, Hacskaylo & Snow (1959) studied the effect of duration and intensity of light upon mycorrhizal development in soils of different nutrient levels. Their results with all three species show that mycorrhizal development is most complete at moderately deficient but not starvation levels of nutrient supply and in high light intensities. Even at the highest light intensities, the application of fertilizers above moderate levels reduced mycorrhizal development.

The results of Boullard (1961) on the effects of light on the infection of *Cedrus atlantica*, *Pinus pinaster*, *P. sylvestris*, and other tree species, showed that an increase of light period, i.e. increase of the duration of the daily photosynthetic period, from 6 hours to 16 hours or even longer, increased the development of the root systems and the number of short roots on the seedlings. It also resulted in an increase in the number and percentage of roots converted to mycorrhizas. Essentially similar results were obtained by Wenger (1955) using *Pinus* and by Harley & Waid (1955) using *Fagus*. In the latter case, a cause of complexity was indicated in that black mycorrhizas caused by *Cenococcum* were more plentifully developed in the lowest light intensities than were other forms—a result which agrees with the field observations of many workers. This might explain some of the cases where light intensity and mycorrhizas are not correlated. For instance, in an extensive study of *Fagus* in woodland soil, Warren Wilson obtained variable results. Some seemed to agree with Björkman's hypothesis but others did not, and in general the hypothesis was not upheld.

In most of the foregoing experiments carbohydrate analyses of the roots were not made, and although in many cases the results would fit well with those of Björkman, their full interpretation in terms of his hypothesis is not possible. Both Harley & Waid (1955) and Warren Wilson (1951) carried out sugar analyses on the roots of *Fagus* seedlings in their experiments; but they reached different conclusions. The former obtained the correlations predicted by the hypothesis of Björkman, but the latter did not do so consistently.

Meyer (1962, 1964, 1966) re-examined the problem with beech and spruce seedlings. Using three soils—eutrophic brown earth, podsolic brown earth, and podsol—he obtained again some results which agreed, and some which did not agree, with Björkman's hypothesis. Taking the three soils together, there was a general positive correlation between his estimates of sugar content of the fine roots and the percentage of short roots converted to mycorrhizas. However, the highest sugar content and the highest mycorrhizal development occurred in the nutrient-rich, eutrophic brown earth. Moreover, on this soil the addition of nitrogenous and phosphatic fertilizers did not reduce mycorrhizal development but sometimes even increased it. Meyer concluded that the sugar content of the roots was more likely to be a resultant of the activity of the fungi than an effect of nutrient supply in the soil. In an experiment in which the root systems of single seedlings were divided, so that part was allowed to grow in a podsol and part in brown earth, a higher sugar content and a higher mycorrhizal frequency were found in the latter.

In considering the ways in which the fungus might influence the sugar content of the root system, Meyer pointed out that auxins have been reported to encourage hydrolysis and that some mycorrhizal fungi are known to produce auxins. In this way the fungi might, he suggested, increase the soluble sugar content of infected tissues. He further suggested that the associated soil microflora might also aid this process and that, if they did so, an explanation of the higher sugar content of roots in biologically active soils and soil-layers might be obtained. Furthermore the effect of nitrogenous fertilizers in diminishing mycorrhizal development might also be explained following Moser's (1959) observation that increased nitrogen supply reduced auxin production by mycorrhizal and other fungi in culture.

The thinking of Handley & Sanders (1962) was on somewhat similar lines. They also suggested that mycorrhizal infection itself might alter the internal carbohydrate régime of the root system. They tested this hypothesis by repeating one of Björkman's experiments—using similar seeds, methods of growth, treatment, and sugar analysis—with one important exception: in their experiments the plants were all grown in a non-mycorrhizal condition. Handley & Sanders showed that the internal concentration of 'easily soluble reducing substances' did not increase with increasing light intensity as they did in Björkman's experiments with mycorrhizal plants. Hence they concluded that the correlated increase of soluble reducing substances observed by Björkman was due to an effect of the fungus and not causative to infection.

Richards & Wilson (1963) and Richards (1965) experimented with *Pinus caribaea* and *P. taeda* on soils of different nitrogen availability. They observed an inverse correlation between the degree of mycorrhizal infection and total nitrogen in the roots of seedlings. Richards & Wilson also estimated the content of soluble reducing substances in the root systems. A positive correlation was observed between the extent of infection and the ratio of carbohydrate to nitrogen in the root system. Richards recalculated Björkman's results on this basis and showed that a positive correlation existed with carbohydrate to nitrogen ratio in his results also. Richards pointed out that, interesting as these relationships are, they do not explain mycorrhizal formation in any simple way, although they may be of value in the practical problems of raising forest nursery stock.

It is now becoming quite clear that this whole question of correlating infection with carbohydrate status is dependent on the quality of the methods used in analyses. These have been greatly improved during recent years, especially owing to the availability of procedures for the elimination of impurities which interfere with the estimation of sugars.

Carbohydrate Analysis of Mycorrhizas

Recent advances in methods of carbohydrate analysis, and in their application by D. H. Lewis (1963) to mycorrhizal material, have thrown doubt on the validity of some previous estimates of carbohydrate content. Lewis gave the tabular comparison of the methods of analysis used by previous workers shown on Table IX, which indicates the variety of methods of extraction, clearing, hydrolysis, and estimation, employed. The methods of estimation themselves are not specific to carbohydrates, as Björkman indeed noted, but measure reducing substances. They are not only sensitive in different degrees to various kinds of reducing substances, but the diverse preparatory treatments remove variable quantities of non-sugar reducing compounds. The interpretation of the analytical results may be difficult.

An examination of the reducing substances in the 80 per cent alcohol extract of beech mycorrhizas is shown in Table X, which is derived from the 1965 papers of D. H. Lewis & Harley. The extract was freed from charged substances by resins, cleared by barium hydroxide, and had the free carbohydrates fermented away with yeast by the method of Yemm (1935). Estimates of reducing power were made at each stage. The reducing sugars correspond to about 50 per cent of the reducing power of the original extract. The other reducing substances were of various kinds, some charged and some neutral. Amongst the substances lost on clearing and deionizing were also compounds of carbohydrates such as phenolic

TABLE IX

METHODS USED IN ESTIMATING SOLUBLE REDUCING SUBSTANCES

Reference	Material	Extraction	Clearing	Hydrolysis	Estimation technique
Björkman (1942)	shoots, roots: *Pinus sylvestris*, *Picea abies*	Dried at 80°C., powdered, powder extracted with water at 80°C.	(*a*) CCl_3COOH (*b*) $HgNO_3+NaCl$	0·2N HCl + 0·2N NaOH	(*a*) Ferricyanide reduction (Hagedorn-Jensen) (*b*) Copper reduction (Schäffer-Harttman)
Handley & Sanders (1962)	shoots, roots: *Pinus sylvestris*	,,			Ferricyanide reduction (Hagedorn-Jensen)
Warren Wilson (1951)	shoots, roots: *Fagus sylvatica*	Preserved in 80% alcohol+Na_2CO_3, extraction 3×30 min. 80% alcohol at 80°C.	Lead acetate + sodium oxalate	6% HCl 4 min. at 100°C+ 20% NaOH	Copper reduction (Somogyi)
Harley & Waid (1955)	roots: *Fagus sylvatica*	3 changes of 3 hrs. 80% alcohol+Na_2CO_3	$Al(OH)_3$	10% H_2SO_4	Copper reduction (Somogyi)
Richards & Wilson (1963)	roots: *Pinus taeda*	Dried at 65°C., extracted 76% alcohol		10% H_2SO_4	Copper reduction (van der Plank)
Meyer (1962)	shoots, roots: *Fagus sylvatica*, *Picea* sp.	(*a*) Dried at 80°C., powdered, powder extracted on boiling water-bath 3 hrs.; stand overnight (*b*) Boiling 80% alcohol	$CuCO_3$ freshly prepared from $CuSO_4+Na_2CO_3$ (0·2N)	3 ml. extract +1 ml. 0·1N HCl + NAOH	Copper reduction Nelson-Somogyi colorimetric

and other glycosides. By using the anthrone method of carbohydrate estimation, the quantity of these was determined. About 17 per cent of the total carbohydrate in the crude extract, as estimated by anthrone, was removed in clearing and deionizing and was presumed to be in these glycosidic compounds, i.e. not in free sugars. A second source of complexity is that an estimation of increase of reducing power after acid hydrolysis does not give a good estimate of non-reducing disaccharides in mycorrhizas. The fungal layer contains in the case of beech and very probably in other forms (for it is the common fungal disaccharide), considerable quantities of trehalose, which is resistant to acid hydrolysis and cannot be estimated by any of the methods given in Table IX. In addition, the polyhydric alcohol mannitol is dresent in mycorrhizas and is neither reducing nor does it react with anthrone or other sugar reagents in the conditions normally used. Trehalose and mannitol are both important carbohydrates in beech mycorrhizas, and increase and decrease according to conditions (*see* Chapter V).

TABLE X

REDUCING SUBSTANCES IN THE WATER-SOLUBLE FRACTION OF THE ALCOHOL EXTRACT OF BEECH MYCORRHIZAS

Reducing substances estimated by Nelson-Somogyi method. Extract reduced to small volume under reduced pressure at 40°C. and taken up in water. Deionized with resins, cleared with barium hydroxide and zinc sulphate, fermented with yeast (2·5 per cent suspension) at room temperature for 3 hours.

Values given as μg. of glucose equivalent. From Lewis & Harley (1965).

	μg. per root-tip	Percentage of total
Total reducing substances	67·5	100·0
Charged reducing substances	14·5	21·5
Neutral reducing substances removed by clearing	8·3	12·3
Neutral reducing substances remaining after fermentation	10·7	15·8
Reducing sugars by difference	34·0	50·4

A further difficulty arises from the fact that carbohydrase enzymes occur on the mycelial surfaces of mycorrhizas (Harley & Jennings, 1958; D. H. Lewis, 1963) as they do on other mycelial surfaces (Harley & Smith, 1956; Walsh & Harley, 1962). These carbohydrases are resistant to destruction by moderate heat. Hence those methods using dried and powdered mycorrhizas or root systems for extraction, may involve at least a partial hydrolysis of sucrose and trehalose in the process. The extent of the error will depend on the actual method of extraction and the degree to which the enzymes are inhibited by the procedure adopted.

We must therefore conclude with D. H. Lewis (1963): 'As far as interpretation of correlations is concerned it is therefore important to realise that "easily soluble reducing substances" contain substances other than sugars which themselves may be related to polysaccharide levels. They can have little meaning as a single variable.'

The subject of the carbohydrate physiology of ectotrophic mycorrhizas will be considered in Chapter V when the nutrition of the causative fungi has been considered. Here it is worth stressing that the whole subject of the correlation of the intensity of infection with the carbohydrate level in the roots needs some re-examination. There seems good reason to accept the findings of both Hatch and Björkman, which have been amply confirmed, that moderate deficiency of one or more of the essential major nutrients is an important factor in promoting infection. A reasonable rate of photosynthesis is also essential. Hatch was inclined to suggest that internal nutrient status was the primary factor affecting susceptibility to infection, while the spread of the fungus and its effect on the branching and development of short roots might be dependent on their carbohydrate status. Such a view would explain, with few exceptions, the available experimental results. It would also not conflict in any way with the recent work on root exudates and their influence on the growth of mycorrhizal fungi which will be considered in the next chapter.

IV

THE FUNGI OF ECTOTROPHIC MYCORRHIZAS

THE knowledge of the identity and cultural behaviour of the fungi of ectotrophic mycorrhizas has been gained mainly through the researches of Professor Elias Melin and his associates. Before Melin published his first researches on this subject, the lists of so-called mycorrhiza-formers were based upon two unsatisfactory forms of investigation. The first was the attempted isolation of fungi from infected roots. Because of imperfect technique, especially of sterilizing the surface of the roots, many kinds of root-inhabiting fungi—unspecialized soil saprophytes as well as perhaps some true mycorrhizal fungi—were included. Anyone who has compared the frequency of isolation of a pathogen, such as *Ophiobolus*, from surface-sterilized and unsterilized roots, will be aware that it is a chancy business to obtain an obligate root-inhabiting organism from imperfectly cleaned material. The second method was based on the common field observation that certain species of larger fungi were regularly found associated with one or a few species of trees growing in their natural habitats. The extensive lists given by Peyronel (1921, 1922) were mainly compiled in this way. Their value in later work will be seen below.

The basis of Melin's methods, which have become the pattern for the verification of mycorrhiza-forming fungi, was described by him in 1936. The final test of whether a given fungus is, or is not, a mycorrhiza-former, is an experimental demonstration of its ability to form mycorrhizas with host roots in aseptic culture. Aseptic conditions are always somewhat abnormal for growth of the host and of the fungus, especially in respect of humidity, aeration, light intensity, and rooting conditions. Hence negative results, or infections abnormal in anatomical structure, do not demonstrate that a fungus is unable to form mycorrhizas. Positive results must, in contrast, be accepted as proof of mycorrhiza-forming ability. The pure cultures tested by this method may be derived from two main sources: either from mycelia isolated from mycorrhizal organs, or from mycelia isolated from fruiting-bodies of known taxonomic identity. In the former case, the identity of mycelia may be established by use of Buller's method (1933) of hyphal fusion with taxonomically known mycelia (*see* M. C. Rayner & Neilson Jones, 1944; Modess, 1941). In any event the first guess as to which known fungus might be used for inoculation or for testing, is made by imaginative use of a list of possible species such as those given by Peyronel and by Trappe (1962).

The use of these methods by many workers* has produced an imposing list of fungi known to be able to produce ectotrophic mycorrhiza with forest trees. The majority of these fungi belong to the Basidiomycetes—particularly to the families Agaricaceae and Boletaceae. A few Gasteromycetes, e.g. *Scleroderma aurantium*, *Rhizopogon roseolus*, and *R. luteolus*, have also been shown to form mycorrhizas with conifers; a few Ascomycetes are possibly mycorrhizal, e.g. *Gyromitra*, while the imperfect type *Cenococcum graniforme* (*Mycelium radicis nigrostrigosum* Hatch) has a wide host-range in gymnosperms and angiosperms. More than seventy species of Basidiomycetes have been proved to form ectotrophic mycorrhizas, and these belong to about twenty genera; but three genera alone, *Amanita*, *Boletus*, and *Tricholoma*, together comprise the majority. It seems highly probable that the genera *Russula*, *Lactarius*, and *Cortinarius*, species of which are difficult to grow in culture, as well as perhaps others, may include many more mycorrhiza-forming species than is at present believed. Melin (1959), Lobanow (1960), and Trappe (1962), give detailed lists. Those of Trappe are extremely valuable because they include both a host index and fungi which are probable but not proved mycorrhiza-formers.

By far the majority of aseptic experiments have been carried out with seedlings of conifers. The seeds of *Larix*, *Pinus* and *Picea*, as well as of other Pinaceae, may be surface-sterilized by suitable treatment. The angiosperm host-trees, especially the large-seeded *Fagus*, *Quercus* and *Castanea*, have seeds which are difficult to sterilize, and within which contaminants are frequently found between the convoluted cotyledons. The lists of proved mycorrhiza-forming fungi therefore contain fewer species which are known to form composite organs with broad-leafed trees. With this reservation, and also bearing in mind the fact that the full host-range of none of the fungi is accurately known, the apparent specificity of the fungi may be discussed.

The most unspecific of all the mycorrhizal fungi of forest trees seems to be *Cenococcum graniforme*, which forms a kind of mycorrhiza that is easily recognized by its dark colour, sheath structure, and radiating hyphae. These mycorrhizas have been found on the root systems of numerous species of conifers and angiospermous trees belonging to about twenty different genera. The greatest degree of specificity is shown by *Boletus elegans*, which is only known to associate with *Larix*. Between these extremes there are many fungal species of intermediate host-range.

The lack of close specificity of most mycorrhizal fungi to a particular host species is paralleled by the reactions of hosts to the fungi. Most host species seem capable of forming mycorrhizas with several fungi, and on any given root system more than one type of mycorrhiza may be

* E.g. Melin (1923, 1923*a*, 1924, 1925); Hatch & Doak (1933); Hatch & Hatch (1933); Modess (1941); M. C. Rayner & Levisohn (1941); Santos (1941); Norkrans (1949); Hacskaylo (1951, 1953); Hacskaylo & Palmer (1955); Lobanow (1960), etc.

observed, i.e. a single host may associate with members of several fungal species at one time. For instance, Melin (1963) in his list of proved mycorrhiza-formers gives more than 40 species capable of infecting *Pinus sylvestris* alone.*

The relative ease with which many of the mycorrhizal fungi can be grown in pure culture indicates that they are not strictly obligate symbionts. It does not indicate, however, whether or not they are obligately dependent on association with roots under natural conditions. Other mycorrhizal fungi are, by contrast, extremely selective in their nutrition; they have only been grown successfully in media to which metabolites of living roots have been added, and so are probably obligately symbiotic. There is indeed a likelihood that many of the terrestrial Agaricaceae and Boletaceae which have so far proved impossible to culture may be amongst the most specific and the most highly adapted mycorrhizal fungi; this is a worthwhile field of research.

The regular association of the fruit-bodies of certain species of Agaricaceae and Boletaceae with particular forest-tree species has already been mentioned. Many of these associates have been proved to be mycorrhiza-formers. At times, when trees are found outside their usual ecological orbit, their mycorrhizal fungi are found with them, producing fruit-bodies around them. Plantations of exotic species, as they develop woodland or forest conditions within the stand, include in their community their usual entourage of fungi, including mycorrhizal fungi, as is shown by the appearance of their fruit-bodies at the proper season.

In 1938 and 1939, L. G. Romell carried out his celebrated trenching experiments which seemed to show that the potentiality of mycelia of certain mycorrhizal fungi to form fruit-bodies depended upon their being in direct connexion with tree roots. In these experiments, trenches were cut to sever the root systems of spruce. In the course of the observations it appeared that mycorrhizal fungi failed to 'fruit' within the trenched areas where there were no active host-roots attached to their parent trees. This dependence of mycorrhizal fungi on association with host-roots is not quite definite and clear-cut.

M. C. Rayner & Neilson Jones (1944) have reported the widespread occurrence of rhizomorphs and hyphal strands of *Boletus bovinus* in heath soil, in the absence or rare occurrence of host trees. In Romell's experiments, *Boletus subtomentosus*, a proved mycorrhiza-former of *Pinus montana* (Modess, 1941), formed fruit-bodies when isolated by trenching from active host-roots. Again, Mikola (1948*a*) showed, as Ferdinansen & Winge had observed before (1925), that *Cenococcum graniforme* is widespread even in soils where no known host for it is living. Clearly here, as we shall find also in other characters, the proved mycorrhiza-forming species show a fairly wide range of behaviour. Even if, as is probable, the majority are greatly dependent on symbiotic living for the production of fruit-bodies, it is important to know exactly the way in

* *See* Appendix §4.

which the vegetative mycelia behave in the soil. If the mycelium is very localized in the root region, or is little developed except on the root surface, we will have to come to different conclusions about its mode of absorption of nutrients from those that we may come to if it is capable of spreading widely.

The difficulty of identifying the mycelia of Basidiomycetes in the soil is one of the obstacles in the way of studying their ecological activity. The only certain ways of deciding whether mycorrhiza-forming species are present is to find that susceptible hosts become infected, or that fruiting-bodies are formed above ground. These two events are usually concurrent. In many cases where trees have been planted in soils believed to be free from mycorrhiza-forming species, it has been observed that mycorrhizal development occurs spontaneously. We can be reasonably confident that the source of infection is not carried in the tissues of the embryo, and propagules are not commonly found in the seed-coat, so that in such cases either vegetative mycelia or dormant spores must have been present in the soil, or wind-blown spores must have infected the young plants (Rosendahl & Wilde, 1942; McComb, 1943; M. C. Rayner & Neilson Jones, 1944; etc.).

A particularly interesting series of observations upon this point was made by Dimbleby (1953), who examined regeneration of birch on moorlands in Britain. Here birch seedlings or saplings were observed to form compact groups, often in irregular lines radiating from a centre. On examination these groups were found to be arranged around old, buried tree stumps, which were shown to be the relics of old pine woodland destroyed some 150 years previously. The birch was rooting chiefly in and around the decayed roots, especially where humus pipes occurred. In these humus pipes, extensive mycorrhizal development was observed. There were three correlated events. First, vigorous birch regeneration occurred only in sites where humus derived from old roots existed; secondly, dense mycorrhiza formation occurred here only; and thirdly, fruit-bodies of *Boletus scaber* and *Amanita muscaria*, both mycorrhiza-formers with birch, were found in these places only. It must be concluded either that the mycelia of these fungi had persisted in the peculiar habitat of decaying roots in an active vegetative state, or that these sites afforded a good place for spore germination as well as for seed germination of birch.

On Wareham Heath, *Boletus bovinus* in a non-virulent form was proved by M. C. Rayner & Levisohn (1941) to be present in the soil before trees were planted. When soil amendments were made by compost addition, mycorrhizal infection occurred, and this was ascribed to a change in virulence of this *Boletus*.

The mycorrhizal Basidiomycetes, like other members of the same group of fungi, have a prolific output of spores. In any region where fruit-bodies are formed, spores carried by air currents are likely to be dispersed everywhere. The most important species are very widely distributed and a study of the Agaricaceae and Boletaceae of Europe, or indeed of the whole north-temperate region, will show how wide a range many species have. This, combined with the lack of close specificity of any species to a single host species, leads to the conclusion that potential mycorrhizal fungi are likely to be available as spores to almost all host-trees.

It is now certain that infection can take place from spores. This has been well demonstrated by Robertson (1954) in a series of experiments with *Pinus sylvestris.* In his first experiments, pots of sterile compost were exposed in pine woodland to the natural spore 'rain' during autumn and winter, whilst others were kept some distance away in a laboratory in Cambridge, England. In the following spring all the pots were placed in similar conditions and sown with seeds. The seedlings were harvested in the following August. Out of 736 seedlings in exposed pots 89 per cent were infected, and out of 596 seedlings in unexposed pots only 22·5 per cent were infected. The significant rise of infection in the pots exposed in woodland to the normal spore 'rain' was tested in a later experiment when especial care was taken to exclude spores from the controls. In this case all the unexposed pots produced 100 per cent uninfected seedlings. In the exposed pots there were 23·7 per cent of infected seedlings.

During the course of these experiments it was noted that seedlings subjected to a spore 'rain' developed local centres of infection on their root systems. As growth proceeded, these local hyphal wefts spread and a more general infection developed. The results are given in Table XI, which shows the numbers of centres of infection developing on seedlings in pots naturally exposed to spore 'rain' as compared with the controls, and leave no doubt as to the efficacy of spores as infective agents.

TABLE XI

DEVELOPMENT OF CENTRES OF MYCORRHIZAL INFECTION IN POTS OF *Pinus sylvestris* SEEDLINGS EXPOSED AND UNEXPOSED TO A 'RAIN' OF *Boletus granulatus* SPORES. Data of Robertson (1954).

	Control	Exposed
Total number of pots	43	43
Number showing centres of infection	9	40
Total number of centres	15	292
Number of centres per pot	0·37	6·8

The ease with which spores may cause mycorrhizal infections contrasts with the difficulty of germinating them in artificial media and is itself evidence of the dependence of mycorrhizal fungi on their hosts.

Robertson's results explain some of the failures which have troubled experimenters who have attempted to raise uninfected and infected seedlings in adjacent nursery beds, and also the difficulties of others who have found that mulching the soil around pine seedlings, with pine needles from woodland, has led to mycorrhizal development. Any treatment which exposes seedlings to a source of suitable spores, may be expected to lead to infection if other conditions are right. Mycorrhizal infection, once established on the root system even as a local area, will spread.

The spread of infection was further examined by Robertson, who compared the relative efficiency of chopped mycorrhizal roots and of infected seedlings interplanted with uninfected stock, as a source of mycorrhizal inoculum. He obtained a clear demonstration that chopped mycorrhizas were hardly more effective, as a source of inoculum, than the casual sources of contamination in an experiment of this kind. Infection, however, spread from the roots of the infected seedlings to give a high percentage of infection in the stock.

The spread of infection from infected seedlings has been observed by Mitchell *et al.* (1937) in their experiments in nursery beds, and had been used by Roeloffs (1930) as a method of ensuring that mycorrhizal development occurred in exotics planted in the Dutch East Indies. Its occurrence raises an important set of problems. The ready spreading of mycorrhizal fungi from living mycorrhizas, in contrast with their slow or negligible rate of spreading from chopped tissue, is a property that they hold in common with obligate root-infecting parasites. It is common to find that the obligate root parasites persist in the remains of host tissues, leading a precarious existence unless a living root grows into contact with them, in which case they infect it. *Fomes annosus*, which causes butt-rot of pines, shows this characteristic, and its period of persistence in old host-roots depends upon habitat conditions and especially upon the type of soil. It is also capable of infecting newly felled stumps by spores. There is a possibility that the apparent persistence of the mycelia of *Boletus scaber* and *Amanita muscaria* observed by Dimbleby is a parallel to that of *Fomes annosus*, but it remains to be determined whether those fungi persisted in the humus pipes as mycelia rather than as spores.

One point has become strongly apparent in considering the available information on the ecology of mycorrhizal fungi, and that is our great ignorance of the state in which they live in the soil. Most of those who have worked on mycorrhizal plants hold definite views on this subject. But these views are largely based on surmise. We know very little indeed about the behaviour of the mycelia of Basidiomycetes in their natural habitats and, apart from the few wood-destroying and coprophilous species that have been studied in detail, their vegetative activities are very little known. It will be seen from the following sections, on the physiology of mycorrhiza-formers, that they have properties which are so characteristic that we may hazard a guess that their vegetative activity is localized in the immediate environs of their host-roots. Here is a whole, almost untouched, subject which is of first importance to the interpretation and explanation of the formation of ectotrophic mycorrhizal organs. Russian workers (e.g. Kljusnik, 1952), in their investigations of oak mycorrhiza, have not only added *Scleroderma aurantium*, *Hebeloma crustuliniforme*, *Pisolithus arenarius*, and *Tricholoma lascivium*, to the number of known mycorrhiza-formers, but have also reported something of their behaviour. The fungi, as observed by the buried-slide technique, grew normally in moist soil, but under dry conditions became broken up into chlamydospores. Somewhat later, Levisohn (1955) showed that the three mycorrhiza-forming fungi, *Boletus bovinus*, *B. scaber*, and *Mycelium radicis nigrostrigosum* (*Cenoccum graniforme*), may be found in soils as hyphal strands or rhizomorphs. She succeeded in obtaining pure cultures of these from soil by placing washed fragments of the rhizomorphs on nutrient agar. Such observations give us a beginning to some knowledge in this area of regrettable ignorance.

The Nutrition of Mycorrhizal Fungi in Culture

The basidiomycetous fungi which have been shown to form ectotrophic mycorrhizas have been studied fairly extensively in pure culture. The results of these investigations show that, although there is considerable variation, the known mycorrhiza-formers do comprise a type of Basidiomycete which contrasts in behaviour with the lignicolous and litter-decaying forms.

In assessing the value of the results obtained in pure culture, several limitations of the methods must be appreciated. It may, as a general rule, be assumed that the results apply in the broad sense to the behaviour of the fungi in association with their hosts; but one may reasonably expect that there will be some deviation in detail, perhaps in important detail, on account of the influence of the host in the provision of a special habitat, special substances, or special conditions. There is, moreover, considerable variation, in both physiological and morphological characters, between strains isolated from different sources. Again, the value of particular cultural idiosyncrasies, in the sense of their being of adaptive significance, cannot really be assessed. It may be that the special root-surface habitat of the mycelium does not select against such features as vitamin or growth-factor dependence. It has been shown that, in general, mycorrhizal fungi have many properties in common with other specialized rhizosphere and root-surface organisms, and we cannot yet see what characteristics they possess which allow them to form their peculiar root-fungus organs.

Carbon Sources

The ectotrophic mycorrhizal fungi as a group are incapable of lignin decomposition, and few are able to cause the breakdown of cellulose.

They therefore form a third group amongst the Basidiomycete species inhabiting the litter layers of forests: the white-rots causing a breakdown of cellulose and lignin, the brown-rots attacking cellulose and hardly affecting lignin, and the mycorrhizal fungi dependent on simple carbohydrates and having little or no effect on cellulose or lignin.

E. Melin in 1925 showed that the mycorrhizal fungi which he had then investigated made their greatest growth in culture upon glucose. More complex carbohydrates were less useful as carbon sources, although some growth was made on the alcohol mannitol and on starch. These results have been generally confirmed, for instance by How (1940) for *Boletus elegans*, by Mikola (1948*a*) and Keller (1950, 1952) for *Cenococcum graniforme*,

TABLE XII

RELATIVE GROWTH OF SEVERAL *Tricholoma* SPECIES UPON VARIOUS CARBOHYDRATES

T. brevipes and *T. nudum* are not mycorrhizal; *T. flavobrunneum*, *T. imbricatum*, *T. pessundatum*, and *T. vaccinum*, are typical mycorrhiza-formers. *T. fumosum* is a mycorrhiza-former with powers of cellulose destruction. Data from Norkrans (1950).

Carbohydrate	*T. brevipes*	*T. nudum*	*T. fumosum*	*T. flavobrunneum*	*T. imbricatum*	*T. pessundatum*	*T. vaccinum*
Glucose	100	100	100	100	100	100	100
Fructose	79	104	86	71	55	120	22
Mannose	85	85	103	73	121	78	74
Galactose	25	30	28	13	7	8	6
Sucrose	18	112	103	47	17	6	20
Maltose	36	104	67	21	12	23	11
Lactose	17	104	15	3	2	2	1
Raffinose	7	94	97	2	2	2	1
Xylose	4	71	7	57	7	22	1
Starch	18	136	93	11	9	72	13
Starch + tr. glucose	54	133	83	93	45	117	50
Inulin	11	92	105	3	2	1	2
Inulin + tr. glucose	44	101	107	49	25	13	17
Cellulose	+	+	+	—	—	—	—

and by Norkrans (1950) for several *Tricholoma* species. The dependence of mycorrhizal fungi upon simple carbohydrates is not so complete as was once supposed. *Cenococcum* has, for instance, the ability to utilize starch in the presence of glucose, and may make good growth on cellobiose and other disaccharides. Similarly, Norkrans's work (1949) has shown a wide range of behaviour amongst the mycorrhizal *Tricholoma* species. Some are capable of using the simple carbohydrates only; *Tricholoma flavobrunneum* utilizes cellobiose; *T. vaccinum* uses cellulose in the presence of glucose; *T. fumosum*, undoubtedly mycorrhizal, can utilize lignin and cellulose to a significant extent (other details in Table XII).

A correlation of the formation of extra-hyphal polyphenol oxidases with the ability to cause breakdown of lignin and cellulose is usually observed in Basidiomycetes. Those forms which are only able to utilize simple carbohydrates usually release no polyphenolase, or, at most, traces of the enzyme system, into the medium. Strong lignin-destroyers such as *Polystictus versicolor* produce the enzyme in great quantity. It is not surprising therefore, that the majority of mycorrhizal fungi do not possess this property of breaking down lignin and cellulose.

Lindeberg (1948) made a survey of the ability of mycorrhizal fungi to produce extra-hyphal polyphenolase and found two particularly striking exceptions. He observed that *Boletus subtomentosus*, a mycorrhiza-former with *Pinus montana*, and *Lactarius deliciosus*, which forms mycorrhizas with *P. sylvestris* and *Picea abies*, produced the enyzme as vigorously as many wood-destroying species. Lindeberg was so struck with the extra-cellular polyphenolase production of *B. subtomentosus* that he confirmed by experiment that it caused a rapid breakdown of leaf-litter with a loss of 49 per cent of the weight of the culture medium in 185 days. It will be remembered that *Boletus subtomentosus* was one of the few mycorrhiza-formers which appear to be able to 'fruit' in the absence of host-roots.

Lyr (1963*a*) examined the ability of six mycorrhizal fungi to produce exo-enzymes in culture, and compared them with non-mycorrhizal litter- and wood-destroying fungi. He was able to demonstrate the presence of cellulase, xylanase, amylase, and a proteinase, but not pectinase, in the culture media of all the species that he used. The quantities were in general very much smaller in the mycorrhizal fungi than in others, but they showed some variation. Lyr concluded that *Boletus subtomentosus*, *B. luteus*, *B. variegatus*, and *Amanita citrina*, might be able to use cellulose to some extent, but that *Boletus badius* and *Amanita muscaria* could not.

The inability of most mycorrhiza-formers to utilize complex polysaccharides shows that we cannot generalize about them in one direction. We cannot assume on present evidence that they have an important influence on their hosts by virtue of their ability to cause rapid breakdown of humus and plant debris, for most of them do not display this property at all. On this point we must, however, keep an open mind, for in all forms the Hartig net penetrates the middle lamella of the cells, and some degree of intracellular penetration occurs in almost all mycorrhizal roots. The so-called ectendotrophic mycorrhizas exhibit a considerable frequency of penetration of hyphae through cell-walls, and it has been suggested by Melin (1948, 1953) and by Norkrans (1950) that these penetrations may be effected by mycorrhizal fungi which are able to produce cellulases that might come into operation when the available glucose in the root is diminished. Other observations make it seem possible also that the dependence of mycorrhizal fungi upon simple sugars in culture results from the artificial conditions of the culture and is not operative when the fungus is attached to the root. The suggestion of Clowes (1954), for instance, that the mycorrhizal sheath of beech digests the root-cap, might imply that the fungus can produce cellulolytic enzymes when growing in the symbiotic state. We have nevertheless no evidence at all that the mycorrhizal sheath and outgoing hyphae which penetrate the humus have as a general rule any powers of digestion of complex carbon compounds.

It is possible, however, that the fungus interferes with the normal synthesis of cell-wall material during the development of the cells of the epidermis-cap complex of the root, and that the simple precursors of cellulose are utilized. Such a mechanism would explain the facts in the case of mycorrhiza, and afford a parallel to the similar explanation of 'leakiness' in lichen algae. This possibility will be considered in the next chapter where it will also be shown that the fungus absorbs photosynthates from its host.

Nitrogen Sources

The nitrogen intake of mycorrhizal fungi does not differentiate them from other Basidiomycetes (*see* Rawald, 1963). In common with many organisms of this group, they make better growth upon ammonium compounds and salts of organic and inorganic acids than upon nitrates. Several scraps of information suggest that some mycorrhizal fungi have limited powers of nitrate reduction both when free and when attached to roots, but this matter needs further study. Amides such as those of glutamic and aspartic acid, amino-acids, urea, nucleic acids, and protein hydrolysates, are suitable sources of nitrogen for some species, as was shown by Melin and his colleagues and pupils. In some cases not only have organic nitrogen compounds been found to be better sources of nitrogen than ammonium salts, but amino-acids in small quantities have been found to exert a growth-stimulating effect. Some evidence is available that the deficiency of amino-acid synthesis, which appears to be present in the metabolism of many forms, is not entirely due to defects in the synthesis of these compounds themselves. Norkrans (1953) not only found, with *Tricholoma* species, that glutamic acid caused stimulation of growth, but that α-ketoglutaric acid could also stimulate; moreover, other acids of the tricarboxylic acid cycle possessed this property. Mikola (1948) and Melin (1953) showed a similar situation with several strains of *Boletus* and with *Cenococcum graniforme* and *Rhizopogon roseolus*. In the presence of ammonium phosphate, both glutamic acid and α-ketoglutaric acid caused stimulation of growth, but excess glutamate caused its inhibition.

These stimulating effects are undoubtedly complex and may be due to deficiency of synthesis of particular amino-acids, as evidenced by the stimulation of growth by casein hydrolysate exceeding all others; or they may be due to the provision of tricarboxylic acids of the Krebs cycle or closely related compounds easily assimilated into the metabolic system. They may also be due to glutamyl radicles which are compounds from which trans-amination and protein synthesis may be accomplished.

Such kinds and degrees of dependence upon the provision of complex metabolites are not the sole property of mycorrhizal fungi but are common in fungi and bacteria. What is remarkable is that they should exist so regularly in a group of species and strains isolated from one set of natural habitats.

Accessory Substances and Growth Factors

Most mycorrhizal fungi are also characterized by being dependent in culture on the presence of one or more vitamins or organic growth-factors. All those then tested, as Melin stated in 1953, were partially or completely dependent on thiamin or one of its constituent moieties, pyramidine or thiazole (Table XIII). But within any one species, grades of dependence may be found—as Melin & Nyman (1940, 1941) have

TABLE XIII

EFFECT OF THIAMIN AND ITS CONSTITUENT MOIETIES ON THE GROWTH OF VARIOUS MYCORRHIZAL FUNGI. RELATIVE GROWTH-RATES ARE GIVEN. Data from Melin (1954).

Fungus	Without addition	+ Thiazole + Pyramidine	+ Thiazole	+ Pyramidine	+ Thiamin
Amanita pantherina	2	10	2	3	11
Boletus granulosus	0·3	10	0·3	1	10
B. luteus	1	10	5	2	8
B. variegatus	0·1	10	0·1	0·2	10
Paxillus panuoides	0·2	10	0·4	0·2	10
Cortinarius glaucopus	0·1	0·1	0·1	0·1	10
Tricholoma albobrunneum	5	10	5	8	12
T. fumosum	2	10	1	1	5
T. imbricatum	2	10	2	5	8
T. pessundatum	2	10	2	10	10
T. vaccinum	1	10	2	5	8

shown for *Boletus granulatus*—and lately there have been reports of a few strains being autotrophic for thiamin (*see* Melin, 1963). Requirment for other vitamins has frequently been reported, such as for pantothenic acid (*Tricholoma imbricatum*), nicotinic acid (*Tricholoma fumosum* and *Lactarius deliciosus*), biotin (*Cenococcum graniforme*), and inositol (*Rhizopogon luteolus*).

Unidentified stimulating substances have also been found in the secretions of germinating pine seeds (Melin, 1925), yeast extract (Melin & Lindeberg, 1939), newly-fallen leaf litter (Melin, 1946), and root secretions (Melin, 1954; Melin & Das, 1954).

A particular study has been made by Melin and his group of the effects of the exudates of living roots upon the growth of mycorrhizal fungi in culture (Melin, 1959, 1963). This is of great importance because it already affords a partial explanation of the reaction of particular fungi to their hosts, and it foreshadows possibilities which may lead to the resolution of the problems of the initiation of infection. It also links mycorrhizal researches with those upon the influence of root exudates on rhizosphere populations of plants, which were discussed in Chapter II.

TABLE XIV

EFFECT OF ROOT EXUDATES OF PINE IN THE PRESENCE AND ABSENCE OF THIAMIN AND OTHER B VITAMINS UPON THE GROWTH OF VARIOUS MYCORRHIZAL FUNGI. RELATIVE GROWTH-RATES ARE GIVEN. Data from Melin (1954).

	+ Thiamin		+ Amino-acids and B vitamins	
Fungus	— root exudate	+ root exudate	— root exudate	+ root exudate
Amanita muscaria	31	56	44	100
A. pantherina	—	—	27	100
Boletus bovinus	—	—	42	100
B. edulis	—	—	26	100
B. elegans	—	—	34	100
B. subtomentosus	41	60	66	100
B. variegatus	14	64	20	100
Cortinarius glaucopus	—	—	7	100
Pholiota caperata	—	—	6	100
Rhizopogon roseolus	8	55	16	100
Russula xerampelina	0·2	100	1	100

Mycorrhizal fungi growing in full nutrient solutions to which B vitamins and amino-acids and other essential substances were added,* were stimulated by aseptic addition to the cultures of living excised roots or their exudates or of exudates of attached pine roots. The roots of pine and of many herbaceous plants (not forming ectotrophic mycorrhizas) brought about this stimulation of growth whether they were present in a free condition or enclosed in celluloid bags in the medium. All the mycorrhizal fungi tested showed some increase of growth, due to the diffusable substances. Certain species showed a relatively small stimulation, while others, such as *Russula xerampelina*, made no growth without the exudate (Table XIV). Melin called the stimulating principle the M factor. Measurements of the progress of growth in the presence of the M factor showed that certain species grew faster than the controls in the

* Thiamin, pantothenic, folic and *p*-aminobenzoic acids, pyridoxin, riboflavin, biotin, choline, inositol, glycocoll, alanine, valine, leucine, isoleucine, phenylalanine, tyrosine, tryptophan, glutamic and aspartic acids, proline, hydroxyproline, serine, threonine, cystine, methionine, arginine, histidine, and lysine.

early stages, but later, as the size of their mycelia increased, both grew at a similar rate. Examples of this kind of fungus were *Rhizopogon roseolus* and *Boletus variegatus*. By contrast, *Russula fragilis* and *R. xerampelina* were totally dependent on the M factor over a period of 80 days, and the control growth-rate was very small throughout that time.

For quantitative work Melin standardized the doses of the M factor in arbitrary units. One unit represented the quantity released into distilled water from 1 mg. of living pine root in 6 days at 4°C. Using *Boletus variegatus* as a test organism, it was observed that although doses of between 5 and 10 units in 50 ml. increased growth by some 50 per cent, further doses up to 20 units caused no further stimulation and indeed doses of 40 units caused an inhibition of growth down to 30 per cent of the control. The explanation of these varied responses to different doses lay in the fact that the exudate contained, in addition to the M factor, a growth inhibitor. Different parts of the root system of *Pinus sylvestris* were observed to release different proportions of stimulatory and inhibitory principles. The main secondarily-thickened axes produced chiefly those that were inhibitory, and the ultimate rootlets those that were stimulatory. This localization of exudation is reminiscent of similar results described in Chapter II during a consideration of the effects of exudates upon rhizosphere organisms.

In addition to a diffusible principle, Melin also showed that indiffusible stimulatory substances existed in pine roots. Not only did hyphae, stimulated to grow in the environs of roots, enter the tissues and colonize the cortex, but also after six days' exudation in water, roots could still yield a stimulatory principle on extraction with hot water. Melin (1959) has reported that both the diffusible and indiffusible M factors are inhibited in their effects by adenine and its derivatives, and this led him to suggest that they are related chemically to such substances, though little is known of their nature. It may be assumed that they are not gibberellins, because Santoro & Casida (1962) report that these inhibit a number of mycorrhizal fungi. It will be of interest to obtain an amplification of the report (Melin, 1963) that H. Nilsson has obtained evidence that the effect of the M factor on hyphal growth can be replaced by diphosphopyridine nucleotide.

The nature of root exudates of *Pinus* seedlings has been studied in some degree by Slankis (1958). *Pinus strobus* seedlings were allowed to assimilate $^{14}CO_2$ in the light and the exudates from their roots, produced under aseptic conditions, were analyzed chromatographically and electrophoretically. Sugars were exuded in relatively small quantities; glucose and arabinose were identified. In addition, relatively large quantities of amino-acids and especially of organic acids were released. Further investigation of these substances and possible changes in their quantity and quality with the cultural conditions under which the seedlings are grown, will yield information that should be valuable for the interpretation of the effect of external factors on mycorrhizal development.

The spore germination of mycorrhiza-forming Basidiomycetes is frequently also dependent upon the presence of factors of an organic nature. The percentage germination which occurs is always very low when the spores are set to develop in distilled water or in pure culture solutions. In this respect, once again, the majority of mycorrhizal fungi differ from the majority of lignin- and cellulose-destroying fungi whose spores germinate more easily. Fries (1941, 1943) has shown that successful germination may be obtained by growing colonies of yeasts, such as *Torulopsis sanguinea*, and other non-spreading micro-organisms, on the agar upon which the spores are set to germinate. In this way he successfully germinated and grew a number of monospore cultures of several species of mycorrhizal Basidiomycetes. The presence of some other fungi had similar effects. Thus *Cladosporium* encouraged the germination of spores of *Lycoperdon umbrinum*, while *Mycelium radicis atrovirens* and *Cenococcum graniforme* stimulated into action the spores of species of *Boletus*. These effects, although they could not be brought about by pure vitamins instead of the growing cultures, could sometimes be produced by the use of fruit-body extracts.

Root exudates, including the M factor, may influence spore germination. Melin (1959) showed how the spores of some species of *Russula*, which fail to germinate in nutrient media, will do so and grow to form colonies in the presence of living roots. The spores of two species, *Russula adusta* and *R. rosacea*, germinated only with pine roots, whereas those of two others were stimulated also by tomato roots. These observations help to explain the ease with which some spores seem to act as sources of inoculum in mycorrhiza formation (Robertson, 1954). Further work is needed to determine why other mycorrhizal species do not germinate in contact with roots in culture.

In order to put these valuable researches on root exudation in perspective, it is perhaps necessary to emphasize that the stimulation of hyphal growth by the living roots of potentially mycotrophic hosts is not specific to mycorrhizal fungi. As Lundeberg (1960, 1963) has shown, other species of fungi may also be so stimulated. Further consideration of these complexities will be given later, when the effect of fungal metabolites on root growth is discussed.

Substances of unknown composition, which stimulate mycorrhizal fungi or alternatively inhibit their growth, have also been shown to be present in extracts of leaf-litter and humus (Melin, 1946; Mikola, 1948). The effects of these extracts are complex because they are complex mixtures of organic and inorganic substances; they contain nutrients, ash compounds, accessory growth-factors, and inhibiting substances. Melin demonstrated something of the effects of various constituents. Incinerated humus extract caused some stimulation of growth, possibly owing to the calcium it contained; for calcium is required in considerable amounts by many Basidiomycetes (*see* Treschow, 1944; Lindeberg, 1944).

Even in the presence of an excess of the ash constituents, a further increase of growth was caused by the addition to the medium of water extracts of humus. This was ascribed to the presence in the extracts of organic substances which stimulated growth, perhaps acting as vitamins or growth factors do. In the presence of high concentrations of extracts of humus there was, in addition, an inhibition of growth. Large doses of cold-water extracts depressed the growth of mycorrhizal fungi but had little effect on free-living litter-decomposing Basidiomycetes.

Extracts of *Calluna* peat have been found by W. R. C. Handley (1963) to be particularly inhibitory to mycorrhizal fungi. *Boletus bovinus*, *B. variegatus*, *B. elegans*, *Amanita muscaria*, *Rhizopogon roseolus*, and *Mycelium radicis sylvestris*—all proved mycorrhiza-formers—were almost totally inhibited by dilute *Calluna* extracts. The inhibiting principle in these peat extracts was destroyed by the action of some saprophytic micro-organisms. Hence if the extracts were inoculated with unsterile peat and incubated for a few weeks, growth of mycorrhizal fungi upon them became possible. Amongst the mycorrhizal fungi used by Handley in these experiments, one, *Boletus scaber*, was relatively insensitive to *Calluna* peat extracts. This fungus is one of those shown by Lindeberg to produce small quantities of extracellular polyphenolase, and it is also capable of forming mycorrhiza with birch on *Calluna* moorland (Dimbleby, 1953).

The explanation of the usual sensitivity of mycorrhizal fungi to peat and humus extracts may lie in the presence in these extracts of tannin and phenolic compounds or their oxidation products. The non-mycorrhizal Basidiomycetes of the litter layer which form extracellular polyphenolases are relatively insensitive to the toxic action of tannin and phenolic compounds that are oxidized by the polyphenolases which they release. Whatever the actual inhibiting action, the sensitivity of mycorrhizal fungi to humus extracts emphasizes again their specialized nature.

Competition with Other Organisms

The sensitivity of mycorrhizal fungi to competition with other soil organisms gives rise to one of the main difficulties in isolating them from root tissues. Where an elaborate medium, reinforced with pure vitamins and amino-acids, and provided with trace elements, is used, it readily supports the growth of a great variety of fungi and bacteria. The mycelia of mycorrhiza-formers are readily overgrown by competitors in such conditions, and they seem to be intolerant of competition from soil saprophytes. In this respect, as in many others, the mycorrhizal fungi have properties similar to obligate root-inhabiting pathogens. It may be reasonably supposed that their behaviour in natural conditions also resembles that of specialized soil-borne parasites. As we have seen, there is some doubt as to whether most of them can even persist as mycelia away from their hosts, and there is no question that their reactions to competition are exactly those expected of obligate root-surface organisms. Even those species such as *Boletus bovinus*, which according to M. C. Rayner & Neilson Jones (1944) exist as rhizomorphs in heath soil in the absence of their hosts, are not virulent enough to infect susceptible hosts when planted in the soil.

In the course of these investigations, Rayner ascribed the lack of virulence to a biologically formed toxic principle in the soil. Later work showed that the antibiotic production of certain species of *Penicillium* in the soil could account for this effect (Brian *et al.*, 1945). Although this was a peculiar case which concerned a special soil site, it is typical of the sensitivity of mycorrhizal fungi to competition. For instance, Levisohn (1957) has tested the effect of *Alternaria tenuis*, a well-known fungal antagonist, upon the growth of several mycorrhizal fungi, viz. *Boletus granulatus*, *B. variegatus*, *B. bovinus*, *B. scaber*, and *Rhizopogon luteolus*. All were inhibited to some extent, but the first particularly so. *Boletus scaber*, now well known to be less restricted in habitat than other mycorrhizal fungi, was only mildly affected.

All these features of sensitivity to competitors and antagonists group the mycorrhizal fungi fairly with other root-inhabiting fungi of a specialized kind which are known to have little effect in the surrounding soil.

The Effect of Temperature and pH on Growth

Temperature has the expected effects upon the growth of most mycorrhizal fungi in culture; their rate of growth falls on an optimum curve. For most species the optimum temperature lies between about 18° and 27° C. Species of *Lactarius*, *Amanita*, some species of *Tricholoma* such as *T. vaccinum* and *T. flavobrunneum*, and some species of *Boletus*, have an optimum not far from 20° C. Others, such as *Cenococcum graniforme*, most species of *Boletus*, and *Tricholoma pessundatum* and *T. imbricatum*, have an optimum below 25° C. Norkrans (1949) reports that *Tricholoma fumosum* has an optimum of 27°–30° C., and Moser (1956) has shown that *Suillus* (*Boletus*) *plorans* grows best between 15° C. and 18° C., yet the optimum temperature for the majority of mycorrhizal fungi is usually about 20° C.

No very valuable generalization about the temperature requirements of mycorrhizal fungi can be made because there is a variation in temperature reaction between the strains of one species and within the same strain on different media (Hacskaylo *et al.*, 1965). Mikola (1948*a*) pointed out that the ambient temperatures in the natural environment of mycorrhizal fungi were often less than their optima observed in culture. For instance for *Cenococcum* the optimum temperature was about 25° C., a temperature not reached at any time in the soils in which it was commonly found.

Moser (1958) determined the ability of a large number of Basidiomycetes, including mycorrhizal fungi, to withstand freezing conditions

(−11° C. to −12° C.) as mycelium. He also measured their growth-rates at temperatures close to zero. There was considerable variation in both respects. Moser pointed out that spring and autumn fungi, and fungi from mountains, may both withstand freezing and be able to grow at low temperatures. Hence in attempting to establish mycotrophic trees in montane regions, low-temperature fungi or strains should be used in artificial inoculation.

The optimum pH for mycelial growth of mycorrhizal fungi is always on the acid side of neutrality. Modess (1941) studied a number of species: *Amanita* spp. and *Paxillus panuoides* had optima between 3·5 and 4·5, most Boleti had optima about 5·0 or a little above, and *Lactarius deliciosus* and *Rhizopogon roseolus* had optima between pH 5·5 and 6·0. How (1941) gives the optimum for *Boletus elegans* as pH 4·0 to 5·5, the actual value being, as expected, somewhat lower on nitrate than on ammonium media. In *Tricholoma*, Norkrans (1949) found some variation of the optimum pH, but most species had optima between 5 and 6, with *T. imbricatum* somewhat lower, while *Cenococcum* was shown by Mikola (1948*a*) to grow best at pH 4·0, with limits at 2·4 and 7·0.

Most of the species are sensitive to aeration and fare best with an adequate oxygen supply. In solid stock cultures, large culture vessels are valuable because of the large oxygen volume provided. In liquid culture, shaking or forced aeration greatly increases the growth-rate.

Problems Arising out of the Physiology of the Fungi

(1) activity in the soil

There is no single, striking physiological property common to all mycorrhiza-formers which characterizes them as a group and separates them from other Basidiomycetes. This ecological category of fungi is defined by a general aggregate of characters, and the individual species within it may depart from the typical by a wide margin in one or more characters. Variation is of course to be expected in any ecological grouping, and is shown by such aggregates as white-rots, brown-rots, root-inhabiting fungi, and so forth. In summary, the characteristic features of ectotrophic mycorrhizal fungi are: dependence on simple carbohydrates; utilization of ammonium and organic nitrogen compounds; requirement for thiamin, other B vitamins, and growth factors including root exudates; inhibition by phenolic products in the humus; sensitivity to competition from saprophytes; possession of pH optima in the acid range (typically about pH 5·0); and possession of low (20° to 25° C.) temperature optima. Neither a single character nor the aggregate of them helps to explain the special ecological behaviour of the fungi. These are, in a broad sense, the characters of all root-infecting fungi: they are characters of fungi which are vegetatively active on living root surfaces only—surfaces from which it has been shown that amino-acids, vitamins, sugars, and other substances, which affect fungal growth, can be secreted. All these characteristics together, fail to indicate in the slightest way why the mycelia involved form a morphologically differentiated organ, the mycorrhiza, in conjunction with a suitable host-root.

The assumption has often been made that the result of the activities of mycorrhizal fungi of forest trees in some way causes the release of soluble nutrients locked in the plant debris in the soil, so that these nutrients become available to the host. In a few cases, e.g. *Boletus subtomentosus*, this is credible, but as a general thesis it lacks conviction; for we have in addition to assume, for the majority of the fungal symbionts, that the host so affects their physiology as to enable them to break down efficiently materials which, in a free state, they are unable to attack at all. The existence of variation in this respect does, however, give some point to the contention of Hacskaylo & Palmer (1955) and of Bergemann (1956) that comparisons should be made of the effects of different fungi on the growth of a single host. But as will be seen in a later chapter, the available experimental analyses of the effects of mycorrhizal fungi in bringing humus materials into solution, suggest that this is not a process of universal importance in the nutrition of mycotrophic trees. In the case of orchidaceous mycorrhizas, on the other hand, the physiologically very different endotrophs do have this kind of effect, as will be seen in Chapter XI.

It was shown in Chapter III that the intensity of mycorrhizal infection is dependent upon the light intensity in which seedlings are grown and hence upon their photosynthesis. Factors such as high mineral supply, which might depress the availability of free carbohydrate, acted against the effect of light. It has therefore been held, following E. Björkman, that the dependence of the majority of mycorrhizal fungi upon simple carbohydrates in culture afforded an explanation of the effect of light. For photosynthesis increased the quantity of sugars in the tissue and so encouraged the growth of the fungi that the intensity of infection was also increased. The prime difficulty of this thesis is the absence of an explanation of how the presence of free sugar in the tissues might be perceived by the fungal mycelia or fungal spores in the soil. The roots must release some stimulus to fungal growth, but this is not likely to be a carbohydrate itself, for this would be completely unspecific and stimulate the growth of many kinds of fungi. It is, moreover, common to find that the ease with which obligate parasites infect their hosts depends upon the availability of carbohydrates in the host tissues (Allen, 1954). Hence, although we have good evidence that active photosynthesis is one of the factors predisposing the hosts to infection, the presence of free sugars in their roots cannot explain, in this or in any other case, the production of composite organs or the morphological pattern of associated growth. The effective principle of root exudates, which has been called the 'M factor' by Melin, has the merit of being released by root systems; but since it is formed by many kinds of root which do not produce mycorrhizas, it again does not alone afford the required explanation of mycorrhiza formation. Nor do we yet know the specificity of the M factor—it is effective in stimulating the growth of many mycorrhiza-formers and some other fungi.

As yet we know too little of the habitat of mycorrhizal fungi. A few, such as *Boletus bovinus*, *B. subtomentosus*, *B. scaber*, and *Cenococcum graniforme*, may perhaps exist as free-living mycelia in the soil, but for most of the known species the evidence suggests that they have a restricted existence apart from living on roots. The onset of infection in N. F. Robertson's experiments, for instance, shows that the beginnings of mycorrhizal development in seedlings are seen as a few centres of infection which arise from spore germination, or by contact with an adjacent infected seedling, or, as in G. W. Dimbleby's experiment, by contact with a particular organic site in the soil which contains spores or vegetative myclia.

After the initial phases of infection have been completed, a considerable development of mycelium may sometimes occur in the root region of the host. Levisohn (1956*a*) has described this in potted seedlings and trees of Scots pine, and Doak (1955) in those of *Quercus macrocarpa*. In the former case, extensive mycelium developed in the root region after the soil had been covered with a heather mulch, and in young trees of spruce and birch in forest conditions after mulching with bracken or heather. Such treatment resulted also in the production of fruit-bodies of various mycorrhizal fungi in the vicinity of the trees. R. H. Thornton and his collaborators (1956) have described a similar development of mycelium in the root regions of *Pinus radiata*. Four types of mycelium were seen in the soil: three with, and one without, clamp connexions. In addition, fruit-bodies of *Rhizopogon*, *Clathrus*, and *Geastrum*, occurred on the soil surface around the trees. It is interesting to note that the sandy soil of the root region, which was permeated by mycelium, was different in structure from the surrounding soil—in the larger size of its pores, and in the greater size and cohesion of its granules.

It will be of the greatest value to obtain more knowledge of the saprophytic survival of mycorrhizal fungi along the lines of investigation carried out by Dr. S. D. Garrett and his colleagues. The work of F. C. Butler (1953) has shown that the obligate root-infecting fungus, *Helminthosporium sativum*, may be partially eliminated from dead host tissues by high nitrogen supply, and Rishbeth (1950, 1951, 1951*a*) has indicated some of the ways in which the parasite, *Fomes annosus*, is affected in the saprophytic phase by soil conditions. These and similar researches might provide a pattern for the study of the saprophytic survival of mycorrhizal fungi. It should be borne in mind that the effect of high rates of manuring, in decreasing mycorrhizal development, could be completely interpreted along the lines of alteration of competition with saprophytic organisms, if the inverse correlations with light intensity and manurial status had not been observed. The two effects might still be additive.

(2) MORPHOGENIC EFFECTS ON THE HOST

The known nutritional requirements of the mycorrhizal fungi afford some explanation of their existence in the specialized habitat of the root surface; but they afford no explanation of their regular association with particular host species, and none again of their effect on the host. Nor have we at present any idea of the stimulus which causes the fungus to form the peculiar pattern of sheath and Hartig net when it grows upon the root; no such structures appear in pure fungal cultures. Whatever the nature of the stimulus needed to cause sheath differentiation, it is not purely a stimulus to rapid growth but is, in particular, a stimulus to morphological differentiation. At present all the active extracts of root tissue, as for example the M factor and secretions such as amino-acids, vitamins, and the like, are growth-promoting: none causes structural changes in the mycelium. The problems concerning the causal morphology of sheath formation are perhaps similar to those concerned with the development of rhizomorphs, sclerotia, and fruit-bodies, which may have a somewhat similar anatomical structure; such problems remain almost uninvestigated. We can, however, be quite clear on an important point which has been greatly misunderstood: the stimulus to sheath formation is not one which merely causes an increase of dry-weight production or greater lateral spread of mycelium in culture, but is morphogenic.

The effect of mycorrhizal fungi upon the morphology and histology of the host root has been widely studied. It has generally been assumed that the histological differences between the host tissue of mycorrhizas and that of homologous uninfected laterals are due solely to the effects of the fungi upon them. This assumption, so readily made by many people, is in need of greater justification than it appears yet to have. The work on the structure and development of mycorrhizas described in Chapter III gives no sure ground for making this assumption with absolute certainty; Clowes and others, it is true, have described the sequence of events during the initial stages of infection and have shown that the fungal hyphae may colonize roots of virtually normal structure. The work of J. Warren Wilson with beech, and the descriptions of Hatch, Doak, and others, of pine, show that many of the lateral short-roots readily abort if not infected. In Warren Wilson's experiments the short roots became progressively more similar in structure to the host cores of mycorrhizas as they ceased growing in length.*

As Clowes has pointed out, many factors which cause a diminution in growth may bring into existence the histological pattern found in the apical regions of mycorrhizas, as the meristem ceases to divide rapidly. It now seems extremely probable that a fungal stimulation of some kind maintains the potential for continued growth and branching in short roots, so that the tendency for them to abort, whether as a result of internal or external influences, is suppressed. The sequence seems to be that branch initials are formed in considerable numbers on mother-roots,

* *See* Appendix §2.

a few of these initials being capable of extensive growth to form long roots. Others, by virtue of their position on the root system or owing to external influences, grow only for a limited period and may soon abort in the uninfected condition. In beech the structure of the apices of roots which cease growing approximates to that of the core of mycorrhizas as growth ceases, and they may become infected in that state. Other roots grow normally until after their surface has been colonized by fungal hyphae, and then mycorrhizal development with differentiation of the core tissues occurs.

Slankis (1948, 1948*a*, 1949, 1951) has investigated the effects of fungi, of fungal culture-medium, and also of synthetic auxins, upon the behaviour of the roots of pine in pure culture. He showed that when mycelia of *Boletus* spp., especially *B. variegatus*, were grown in the same medium as excised pine roots, an unusually large development of short laterals, which often branched dichotomously, occurred even when the hyphae were not in physical contact with the roots. Filtrates of the medium in which *B. variegatus* had been grown exerted a similar effect, proving that a substance had been released by the mycelium into the medium and that this substance stimulated short-root development.

Ulrich (1960) tested the ability of several species of mycorrhizal Basidiomycetes to form auxins in the presence or absence of tryptophan as an auxin precursor. Many species of *Boletus*, especially those forming nodule mycorrhiza with pine, formed indolyl acetic acid in the absence of tryptophan, but *B. badius* did not. Other fungi required tryptophan to be added to the culture media and some of these only produced auxin as active young mycelium. Species of *Amanita* were observed to produce a variety of indolyl compounds from tryptophan. In auxin production as in other characteristics mycorrhizal fungi vary.

Slankis further tested the effects upon pine roots of concentrations of β-indolylacetic, β-indolylpropionic, β-indolylbutyric, and α-naphthalene-acetic acids, in concentrations between 0·001 μg. and 10 μg. per 20 ml. Somewhat similar effects to those observed with filtrates of culture medium were obtained. In low concentrations (0·001 μg. to 1·0 μg. per 20 ml.) of β-indolylacetic acid (heteroauxin), the formation and growth of long laterals was stimulated; but at higher concentrations the number of short branches increased whilst long-root development was inhibited. The short branches forked dichotomously and bore a superficial resemblance to mycorrhizal roots. Where dichotomy was stimulated, inhibition of root-hair formation occurred. In very high concentrations, branching was inhibited. The other compounds were also active; α-naphthalene-acetic acid was more so than heteroauxin, β-indolylpropionic acid and the others were less so. Slankis was able to produce, by appropriate adjustment of auxin concentration in aseptic conditions, long roots bearing short dichotomous systems similar to coralloid bunches of mycorrhizas, or massed short-roots similar to tuberculate mycorrhizas.

In subsequent work Slankis enclosed excised long roots of pine in three compartments, so that it was possible to apply different auxin concentrations to the different zones. By this means he was able to show that auxins were translocated both distally and proximally within the roots. Hence fungal auxins originating from a region of infection might be expected to affect the growth of the long roots, and to determine the production of short roots and the extent of their branching and development. Slankis concluded (1958) that the fungal auxin has a threefold effect: it causes the morphological deviations of short roots on which mycorrhizal development depends, it influences the whole root system by being translocated, and it determines the frequency and sequence of long- and short-root production. In 1963 he also published photographs of longitudinal sections of short roots produced in culture with 2·5 mg. of indolylacetic acid per litre. In this high concentration the histological pattern resembled that of mycorrhizal roots.

It is of especial interest that recent experimentation on the effect of fungal auxins on tree roots has indicated a connexion with those factors which may affect photosynthesis. Slankis (1961, 1963) has reported that the extent of the effect of auxin in pine seedlings is correlated with light intensity. At low intensities, which in Björkman's experiments would have led to little mycorrhizal infection, auxins were observed to bring about only slight morphological changes in the roots even if applied in high concentrations, whereas in higher light intensities or in the presence of adequate sugar they caused expected alterations in form and structure. Slankis propounds the view that the relative lack of mycorrhiza formation in conditions of shade can most adequately be explained by changes in the physiological state of the host-roots. At the same time Moser (1959), in a study of auxin production by mycorrhizal fungi, has shown that it is diminished when external nitrogen supplies are high.

Putting these points together with the work of Melin on the M factor and the cognate work of Rovira and others on root exudates (*see* Chapter II), a working hypothesis can be formulated concerning the factors which affect mycorrhiza development. Light intensity and external nitrogen supply affect the metabolism of the root system in such a way that the internal concentration of soluble substances, including sugars, and the kind and quantity of exudates, are altered.* These changes have a direct effect, externally to the root, on the growth and development of the mycorrhizal fungi. They also have an effect upon the sensitivity of the root tissues to fungal auxins. The stimulation of the fungi may increase their potential for auxin production, but they are also directly affected metabolically by inorganic nutrients in the soil. There is therefore a complex constellation of interacting factors which influence mycorrhiza development, and it will require much detailed work to test and elaborate this hypothesis. The local development of fungi around particular sites in the root system, and the local development of bunched mycorrhizas when other parts of the

* *See* Appendix §2.

root system do not react so intensively, may be compared with similar phenomena in the development of leguminous nodules (Nutman, 1958). It remains to be determined whether intensive mycorrhiza development at one locale may inhibit further mycorrhiza formation in its immediate environs in the manner in which nodulation at one site affects further development in its neighbourhood.

There are, however, yet further complexities. It is certain that other factors besides auxins influence the development of the rootlets both morphologically and histologically. The growth and dichotomy of the short roots of pine may occur in aseptic culture, without the addition of auxin, as Clowes and L. Leyton have reported. Barnes & Naylor (1959) have also observed forking of excised roots of *Pinus serotina* in aseptic culture containing nitrogen compounds such as various amino-acids. Goss (1960) made a study of this subject in *Pinus ponderosa* growing in natural soil in Nebraska. He observed forking of short roots both in adult pine in woodland soils and in seedlings growing in mycorrhizal and non-mycorrhizal soils. He concluded that forking was an inherent factor in *Pinus ponderosa*, which becomes more pronounced as a result of the fungus association.

In addition, those species forming ectotrophic mycorrhizas are typically heterorrhizic even in the uninfected state. The ultimate branches of their root systems, the short roots, may be extremely short-lived, especially if uninfected. A worthwhile problem, greatly in need of solution, is to discover the processes initiating development and senescence of short roots in the uninfected and mycorrhizal state under different conditions for growth.

An interesting development in the experimental technique of root culture which may be of great future importance in this regard, has been made by Fortin (1966). He used as inocula excised roots of *Pinus sylvestris* with a portion of hypocotyl attached to them. With these he obtained good root growth in defined media into which he also introduced mycorrhizal fungi as hyphal suspensions. In this way infection was obtained and mycorrhizal short roots equipped with sheaths and Hartig nets were developed. The production of mycorrhizas in this way seems to afford a means not only of investigating the factors affecting infection, but also the physiological processes of the mycorrhizal organs.

(3) PSEUDOMYCORRHIZAS

The term pseudomycorrhiza was first applied by E. Melin to short roots attacked by certain root parasites and other fungi in the absence of mycorrhizal fungi, so that the cells were penetrated, growth was slow, and branching was rare or absent. The special fungi particularly involved were called *Mycelium radicis atrovirens* and *Rhizoctonia sylvestris*, which were isolated repeatedly from the pseudomycorrhizal roots. Later observation has shown that sterile mycelia, answering to the description of these pseudomycorrhizal fungi of Melin, are very common members of the root-surface population of many plants. Robertson (1954) has isolated them from the long roots of *Pinus sylvestris*, and Harley & Waid (1955) have reported them to occur as mycelia on the long roots of beech. Such mycelia are most easily isolated from the subapical regions, for on the more mature parts they are associated with sporing fungi which grow more strongly on culture media than they do themselves.

The parasitic potential of these fungi seems to be low, though short laterals are parasitized by them when mycorrhizal fungi are absent, where seedlings are grown in very low light intensity or in near-starvation conditions, and where seedlings are raised in old forest nurseries on agricultural soil. It may be assumed that these weak pathogens do not compete successfully with mycorrhizal fungi for the colonization of the surface tissues of the roots in normal conditions. But they do attack the sheaths of old mycorrhizas when these go into senescence, and may then form an outer secondary sheath-like layer as described in Chapter III.

M. C. Rayner & Levisohn (1941), and Levisohn (1954), have made a study of these kinds of parasitic attack. Usually the root becomes invested to some extent with a kind of sheath of fungal tissue and the cortex is permeated by hyphae which penetrate the cells to form structures called haustoria. When seedlings of pine and spruce are infected, some decrease in vigour of growth, which varies with the species, occurs. Levisohn takes the view that the primary predisposing causes of this kind of infection are factors which prevent the colonization of the root by mycorrhiza-forming fungi, and that in the presence of active mycorrhizas, under the right conditions, *Rhizoctonia sylvestris* and *Mycelium radicis atrovirens* do not cause much damage to the hosts. Björkman (1942), in experiments with conifers on peaty soils, showed that pseudomycorrhiza formation was stimulated in conditions where mycorrhizas did not form.

Mycelium radicis atrovirens and *Rhizoctonia sylvestris* have been studied in culture and have been found to have a somewhat less exacting nutrition than mycorrhizal fungi. A comparison is given in Table XV, based on the data of Schelling and Keller who worked in Professor E. Gäumann's laboratory, of the growth made on various 'carbon sources' by *Mycelium radicis atrovirens* and one of the least discriminating of mycorrhiza-formers, *Cenococcum graniforme*. It can be seen that *Mycelium radicis atrovirens* makes good growth on a great variety of polysaccharides and simple sugars, whereas *Cenococcum* is much more restricted (*see also* Mikola, 1948*a*). It is noteworthy that although Schelling (1950) records *M.r. atrovirens* to be unable to grow on cellulose, Gams (1963) describes it as cellulolytic. He ascribes the difference to the fact that Schelling omitted trace elements and growth factors from his media. In any event *M.r. atrovirens* was shown by Gams to be composed of many strains, and so it may be variable in cellulolytic potential.

We can at the moment view these fungi which form somewhat abnormal morphological patterns with host root-systems and at the same time often reduce their vigour, as one extreme variation of the mycorrhizal fungi, which may be contrasted with very exacting species such as *Boletus elegans* or *Russula xerampelina* as the other extreme.

TABLE XV

THE GROWTH IN mg. DRY-WEIGHT PER DAY OF *Cenococcum graniforme* AND *Mycelium r. atrovirens* ON VARIOUS CARBON SOURCES. (After Schelling, 1950; Keller, 1950.)

C source	*M. r. atrovirens* mg./day	*Cenococcum* mg./day
Glucose	8·7	11·2
Fructose	5·7	6·2
Mannose	8·0	10·4
Galactose	3·0	0·14
Arabinose	5·1	0·006
Xylose	5·4	0·003
Sucrose	10·4	7·5
Maltose	8·6	6·0
Cellobiose	9·4	—
Lactose	3·6	0·0015
Starch	9·1	0·16
Inulin	6·2	1·2
Cellulose	0	0
Mannitol	4·0	5·9
Erythritol	3·6	0·006
Raffinose	6·0	1·7

The name pseudomycorrhiza must be used with some care; there has been a tendency for it to be applied in a loose manner. Those who held a pseudoreligious belief that mycorrhizas 'benefited' their hosts under all conditions by some perhaps magical process, would refute, by the suggestion that the state being observed was pseudomycorrhizal rather than mycorrhizal, any observation which looked contrary to their own belief. The sensible view to take is one well founded on experiment and observation; that there is a considerable range of behaviour amongst the fungi which are capable of infecting tree roots and of forming special morphological patterns with these roots. Many of these fungi set up a composite organ whose life period is as long as or longer than the uninfected axes. It is the usual condition of many forest trees to possess root systems that are heavily infected in this way. Fungi capable of forming these mycorrhizal organs may, under some conditions, cause considerable damage to root systems, and some of the fungi, although not virulent killing parasites, normally cause premature death of their hosts or do not set up physiologically active composite organs with them. We may perhaps compare this range of behaviour with the behaviour of strains of *Rhizobium* on leguminous plants, where the genetic nature of the host and of the *Rhizobium*, in addition to the external conditions, may all play their part in determining whether the nodule is a physiologically functional entity or an 'ineffective' gall.

It is not, therefore, of any value to attempt to classify the joint composite organs of fungus and root into sharply defined categories such as mycorrhizas and pseudomycorrhizas. A number of workers are currently investigating the nature and variation of the physiological properties of the composite organs produced by different fungi on any one host, because this is both of great ecological interest and of practical importance in forestry. In this context the work of Moser, Bowen, Mikola, and Laiho and others, will be discussed in a subsequent chapter. It would also be of great interest to have a physiological comparison of the fungi which mediate the epiparasitism of 'saprophytes' such as *Monotropa* and *Gastrodia minor* on ectotrophic organs of woody plants (*see* p. 189 and p. 226, ref. Björkman, 1960; Campbell, 1963) with other mycorrhizal fungi.*

* *See* Appendix §8.

V

CARBOHYDRATE PHYSIOLOGY OF ECTOTROPHIC MYCORRHIZAS

Introduction

A knowledge of the carbohydrate physiology and carbon metabolism of ectotrophic mycorrhizal organs is essential for the understanding of their functioning. Not only is there a connexion between the photosynthetic process and intensity of mycorrhizal infection (cf. Chapter III), but also the majority of the causative fungi are dependent upon simple carbohydrates for growth in culture (Chapter IV). Furthermore, for the nutrient absorptive activities of mycorrhizas, like those of roots and other plant organs, respiratory substrates and carbon-containing acceptor radicals (e.g. for nitrogen assimilation) are needed. It is therefore surprising that there had been virtually no detailed study of carbon metabolism of ectotrophic mycorrhizas, as opposed to the simple analysis of carbohydrates in them, until recently. Much of the information in this chapter is based upon the work with *Fagus sylvatica* which provides the only available corpus of observations with which to attempt the erection of working hypotheses concerning the carbon nutrition of the fungi in symbiotic state and the respiration and nutrient absorption of mycorrhizal organs (*see* Harley & Jennings, 1958; D. H. Lewis, 1963; D. H. Lewis & Harley, 1965, 1965*a*, 1965*b*).

The dependence of most of the fungi of ectotrophic mycorrhiza on relatively simple carbohydrates for growth in culture, has led to the assumption that they probably rely upon their hosts for carbon nutrition, and although there is a range of behaviour amongst them, the majority of those that have been examined have restricted ability to break down and utilize the complex carbon sources that usually occur in the soil. The opposing view, that carbon compounds are indeed absorbed by mycorrhizal fungi from the soil, has also been put forward. MacDougal & Dufrenoy (1944, 1946) described the production of new mycorrhizas by detached pine roots in the soil and their growth during a period of months. This did not, of course, prove that carbohydrates were absorbed by them from the soil, because of the probable existence of stored carbohydrate in the axes of the mother-root; but Young (1947) has shown that *Pinus taeda* could absorb carbohydrate from maltose and grow in the absence of photosynthesis, and also that a species of *Boletus* with which it formed mycorrhizas could digest cellulose to yield soluble reducing substances. This might indicate a self-sufficiency of this species of mycotrophic fungus for carbon compounds, and a partial reliance of its host upon carbohydrate released by the fungus. But as it is unlikely that mycorrhizal fungi in the symbiotic state develop enzymic activities that are not exhibited by them in culture, a general hypothesis on the basis of these observations is untenable. W. F. Kuprewitsch (quoted by Lobanow, 1960) has also put forward a view that mycorrhizal roots produce external enzymes by the activity of which organic substances are made available for absorption by the root systems, so that they might therefore be partially carbon-heterotrophic. The subject of external digestion of carbohydrate will be considered later in this chapter.

Direct evidence of the movement of carbon compounds, formed in photosynthesis, from the host to the fungus, was obtained by Melin & Nilsson (1957) in a series of experiments using $^{14}CO_2$. Artificially produced mycorrhizal seedlings of *Pinus sylvestris*, in two-member cultures with either *Boletus variegatus* or *Rhizopogon roseolus*, were allowed to assimilate $^{14}CO_2$ in the light. Decapitated plants were used as controls. The results showed that, during photosynthesis, radioactive carbon dioxide was assimilated and the products were transported to the mycorrhizas where considerable activity was found in the fungal sheath. The decapitated controls also showed some assimilation of radioactivity. D. H. Lewis (1963) has recalculated some of the results of Melin & Nilsson to show that this non-photosynthetic, dark assimilation of $^{14}CO_2$ is particularly active in the fungal sheaths of the mycorrhizas of the controls. As will be seen later, mycorrhizas of *Fagus sylvatica* have the property of dark fixation of CO_2 into organic acids—a property that they hold in common with many plant organs, including other salt-absorbing organs. The presence, in the experiment of Melin & Nilsson, of considerable $^{14}CO_2$ fixation in the controls, is therefore to be expected, and does not in any way weaken their conclusions that a very significant transport of photosynthate from host to fungus occurs in normal seedlings.

A further example of movement of carbon from host to fungus is given by Björkman (1960), who also used ^{14}C as a tracer. He injected [^{14}C] glucose into the trunk of *Picea excelsa* and observed the movement of radioactivity not only into its mycorrhizal fungus but also into *Monotropa*, an epiparasite sharing the same mycorrhizal mycelium (*see* pp. 188–9).

It is of interest also that Shiroya *et al.* (1962) and Nelson (1964) measured the rate of translocation of ^{14}C fixed in photosynthesis and transported to the roots of *Pinus resinosa* and *P. strobus*. More was transported in a given time to well-developed mycorrhizal root-systems than to others. Again, Tranquillini (1959, 1964) estimated the seasonal carbon balance of *Pinus cembra* at the tree-line in the Alps. He was of the opinion that, of the carbon fixed (as calculated from gas exchange), about 40 per cent was released as CO_2 by the host, while of the rest only about one-third was accounted for by dry-weight increase, so that the remaining two-thirds were utilized by the mycorrhizal root systems and by secretion from them.

Neither the experiments of Melin & Nilsson nor those of Björkman have attempted to determine the nature of the carbon compounds passing to the fungus. They provide, moreover, no knowledge of the mechanism of transfer. For this purpose much information is required about the carbohydrate physiology of mycorrhizal organs.

Carbohydrate Absorption and Metabolism

As an approach to carbohydrate metabolism of mycorrhizas, their processes of sugar absorption and assimilation were examined first by Harley & Jennings (1958) using excised mycorrhizas of beech. Carbohydrate was readily absorbed from dilute solutions of glucose, fructose, and sucrose. This in itself shows that, if the unlikely ecological event of the existence of significant amounts of simple sugars in the root region occurred, the views of MacDougal & Dufrenoy and of Young would be tenable and carbohydrate could be absorbed by that route into the fungal layer.

The monosaccharides were absorbed by mechanisms which were sensitive to concentration, so that the curves for uptake against concentration approximated to rectangular hyperbolae. The calculated concentration at which the half-maximum rates of uptake (K_m) occurred, was about ten times larger for fructose than glucose. Hence the absorptive system appeared to have a greater avidity for glucose than fructose. This was indeed confirmed because the two sugars competed for uptake. Fructose uptake was reduced to small proportions in equimolar mixtures of glucose and fructose, whereas glucose uptake was almost unaffected. The uptake mechanism was associated with metabolic turnover and probably with phosphorylation of the hexose. It was sensitive to temperature, oxygen supply, and pH, it occasioned an increase of oxygen uptake during sugar absorption, and it was inhibited by dinitrophenol and by silver nitrate.

Absorption of carbohydrate from sucrose solutions was somewhat more complicated, because it involved an external breakdown of sucrose by a surface-attached heat-resistant enzyme system. There was a preferential selection of glucose from the products of breakdown, and a release of fructose into the medium. Factors, such as metabolic inhibitors, which eliminated absorption, did not necessarily affect hydrolytic breakdown of sucrose. Hence inhibited mycorrhizal roots broke sucrose down to its constituent glucose and fructose in the absence of absorption.

Later work (D. H. Lewis & Harley, 1965*a*) has shown that uninfected beech roots also have the property of absorbing glucose and fructose at rates between one-third and one-half as fast as mycorrhizal roots—calculated on a fresh-weight basis. Moreover they exhibited the properties of causing an external breakdown of sucrose into its constituent glucose and fructose moieties, and a selective uptake of glucose from the products.

All these features are also held in common with many other kinds of plant material. They have been observed in lichens by Harley & Smith (1956), in *Chaetomium globosum* by Walsh & Harley (1962), and in *Neocosmospora* by K. Budd (unpublished). Rather similar mechanisms occur in higher plant tissues, e.g. in tomato roots as shown by Street & Lowe (1950).

The ability of mycorrhizal and uninfected roots to absorb trehalose and mannitol was also examined by D. H. Lewis & Harley (1965, 1965*a*). Both these substances were utilized readily by mycorrhizal roots but only slowly or negligibly by uninfected roots. Trehalose absorption was similar to that of sucrose, in that an external breakdown of this disaccharide into its constituent moieties of glucose occurred before absorption. The contrast in behaviour of uninfected and mycorrhizal roots in respect of trehalose utilization is important and will be considered again later.

TABLE XVI

PERCENTAGE DISTRIBUTION OF RADIOACTIVITY IN SOLUBLE SUGARS OF MYCORRHIZAL ROOTS, AND THE PERCENTAGE OF TOTAL COUNTS IN INSOLUBLE CARBOHYDRATES, AFTER FEEDING [^{14}C] GLUCOSE, -FRUCTOSE, OR -SUCROSE. (D. H. Lewis & Harley, 1965*a*.)

	Percentage of total carbohydrate:		Percentage within soluble fraction due to:				
Sugar supplied	Soluble	Insoluble	Trehalose	Sucrose	Glucose	Mannitol	Fructose
Glucose	41·2	58·8	66·4	16·9	5·1	11·6	0·0
Fructose	87·2	12·8	3·2	13·6	2·4	70·3	10·5
Sucrose	46·0	54·0	37·3	5·8	1·9	53·1	1·9

However, the general similarity of the behaviour of mycorrhizas, uninfected roots, and other plant tissues of many sorts, considerably weakens any support that this work might appear to give to the views of MacDougal & Dufrenoy, and of Young, that carbohydrate absorption is a special feature of ectotrophic mycorrhizas.

Harley & Jennings (1958) showed that, during the absorption of monosaccharides by beech mycorrhizas, there was only a small change of internal concentration of reducing sugars. The monosaccharides were assimilated into disaccharides and insoluble carbohydrates. This problem was further examined by D. H. Lewis & Harley (1965), using ^{14}C-labelled carbohydrates as well as by straight analysis. The main accumulation was to be found in insoluble carbohydrates (identified as glycogen), in trehalose, and in the sugar-alcohol mannitol (*see* Table XVI). This emphasizes the association of the process of absorption with assimilation, and perhaps explains why a phosphorylation step, as evidenced by the effect of the phosphorylation inhibitor DNP on uptake, is likely to be involved.

Comparison of the absorption of carbohydrate by uninfected and mycorrhizal beech roots showed important differences to exist. The core tissue of mycorrhizal roots which had been stripped of their sheaths had much in common with uninfected roots, but the fungal sheath differed from both (D. H. Lewis & Harley 1965*a*). Uninfected roots synthesized sucrose in their tissues from glucose, fructose, and sucrose, and at the same time formed little insoluble carbohydrate. The mycorrhizal sheath by contrast did not produce sucrose, but instead, trehalose, mannitol, and insoluble glycogen. Any departure by the core from the pattern of uninfected roots was explicable by the presence of the Hartig net in the cortex or by possible wound effects.

The absorbed carbohydrate was not distributed evenly between the constituent sheath and core tissue of mycorrhizal roots. About 70 per cent of the accumulated carbohydrate was found in the sheath, and only some 30 per cent in the core. Of the latter at least some must have been in the Hartig net; hence by far the greater accumulation was in the fungus.

TABLE XVII

CARBOHYDRATE CONTENT OF EXCISED BEECH MYCORRHIZAS AND CHANGES AFTER STORAGE IN WATER OR VARIOUS CARBOHYDRATES. Temperature 20° C. Concentration 0·5 per cent w/v. Mg. per gm. fresh-weight. D. H. Lewis & Harley (1965).

	Control	Changes after storage in:					
		Water	Mannitol	Fructose	Sucrose	Trehalose	Glucose
Soluble							
Total sugars	14·69	−5·36	−5·66	−1·53	−0·10	+2·02	+2·91
Total reducing	4·88	−1·02	−0·22	+0·34	−0·43	−0·69	+0·06
Sucrose	5·07	−2·34	−3·20	−0·99	−0·80	−0·06	+0·28
Trehalose	4·74	−2·00	−2·24	−0·88	+1·13	+2·77	+2·57
Mannitol	Nil	Nil	+7·69	+6·36	+3·86	+2·27	+3·86
Insoluble carbohydrate[1]	25·94	−2·84	+3·87	+6·83	+5·20	+8·10	+11·14

[1] Estimated, after hydrolysis, as glucose equivalents.

A point of great interest and importance is the occurrence and formation of the sugar-alcohol mannitol in the fungal tissue of mycorrhizas. Analysis of beech mycorrhizas, as obtained directly from beech woodland, shows that this polyol is present in variable quantity, while sometimes indeed it may be absent. This variation is likely to be seasonal and requires further study. In the feeding experiments described, D. H. Lewis & Harley observed that mannitol was unquestionably the main destination of absorbed fructose (Tables XVI and XVII), often to the exclusion of trehalose or insoluble glycogen. On the other hand, glucose absorption yielded trehalose and glycogen, with little mannitol formation. In spite of the separation in destinations of glucose and fructose, the fact that they compete in uptake is evidence that there is at least one common stage in their processes of absorption.

D. H. Lewis & Harley (1965*a*) suggested possible pathways of hexose utilization by beech mycorrhizas, which they summarized diagramatically to provide a working hypothesis. One of the possibilities which they put forward to explain the separate destination of glucose and fructose was that the phosphohexoisomerase stage (glucose-6-P to fructose-6-P) might be slow. Harley & Loughman (1966) tested this point and obtained evidence that the enzyme was present and as active as is usual in plant extracts. Moreover it appeared to be equally active in both sheath and host tissue. It may be presumed, therefore, that the other reactions utilizing hexose phosphates must compete relatively effectively with the isomerase in the sheath tissue.

Mechanisms of Movement of Carbohydrate from Host to Fungus

The work described above allows one to recognize in any analysis, which of the sugars are from the fungus and which from the host (Table XX). From it followed a method for studying the process and the mechanism by which carbohydrate moves from host to fungus in the manner demonstrated by Melin & Nilsson (1957).

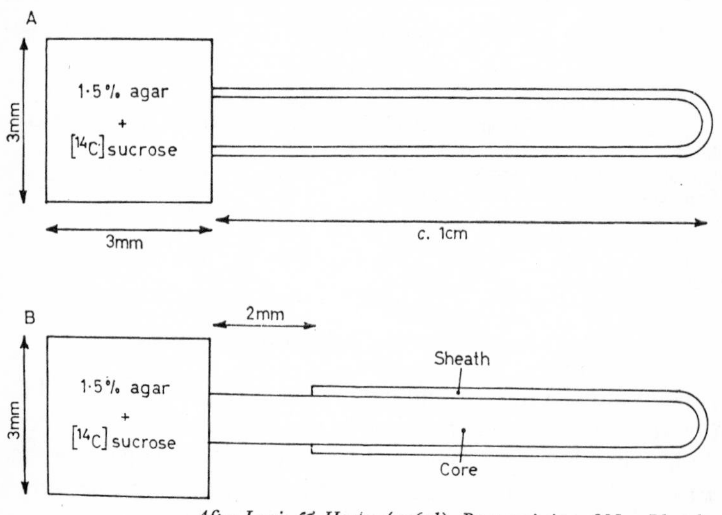

After Lewis & Harley (1965b). By permission of New Phytol.

Fig. 8.—Diagrams showing the method of application of [^{14}C] sucrose to mycorrhizas in translocation experiments. Above A. Application to intact mycorrhiza. Below B. Application to host tissue.

Sucrose is, according to Zimmermann (1961), the sugar most usually translocated in the angiosperms and it was, indeed, the only sugar detected by Ziegler (1956) in sieve-tube exudate of beech. For this reason blocks of 1·5 per cent agar containing [^{14}C] sucrose were applied to excised beech mycorrhizas as shown in the diagram (Fig. 8).

By permission Forstwiss. Zbl.

Results of one of the earliest experiments on the effects of mycorrhizal infection on the growth of trees. Photograph by Frank (1894) of three-year-old pine seedlings in normal, sterile, and reinfected soil.

PLATE I

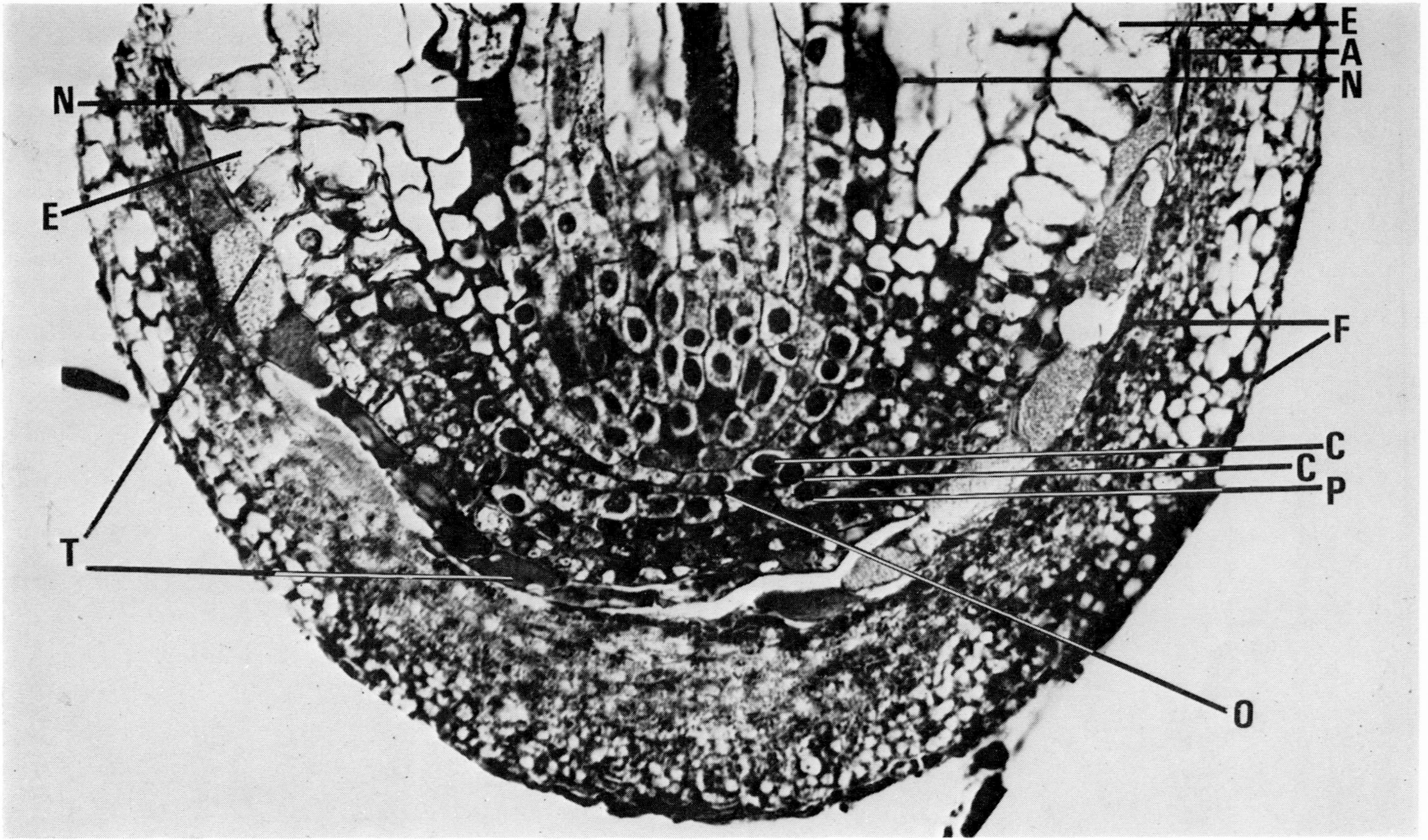

After Clowes (1954). By permission New Phytol.

Median L.S. of the apical region of beech mycorrhiza, ×625. E, epidermis; N, endodermis; A, partly disorganized cap cells; T, cap cells full of brown substance; F, fungal sheath; C, P, and O, approximate sites of the initials of cortex, epidermis-cap complex, and columella.

PLATE 2

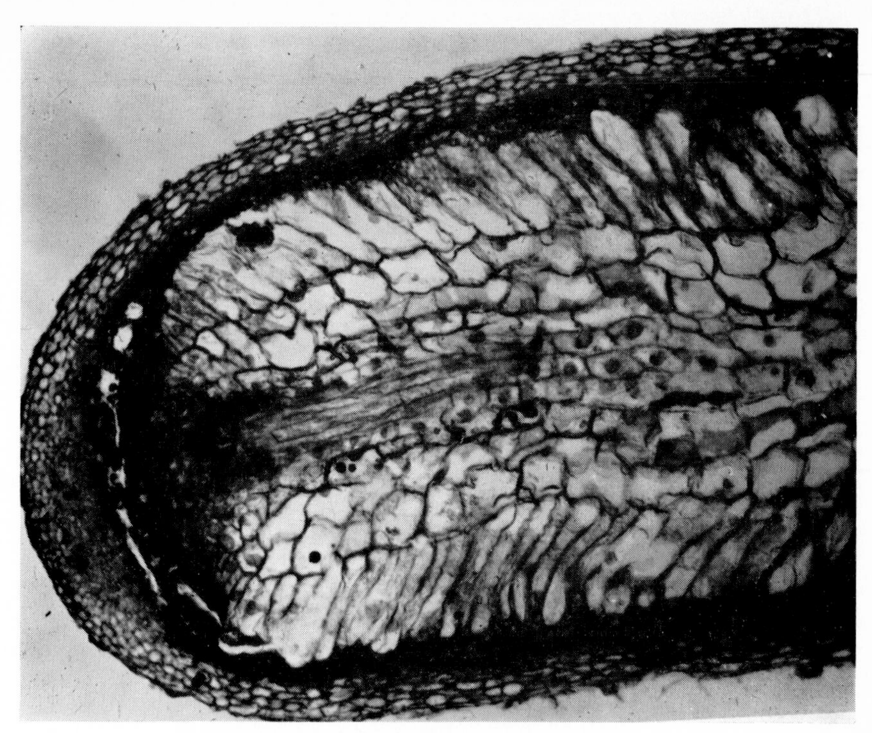

(*a*) L.S. mycorrhizal root of beech, × 250.

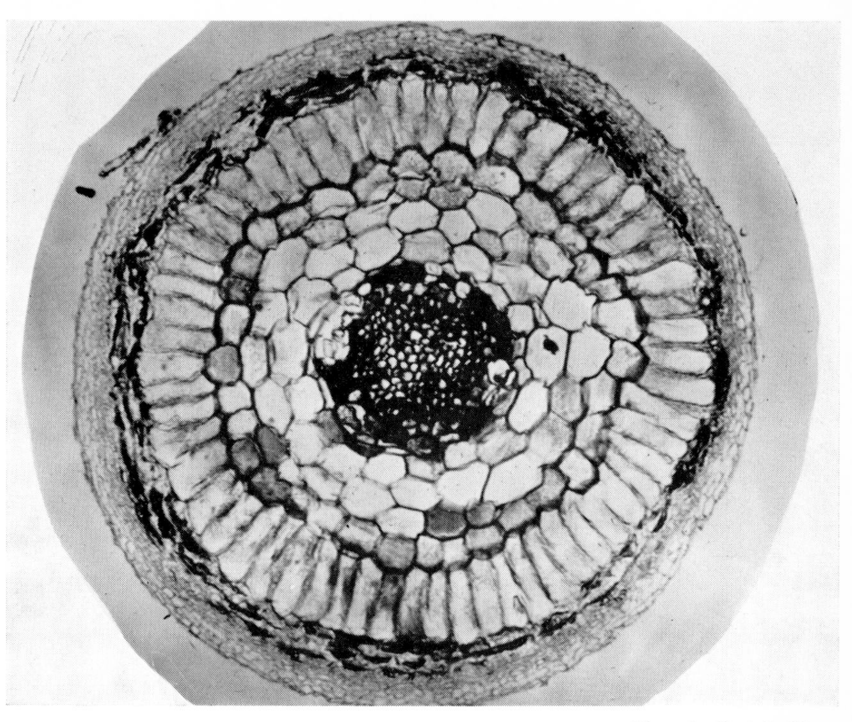

Photos by F. A. L. Clowes

(*b*) T.S. mycorrhizal root of beech, × 250.

PLATE 3

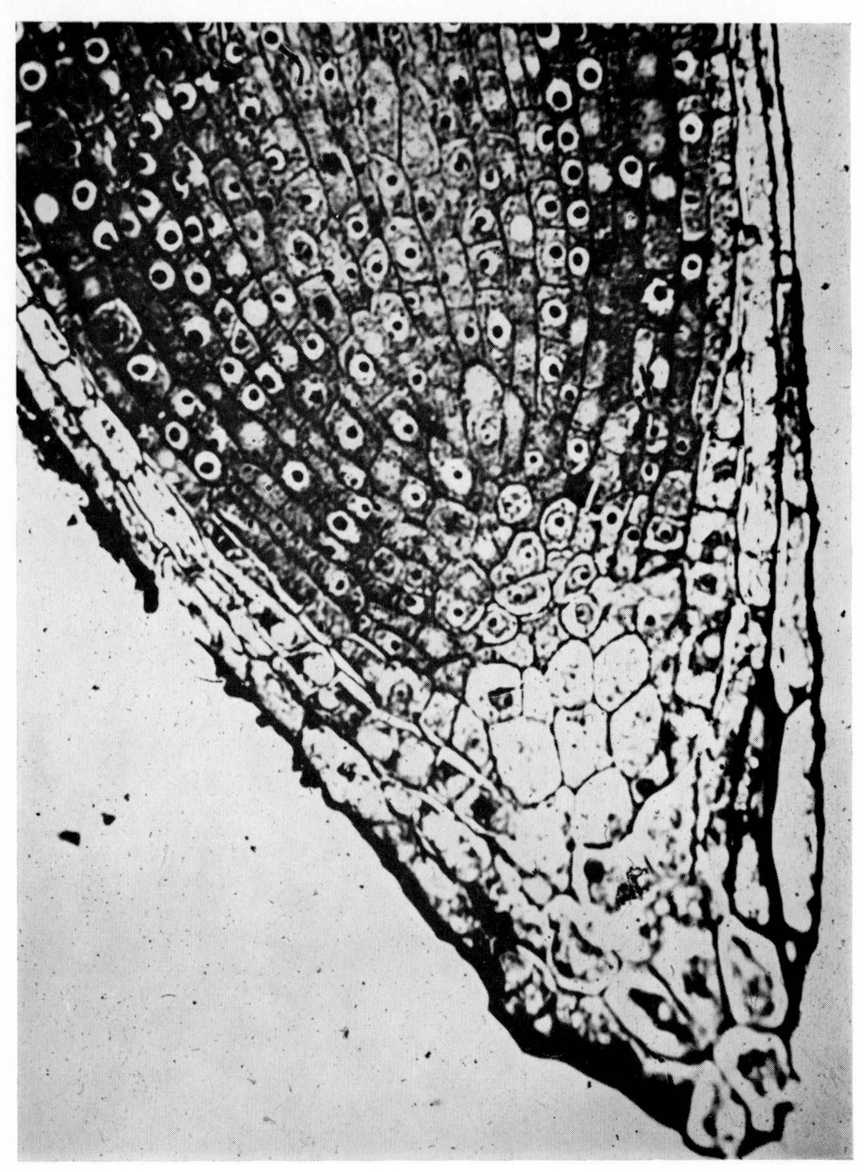

Photo by F. A. L. Clowes

(*c*) L.S. uninfected beech root, ×640.

PLATE 3

Three kinds of arrangement were used. First, the direct application of the agar block to the cut end of the mycorrhiza (Fig. 8A), secondly, the application of the block only to the host core of the mycorrhiza (Fig. 8B), and thirdly, the mycorrhizas treated in the above ways were banded with grease to reduce movement of soluble substances by diffusion in a watery film on their surfaces. After a period of about 24 hours, the tips were

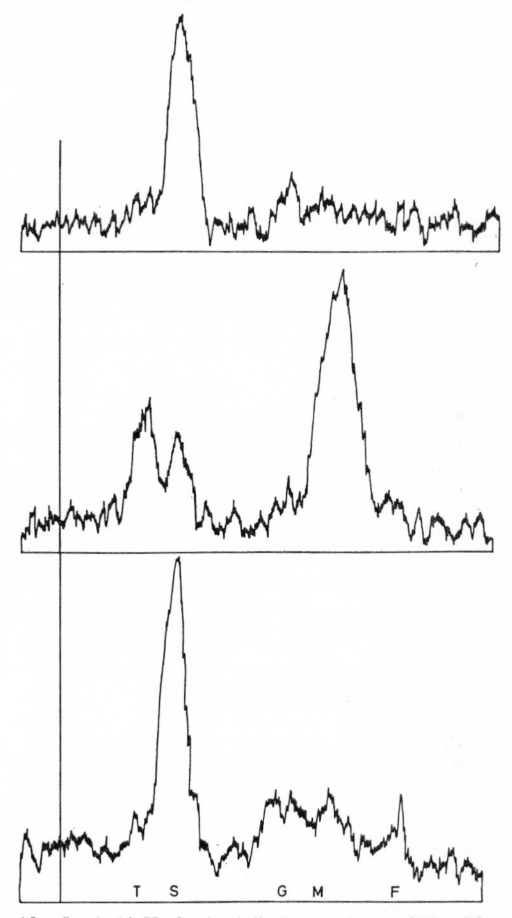

FIG. 9.—Chromatograms showing the distribution of ^{14}C within the soluble fractions of beech mycorrhizas following translocation of [^{14}C] sucrose. Above, basal region on which the block was applied; middle, the sheath of the apex; below, the core of the apex. T, S, G, M, F: trehalose, sucrose, glucose, mannitol, and fructose positions—solvent, ethylacetate, acetic acid, water.

After Lewis & Harley (1965b). By permission of New Phytol.

assayed to determine the quantities of sucrose that had entered the plant material, its distribution in the tissues, and its biochemical destination. It was shown that sucrose was absorbed from the block and translocated through the tissues. Translocation occurred dominantly in the host tissue and to a much lesser extent through the fungus and on its surface. Sucrose moved from the host tissue into the fungal layer in considerable quantities, and accumulated there as mannitol, trehalose, and glycogen (*see* Fig. 9). The magnitude of this transfer may be judged from Table XVIII, which shows the distribution of radioactivity in fungal and core tissue of the apical region remote from the point of application of agar containing [^{14}C] sucrose.

The percentage found in the sheath in all treatments shows how readily the fungus absorbs and assimilates sugars from the host.* Indeed the distribution is not greatly different in kind or proportions from that found when sugars are fed exogenously to the sheath of beech mycorrhizas.

TABLE XVIII

DISTRIBUTION OF RADIOACTIVITY IN FUNGAL SHEATH AND HOST CORE AFTER APPLICATION OF [^{14}C] SUCROSE TO THE CUT END OF BEECH MYCORRHIZAS IN AGAR. (*See* Fig. 8 and text for method.)

Experiment		Treatment	Sheath	Core
I		Agar to host and fungus	76	24
		Agar to host only	68	32
II	agar to host and fungus	Ungreased	55	45
		Greased	66	34

TABLE XIX

RELATIVE UPTAKE OF, AND $^{14}CO_2$ PRODUCTION FROM, [^{14}C] TREHALOSE AND [^{14}C] MANNITOL BY MYCORRHIZAL (M) AND UNINFECTED (U) BEECH ROOTS. 20° C. for (*a*) 21 hrs., (*b*) 18 hrs. (D. H. Lewis & Harley, 1965.)

	Uptake		$^{14}CO_2$ Production		$^{14}CO_2$ Production as per cent of uptake	
	M	U	M	U	M	U
Trehalose (*a*)	100	8·5	—	—	—	—
(*b*)	100	9·2	100	5·3	20·2	17·6
Mannitol (*a*)	100	5·1	—	—	—	—
(*b*)	100	4·2	100	0·3	12·2	1·2

It seems possible, therefore, that the fungal sugars constitute a sink into which carbohydrates are passed and from which, although they may be taken and utilized in fungal metabolism, they are not available for reabsorption by the host. The absence of reciprocal flow from fungus to host was confirmed in experiments by D. H. Lewis (1963). In these, beech mycorrhizas were allowed to absorb exogenously supplied [^{14}C] glucose for 23 hours and then transferred to sugar-free media for a similar period. Although fungal sugars decreased in quantity during the period in the sugar-free medium, there was no evidence of their translocation back to the host, for at the end of the second period no labelled sugars were present in the host tissue. Furthermore, D. H. Lewis & Harley (1965*b*) compared the ability of mycorrhizal and uninfected roots of beech to absorb and utilize exogenously supplied mannitol and trehalose. The results, shown in Table XIX, indicate that uninfected roots have very restricted ability to absorb mannitol, and the very small radioactivity in the carbon dioxide

* *See* Appendix §5.

evolved from it further shows that it is hardly utilized. Trehalose is absorbed only slightly more readily by uninfected roots, but that which is absorbed is fairly readily utilized.

The Absence of Root-cap Tissue in Mycorrhizas

As noted above in Chapter III, Clowes (1950, 1951, 1954) has shown that the architecture of the meristem of mycorrhizal roots of beech is similar to that of uninfected roots. Clowes (1954) pointed out that, although divisions occur which provide root-cap cells, the magnitude of the epidermis and root-cap tissues is much less than that of uninfected roots. He suggested that a continued dissolution of the cap tissue occurs as growth and differentiation proceed.*

It is possible now to view this problem, which was mentioned on pages 37 and 72, in a new light. D. C. Smith and his colleagues (*see*, for instance, Richardson *et al.*, 1967), in their study of lichens, have shown that, following photosynthesis in the algae partner, carbon assimilates move from the alga to the lichen fungus. It is of interest to note that in many of these fungi the assimilates are trapped as mannitol in the fungal layer. Extraction of the algae from the lichen thalli shows that they are, for a considerable time-period after isolation, 'leaky' of carbohydrate, and as they assimilate carbohydrates in photosynthesis, they lose a proportion of the products into the surrounding medium. In this respect they differ from their free-living counterparts which do not leak and which moreover produce, in the case of *Nostoc*, extracellular slime. The algae extracted from the lichen thalli become progressively less 'leaky' and more normal in a short time, if they are kept in culture. It is therefore a reasonable hypothesis that one of the effects of the fungus upon its host is to interfere with the normal enzymatic system which synthesises the cell membranes and external slime, so that the simple carbohydrate precursors, which may often yield glucose, may be directly available to be absorbed by the fungus.

If a similar mechanism occurred in ectotrophic mycorrhizas, it would simplify the explanations of the absence of a root-cap and the penetration of the fungi between the cells to form a Hartig net. The fungi might prevent the normal development of the cell-wall material, and so make available the carbohydrates normally destined to build it. This problem requires further investigation; but it clearly has a relevance not only to lichens and mycorrhizas but also to the penetration and exploitation of tissues by pathogens.

Respiration of Ectotrophic Mycorrhizas

It has, on occasion, been claimed that mycorrhizal roots have a higher respiration rate than uninfected roots. For instance McComb (1943) suggested that the increased absorption of nutrients, particularly of phosphorus, by mycorrhizal roots, might be explicable by a high

* *See* Appendix §4.

respiration rate. The work of Routien & Dawson (1943) is often quoted in this context; but their measurements, which are claimed to show greater respiration rates of mycorrhizas, relate to the respiration rates of root systems rather than to infected roots and mycorrhizal tissues. They indeed assumed that the net linear extent of mycorrhizal root systems of the pine seedlings which they used, were similar whether they were mycotrophic or not. They confirmed that an equal number of short roots were present per unit length on the long roots of mycorrhizal and uninfected plants. They went on to show that the respiration of mycorrhizal roots was between 1·4 and 4·7 times greater per unit length than that of non-mycorrhizal roots. This method of course ignores the elaboration by branching of the short roots and the consequent increase in the quantity of plant material in the mycorrhizal samples. The measurements of Routien & Dawson, although well-suited to their own purpose, give a false impression if quoted out of context.

Kramer & Hodgson (1954) compared the respiration of *Pinus taeda* roots on a number of bases. In this species, uninfected roots had a lower dry-matter content per unit fresh-weight than had mycorrhizal roots. The oxygen absorption of mycorrhizal roots was about 1·4 times as great as that of non-mycorrhizal roots on a fresh-weight basis, but it was considerably smaller on a dry-weight or surface-area basis. Similar results have been observed with *Fagus sylvatica*. On a dry-weight basis, uninfected and mycorrhizal roots respire at much the same sort of rate, namely 300–600 μl. O_2 per hour per 100 mg. dry-weight, and no explanation of function or difference of behaviour can really be linked with their rates of respiration.

As Lobanow (1960) has pointed out, not enough is known of the respiratory metabolism of ectotrophic mycorrhizas: the following account leans heavily on work with excised beech mycorrhizas, and so lacks an adequate comparison with other mycorrhizas.

As might be expected, the rate of respiration is sensitive to temperature. Harley *et al.* (1956) compared the rate of oxygen uptake in a wide range of oxygen concentrations at 10° and 20° C. They recorded temperature coefficients (Q_{10}) of 2–2·5 over the range of 10–100 per cent oxygen in the gas phase. The temperature coefficients have also been measured over various temperature ranges between 15° and 35° C. in air. The Q_{10} values for respiration decreased over temperature ranges above about 15 °C. The decrease was not so great as that shown by Q_{10} values of nutrient uptake—a point which will be reconsidered later, in dealing with nutrient absorption.

The dependence of oxygen uptake on oxygen supply again follows expectation. Fig. 10 shows the effect of varying the oxygen tension in the gas phase above respiring mycorrhizas in a Warburg apparatus. Oxygen uptake rises from zero at zero oxygen supply, to a high value in air. Carbon dioxide evolution almost equals the oxygen uptake between about 5 per cent of oxygen and air, so that over this range the respiratory quotient (RQ) lies close to unity. However, below 5 per cent of oxygen the RQ rises steeply. It will be seen later that nutrient absorption falls to a low value at low oxygen tensions, and indeed potassium uptake may cease at a little below 5 per cent of oxygen.

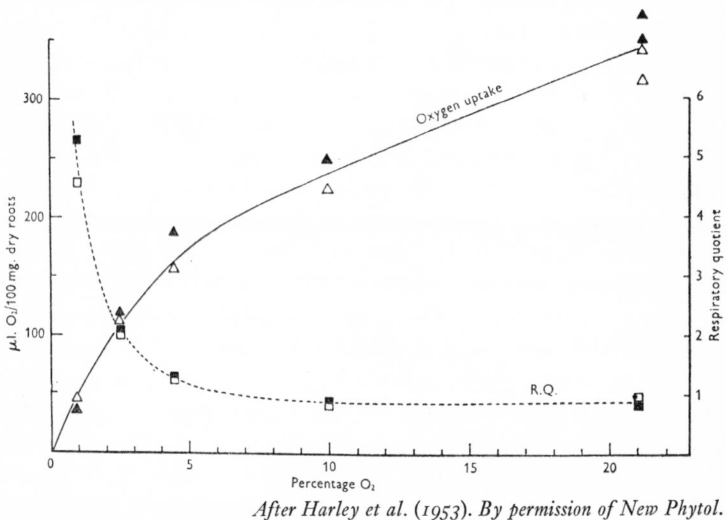

After Harley et al. (1953). By permission of New Phytol.

FIG. 10.—Graph indicating the respiratory behaviour of beech mycorrhizas in various oxygen concentrations. △, O_2 uptake; □, RQ.

The fall in oxygen uptake over the whole range of oxygen tensions below that of air, and especially below 10 per cent in the gas phase, posed the question of whether it was possible that an oxidase system with a low affinity for oxygen could be mediating the respiration in these beech mycorrhizas in whole or in part. The presence of a polyphenol oxidase system in the tissues (Harley *et al.*, 1956) seemed to give added point to this question, as also did the effects of certain inhibitors. The answer to it as shown below was, however, found to be in the negative.

Sodium azide, when applied at concentrations of 1 mM. or less, causes a rapid stimulation of both oxygen uptake and CO_2 production, with a concurrent increase of RQ (Harley & McCready, 1953). This effect of azide was observed to be closely similar to that of dinitrophenol, and in addition it was shown by Jennings (1958) that azide-treated roots exhibited the symptoms of uncoupling of phosphorylation and respiration. Hence azide appears to behave as an uncoupling agent rather than as an inhibitor of a metal-containing oxidase system. Indeed its action here on beech mycorrhizas contrasts with its usual effect on plant tissues that are known to contain cytochrome oxidase as the terminal oxidase in respiration (*see* Beevers, 1961, p. 96). Again, potassium cyanide and carbon monoxide, which are inhibitors of cytochrome oxidase in suitable conditions, were found either to stimulate or to leave unaffected the oxygen uptake of beech mycorrhizas (Harley *et al.*, 1956). Moreover diethyldithiocarbonate (Dieca), regarded as an inhibitor of copper-containing oxidase systems such as polyphenol oxidase, did not inhibit the respiration of beech mycorrhizas but slightly stimulated it. Thus doubt was cast on the possibility that any part of the oxygen uptake in respiration was mediated by the polyphenol system.

This whole question was further investigated by ap Rees (1957) and Harley & ap Rees (1959), who showed that the reaction of the respiration of mycorrhizal roots of beech to inhibitors was largely the reaction of the sheath tissue. The respiration of core tissue of mycorrhizas and uninfected roots was inhibited by azide and cyanide. Moreover, their oxygen uptake was also inhibited by carbon monoxide in the dark, and the inhibition was reversed in the light. This demonstrated that the host tissue in all probability contained a normal cytochrome system which mediated oxygen uptake. In a further examination, the sheath tissue was dissected away from the host and homogenized. The homogenates were shown to contain a functional succinic dehydrogenase and a cytochrome C oxidase system in sufficient quantities to account for a large part of the oxygen uptake of intact organs. The uncoupling effects of azide, and the stimulatory effects of other inhibitors, may therefore be reasonably ascribed to the presence, in the tissues of beech mycorrhiza, of an inhibitor-resistant oxidase bypass which comes into operation when the cytochrome C oxidase system is inhibited, and which is not coupled with oxidative phosphorylation. Rather similar features have been described in a few other plant tissues and have received similar explanations (e.g. Lips & Biale, 1966).

There are no grounds, therefore, for the view that beech mycorrhizas constitute a type of nutrient absorbing organ in which there occur respiratory and electron transport systems that are not mediated by cytochrome oxidase. They do indeed exhibit, as do some other tissues, peculiar complexities which may merit investigation; but these may not be relevant to the general problems of the functioning of ectotrophic mycorrhizas. Added point is given to this conclusion by the work of Kramer & Hodgson with pine mycorrhizas (1954). They have shown that these react in a more expected fashion to respiratory inhibitors, and that their respiration rates are diminished by such inhibitors as are those of uninfected roots.

The observation that the reactions of beech mycorrhizas to inhibitors are primarily a property of their fungal sheath, poses the question of whether the sheath tissue of intact mycorrhizas is responsible for the greater part of the uptake of oxygen and the release of carbon dioxide in respiration. The examination of the respiration rates of isolated sheath and core tissue does not entirely settle this question. Harley *et al.* (1956) reported two experiments in which oxygen uptake of free cores was not greatly different from that of their free sheath tissue in air, although it was significantly lower in lower oxygen tensions. This demonstrates that the fungal sheath on a weight basis respires much faster than the core, because it comprises only about 39 per cent of the total dry-weight of mycorrhizal apices. This type of experiment is, however, not entirely satisfactory, because there can be little doubt that the wounding occasioned by dissection gives rise to stimulation of respiration by both components. However, even allowing for this, and also for the fact that oxygen supply to the host is increased when it is freed from the sheath, there can be no doubt that the fungal respiration comprises at least half of the gas exchange in the intact state. Hence the utilization of carbohydrate and other respiratory substrates by mycorrhizas must be in large part a utilization by the fungal sheath.

When beech mycorrhizas are kept in water or buffer over a period of time their respiration rate falls to a low value. This fall is not due to exhaustion of carbohydrate. Certainly, as shown by Harley & Jennings (1958) and by D. H. Lewis & Harley (1965), carbohydrates do diminish in quantity during this period; but the rate of respiration is not dependent on their concentration. Mycorrhizas stored in this way not only react, by increased respiration, to the application of nutrient salts, but they also do so to the application of the uncoupling agent dinitrophenol, and to azide and cyanide (Harley *et al.*, 1956). In these cases the increased respiration involves a stimulation of both oxygen uptake and carbon dioxide emission. Moreover, a fall in the rate of respiration takes place during storage, whether sugar is present in the external solution or not. When it is present, for instance as sucrose or glucose, the fall is somewhat less steep than when it is absent, and sugar is absorbed in quantities far in excess of those of which the carbon is lost as CO_2 (Harley & Jennings, 1958). The subject will be further considered in Chapter VI, when the association of nutrient absorption and respiration is discussed.

Acid Metabolism and Dark Carbon-dioxide Fixation

Beech mycorrhizas contain the expected acids of the tricarboxylic acid cycle in their tissues. Extraction of them with alcohol, followed by separation on resins, gives acidic, basic, and neutral fractions. If the acid fraction is separated by gradient elution from resin, the quantities of non-volatile acids can be estimated. The relative quantities of these are normal for plant tissue (Carrodus, 1965). If fresh mycorrhizas are fed with [^{14}C] acetate, these acids become labelled in the sequence expected by the operation of the TCA cycle (Carrodus, 1965).* It is moreover to be observed that, although the neutral fraction which contains sugars and the ether-soluble fraction which contains lipids become relatively little-labelled, the basic fraction which contains amino-acids rapidly becomes heavily labelled (Fig. 11). This is in line with the results obtained with other tissues such as those of maize (*see* Harley & Beevers, 1963; MacLennan *et al.*, 1963). In beech mycorrhiza, as in the plant material used by the

* *See also* Carrodus & Harley (1968).

authors quoted, it is glutamate and glutamine that account for most of the label in the amino fraction—presumably by equilibration with the α-ketoglutaric acid of the TCA cycle. These results serve to show that it is likely that the tricarboxylic acid cycle functions normally in beech mycorrhizas, and that the synthesis of amino compounds may occur by reductive amination of α-ketoacids of the cycle.

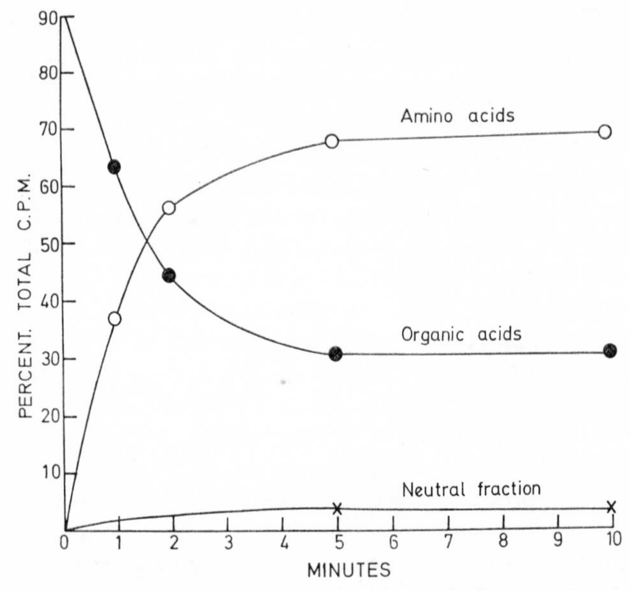

FIG. 11.—Graph showing the relative radioactivity of various soluble fractions during absorption of [^{14}C] acetate by beech mycorrhizas in short time-periods.

Dark fixation of CO_2 serves to replenish the acids of the cycle. If mycorrhizas are treated with [^{14}C] bicarbonate, carbon dioxide is absorbed and radioactivity is found—especially in malate amongst the acids and in glutamate amongst the basic substances. In the presence of ammonium salts in the bicarbonate medium, the CO_2 fixation rate is greatly stimulated. At the same time ammonia is absorbed and glutamine is synthesized. The increase in fixation results particularly in an increase of radioactivity in the glutamine (Harley, 1964). This process of dark carbon-dioxide assimilation is important in the absorption of nitrogenous substances and cations, and will be discussed again in that context (pp. 134–136).

Summary of Carbon Metabolism

Ectotrophic mycorrhizas, consisting as they do of two very different biological components—fungus and higher plant—exhibit features of the carbon metabolism of both. The two kinds of tissue have very different internal carbohydrate substances. It is of special significance that mannitol, trehalose, and glycogen, which are the important storage carbohydrates of the fungal layer, are not found in the host-plant and are not readily absorbed by it if made available. By contrast the host carbohydrates, sucrose, glucose, and fructose, are readily absorbed and utilized by the fungus. This allows the hypothesis to be formulated that the fungus can absorb carbohydrate from the host and convert it into forms that are unavailable for reciprocal flow. This concept of a 'sink' in the fungal layer explains the rapid translocation of photosynthate to the fungal sheath which has been observed in whole pine seedlings by Melin & Nilsson (1957) and in excised mycorrhizas of beech by D. H. Lewis & Harley (1965*b*).*

The complexity of the storage carbohydrates found in mycorrhizas (*see* Table XX) emphasizes once again the care that must be exercised in drawing conclusions from carbohydrate analyses.

TABLE XX

SUGARS IDENTIFIED AS PRESENT IN THE 80 PER CENT ALCOHOL-SOLUBLE, AND IN THE HYDROLYZED INSOLUBLE, FRACTIONS OF BEECH MYCORRHIZAS. Hydrolysis by refluxing 2 hrs. with 1·5 N H_2SO_4. (Data from D. H. Lewis & Harley, 1965.)

Compound	Soluble				Hydrolyzed Insoluble			
	Myc.	Sheath	Core	Uninf.	Myc.	Sheath	Core	Uninf.
Inositol 1	+	+	+	+	—	—	—	—
Inositol 2	+	—	+	+	—	—	—	—
Trehalose	+	+	*	—	—	—	—	—
Sucrose	+	—	+	+	—	—	—	—
Glucose	+	+	+	+	+	+	+	+
Fructose	+	?	+	+	—	—	—	—
Mannitol	+	+	*	—	—	—	—	—
Galactose	—	—	—	—	+	+	+	+
Mannose	—	—	—	—	+	+	*	tr.
Arabinose	—	—	—	—	+	+	+	+
Xylose	—	—	—	—	+	+	+	+
Fucose	—	—	—	—	+	—	+	+
Ribose	—	—	—	—	+	+	+	+
Rhamnose	—	—	—	—	+	—	+	+
Deoxyribose	—	—	—	—	+	+	*	—

+ = Present.
* = Present wholly or mostly in the fungus of the Hartig net.
? = Doubtfully present in fungal tissue.
tr. = Trace present.
— = absent.

The results of the study of respiratory metabolism show that, although the part played by the fungal layer in oxygen absorption and CO_2 emission is a large part of the total gas exchange, there may be no great peculiarities of respiration in mycorrhizas. The behaviour of beech mycorrhizas in respect of reaction to inhibitors is of much general interest; nevertheless it may not be important to the peculiar functioning of mycorrhizas, because pine mycorrhizas do not behave in the same way. But although oxygen

* *See* Appendix §5.

uptake, electron transport, and also the operation of the TCA cycle, seem at present to be normal in ectotrophic mycorrhizas, many problems in this field still await solution.

Beech mycorrhizas and also pine mycorrhizas exhibit dark fixation of CO_2 which, in the case of beech, results in the formation of malate and may be held *inter alia* to be a mechanism of replenishment for the acids of the TCA cycle when these are utilized in synthesis.

Although so much of the information on carbohydrate metabolism of ectotrophic mycorrhizas is derived from work on a single species, *Fagus sylvatica*, it is most probably widely applicable. This is made all the more likely because recent comparative studies of many kinds of mutualistic and parasitic symbioses have shown them to exhibit much similarity of carbon metabolism. The symbioses in question which involve photosynthetic and carbon-heterotrophic partners include lichens, ectotrophic mycorrhizas, rust and powdery-mildew diseases, and symbioses of invertebrates with algae. In all these, the autotrophic partner releases to the heterotroph its normal translocatory substance or a special stable product of the photosynthetic process. Movement of carbon compounds to and into the heterotroph is a movement from source to sink, and often involves the production in the heterotroph of carbohydrates which are not utilizable by the autotroph. In many cases hormones or other special substances may be involved in the process. These studies, which have received comparative comment by D. C. Smith *et al.* (*Biol. Rev.*, 1969, **44,** 17–90), by D. H. Lewis (*Lichenologist*, **4,** 1969, in press), and by J. L. Harley (*Trans. Brit. Mycol. Soc.*, **51,** 1–11, 1968), are not only of great intrinsic interest but also emphasize the need for further work on the carbon metabolism of ectotrophic mycorrhizas. These are examples of symbiosis between fungus and autotroph in which there is a prolonged steady-state condition that may be experimentally exploited more readily than parasitic symbioses, where the tissues exhibit a more rapid continual trend to senescence and death.

VI

THE PHYSIOLOGY OF SALT ABSORPTION BY ECTOTROPHIC MYCORRHIZAS

THE correlation of intense mycorrhizal infection with unbalanced or deficient nutrient supplies in the soil has been described. A greater degree of infection occurs in raw-humus soils than in soils that are well furnished with nutrients in available form. This correlation of intense infection with mineral deficiency has been held by some to prove that mycorrhizal organs must increase the rate of nutrient absorption from the soil. This argument is, of course, specious. Mycorrhizal infection may be expected to exert its greatest effect on the activities of the host where it is developed most intensely, but the nature of its influence, if any, can only be determined by experiment. In this chapter we ask the question 'Are mycorrhizas capable of absorbing substances from the soil and, if so, what is the magnitude of their power of absorption and what factors affect the rate of absorption?'

A. B. Frank's original observations, which have been amply confirmed by later observers, showed that the root systems of Betulaceae, Fagaceae, and Pinaceae were normally infected under natural conditions. Trees deficient in or lacking mycorrhizal development are only found growing under special conditions: where within their normal geographical range, they are grown in heavily manured, usually garden, sites; where these trees occur outside their geographical or habitat range; or where the fungi concerned are absent or inhibited by local conditions. The first useful indications of the possible mode of functioning of mycorrhizal rootlets were accordingly obtained, in large measure, from a comparison of the behaviour of trees planted in unusual places. In 1917, Elias Melin, early in his extensive researches on ectotrophic mycorrhizas, observed that self-seeded pine and spruce seedlings on recently drained peat bogs usually developed very poorly. The exceptions to this rule—seedlings developing normally—were found to have formed mycorrhizas, presumably through infection by wind-borne spores newly brought by natural agencies to the peat soil. The uninfected seedlings showed signs of nutrient deficiency such as are usually associated with nitrogen starvation. Since that time, very numerous examples of a similar nature have come to light, especially where foresters have attempted to introduce exotic trees into new localities. Thus Kessell (1927) described numerous failures of coniferous nursery stock in Western Australia in which the seedlings, mostly of pines, germinated well and at first grew in a healthy manner but later developed deficiency symptoms. Such simple applications of nutrients as were made were not successful in preventing death of the stock, but the introduction of small quantities of soil which had grown pines successfully elsewhere was effective in causing the seedling pines to recover and develop normally. In each case, failure was associated with the absence of mycorrhizal development, and success was correlated with its occurrence.

Roeloffs (1930), who worked in the former Dutch East Indies, described how the interplanting of unhealthy seedlings of *Pinus merkusii* and *P. khasya* with fully mycorrhizal pine seedlings from elsewhere, caused the spread of mycorrhizal development to the other seedlings in his nursery beds and their recovery of health. Similar experience has been obtained with pines in Southern Rhodesia (Anon., 1931), in the Philippines (Oliveros, 1932), in Malaya (Griffiths, 1965), and elsewhere. M. C. Rayner (1938) summarized the answers to a questionnaire sent by the Director of the then Imperial Forestry Institute, Oxford, to many forestry departments in the British Commonwealth; they had had similar experiences to those cited. It is also a common experience to find that, on many 'difficult' sites, the surviving self-sown seedlings or saplings are all mycorrhizal. This was true of the surviving birch seedlings studied by Dimbleby (1953) on *Calluna* moorland.

Although most of the experiments involving mycorrhizal inoculation have been performed with conifers, similar results have been obtained with some angiosperm trees. Lobanow (1960) gives examples of comparisons between mycorrhizal and non-mycorrhizal oaks made in the treeless areas of Russia. Here again, mycorrhizal development is associated with successful establishment and increased growth of young trees. The results are usually expressed in terms of growth of shoot and root in length or weight, but some workers also give estimates of nutrients absorbed from the soil.

Details of the nature of the differences between the non-mycorrhizal seedlings and those infected by chance in nursery sites are given by McComb (1938, 1943). A new nursery site in Iowa, on which a number of species of coniferous seeds had been sown, was mulched with pine needles; soon after mulching, patches of seedlings were observed to be developing vigorously, having deep-green needles, although the remainder were unhealthy and had purplish foliage. The healthy seedlings were found to have become infected with mycorrhizas, and comparative analyses of these and the remainder of the seedlings were made. The results with *Pinus virginiana* are given in Tables XXI and XXII. Not only were the plants which had become infected healthier, taller, and heavier, but a remarkable difference was observable in the numbers of short roots on the seedlings. The mycorrhizal plants possessed more than double the number of short roots—a feature which could be explained either by a stimulation of short-root production or by a prevention of abortion of the short laterals, or again, by the great increase of total root size. More striking still were the results of analyses (Table XXII), which showed that not only were the quantities of nitrogen, phosphorus and potassium present in each mycorrhizal seedling greater than in uninfected seedlings, but the quantity of phosphate was disproportionately greater on a dry-weight basis.

TABLE XXI

COMPARISON OF HEIGHTS AND NUMBERS OF SHORT ROOTS OF SPONTANEOUS MYCORRHIZAL AND NON-MYCORRHIZAL SEEDLINGS OF *Pinus virginiana*. Results of McComb (1938). (Samples of twenty seedlings.)

Type of seedling	Height cm.	Total short roots	Mycorrhizal short roots	Uninfected short roots
Green, healthy	2·7	672	350	322
Purplish foliage	2·0	297	7	290
Significance	P > 0·05	P > 0·01	P > 0·01	—

TABLE XXII

ANALYSES OF SAMPLES OF SPONTANEOUS MYCORRHIZAL AND NON-MYCORRHIZAL SEEDLINGS OF *Pinus virginiana*. Results of McComb (1938). (Means of four samples of twenty seedlings each.) Weights in mg.

	Mycorrhizal seedlings	Non-mycorrhizal seedlings	Significance of difference
Fresh-weight	1,230	592	P > 0·01
Dry-weight	323	152	P > 0·01
Nitrogen per cent dry-wt.	1·78	1·88	P > 0·05
Nitrogen per seedling	5·75	2·87	P > 0·01
Phosphorus per cent dry-wt.	0·185	0·097	P > 0·01
Phosphorus per seedling	0·60	0·15	P > 0·01
Potassium per cent dry-wt.	0·66	0·62	—
Potassium per seedling	2·17	0·96	P > 0·01

Similar results were obtained for *Pinus strobus* and *Picea rubra* by Mitchell *et al.* (1937). In this case a complicated experimental arrangement was set up in carefully prepared beds in which some of the treatments were inoculated with mycorrhizal material and some were left as controls. In some of the uninoculated and unfertilized beds, sporadic airborne infection occurred. The infected plants were soon recognizably different from their neighbours in growth and in the green colour of their needles.

Table XXIII shows the time-sequence of spread of mycorrhizal development from the infection centres, which was similar to the spread of infection described by Roeloffs (1930) and Robertson (1954). In the other unfertilized beds, which had been inoculated by addition of soil containing mycorrhizal fungi, full infection developed during the first year (1935). This indicated that the slow development and spread of mycorrhizas in the uninoculated controls was due to the arrival of suitable fungi by casual means, probably as wind-blown spores, and that the conditions of soil and seedlings were not the determining factors. On the other hand, the beds receiving treatments with fertilizers showed no signs of casual infection, nor did mycorrhizal development occur on them even when they had been inoculated during the first year. In the second year, when considerable growth of the seedlings had taken place, some mycorrhizal development occurred on the fertilized beds. Indeed, the whole sequence exactly agrees with the conclusions drawn from the experimental findings of Hatch (1937), Björkman (1949), and others, concerning the effects of mineral nutrition on susceptibility to mycorrhizal infection.

TABLE XXIII

SPONTANEOUS DEVELOPMENT OF MYCORRHIZAL INFECTION IN *Pinus strobus* ON UNFERTILIZED, UNINOCULATED BEDS OF NUTRIENT-DEFICIENT SOIL. Seeded 29 April 1935. Observations of Mitchell *et al.* (1937).

Date	1 Oct. 1935	15 April 1936	10 June 1936	20 Aug. 1936	20 Oct. 1936
Mycorrhizal development	nil	scattered groups	10 %	75 %	100 %

The results of analysis of the amounts of nitrogen, phosphate and potassium present in the seedlings, are shown in Table XXIV. These are remarkably similar to those of McComb (*see* Table XXII) in indicating, for mycorrhizal seedlings, a disproportionately large mineral absorption as compared with dry-weight. Indeed, between the two dates of sampling, no significant absorption of any of the elements N, P, or K, into the uninfected seedlings took place, whereas the infected seedlings increased their content as regards both the amounts per plant and the amounts as a percentage of dry-weight.

No particular advantage would be gained by giving a greater number of examples of this kind. It must be borne in mind that the sampling technique adopted here is essentially faulty. It cannot be assumed that the seedlings which become infected by casual means are a random sample of the populations of seedlings under observation. We have already seen in a previous chapter that infection occurs only after a given stage of development is reached, i.e. after the first foliage leaves have spread. It is at least likely therefore, that early germinators and rapid growers in the populations will be disproportionately numerous amongst the plants in the mycorrhizal samples. It is only fair to state however, that in the last-described example, graded seeds of fairly uniform weight were used and that the seedlings developed with remarkable uniformity.

TABLE XXIV

ANALYSES OF SPONTANEOUSLY INFECTED AND UNINFECTED SEEDLINGS OF *Pinus strobus* (cf. Table XXIII). Results of Mitchell *et al.* (1937). Weights in mg.

Date	10th July 1936		20th August 1936	
Infection	Mycorrhizal	Non-mycorrhizal	Mycorrhizal	Non-mycorrhizal
Per cent short roots infected	68·6	1·2	70·2	9·5
Total dry-wt.	207·9	160·2	337·2	180·6
Root dry-wt.	72·3	73·0	127·4	74·6
Shoot dry-wt.	135·6	87·2	209·8	106·0
Root/shoot ratio	0·53	0·84	0·61	0·71
N per cent dry-wt.	1·49	1·23	1·60	1·20
N per seedling	3·1	2·0	5·4	2·2
P per cent dry wt.	0·13	0·08	0·21	0·07
P per seedling	0·27	0·13	0·72	0·13
K per cent dry-wt.	0·45	0·55	0·63	0·45
K per seedling	1·03	0·88	2·12	0·81

The observations quoted above constitute a sample of those of which results have been published. One must realize that there have, no doubt, been many cases where no striking differences were observed between different sets of plants which had or had not become spontaneously mycorrhizal. For obvious reasons these would not be published. One must not, from these results, come to the conclusion that mycorrhizal infection always leads to seedling improvement in poor soils. Improvement in growth is often correlated with the onset of mycorrhizal infection, and we shall see that there is good reason to believe that these occurrences may be causally related. The danger of assuming that all defective seedling growth is in some way related to poor or inadequate mycorrhizal development seems self-evident; nevertheless the assumption has often been made, to the great detriment and discredit of the study of mycorrhizal phenomena.

A series of experiments in which there was very much more control of conditions than in those cited above has been described by Routien & Dawson (1943), using *Pinus echinata*. As a substrate, purified colloidal clay mixed with pure quartz sand was employed. The clay was suspended in hydrochloric acid and washed in distilled water, so that the adsorbed bases were exchanged for hydrogen ions. The final material had a base exchange capacity of 32·4 mE at pH 6·2 per 100 gm. of dry material and contained no soluble or exchangeable metallic ions. The hydrogen was replaced in varying proportions by calcium and barium, calcium and iron, or calcium alone, and the clay was then mixed with pure quartz sand. Mycorrhizal inoculation was made where required, by addition of small amounts of infected soil, and seedlings were grown upon the artificial soils with or without addition of salts.

The results of estimates of dry-weight, mycorrhizal development, and the quantity of nutrients absorbed, give a convincing demonstration that nutrients were more readily absorbed by mycorrhizal roots. The amounts of total bases, of potassium, calcium, magnesium, and iron, and also of phosphorus, which were absorbed by the mycorrhizal plants from soils having a low level of base saturation, greatly exceeded those absorbed by non-mycorrhizal plants. Mycorrhizal plants alone grew normally and absorbed salts from soil in which the amount of adsorbed hydrogen corresponded to 37 per cent of the base-exchange capacity.

Comparisons of infected and uninfected seedlings were made by Hatch (1937), Young (1940), and Finn (1942), using pure cultures of fungi as the source of inoculum for the mycorrhizal plants. Hatch used a soil from Cheyenne, Wyoming, as his basic substrate, on the assumption that this naturally occurring treeless soil would be deficient in mycorrhizal fungi. He grew *Pinus strobus* as his test organism, and compared inoculated and uninoculated seedlings after a year's growth. The results are given in Table XXV. In a remarkable way they agree with the results

TABLE XXV

COMPARISON OF *Pinus strobus* SEEDLINGS GROWN IN 'PRAIRIE' SOIL WITH AND WITHOUT PURE-CULTURE INOCULATION. Results of Hatch (1937). Weights in mg.

				Nitrogen		Phosphorus		Potassium	
Treatment	Sample	Dry-wt.	Root/ shoot ratio	% dry-wt.	per seedling	% dry-wt.	per seedling	% dry-wt.	per seedling
Inoculated	1	448·4	0·67	1·20	5·39	0·189	0·849	0·775	3·47
	2	360·9	0·89	1·28	4·62	0 202	0·729	0·713	2·57
	mean	404·6	0·78	1·24	5·00	0·196	0·789	0·744	3·02
Uninoculated	1	360 7	1·02	0·70	2·51	0·074	0·268	0·539	1·94
	2	300·0	1·37	1·06	3·16	0·076	0·229	0·347	1·04
	3	301·4	1·02	0·80	2·40	0·070	0·211	0·390	1·17
	Mean	302·7	1·04	0·85	2·69	0·074	0·236	0·425	1·38

of the casual infections given above, and show that the errors due to the faults in sampling were not excessive in those cases. Young's results, although they do not include such detailed analyses, are similar to those of Hatch. In his experiments the uninoculated seedlings showed symptoms of nutrient deficiency and soon died, whereas those inoculated with mycorrhizal fungi thrived. Finn used *Pinus strobus* and again obtained results within one growing-season which showed full agreement with those of Hatch—a considerable increase of dry-weight, decrease of root/shoot ratio, and a greater absorption of nitrogen, phosphorus and potassium in the inoculated than in the uninoculated samples. There was, however, a much less striking difference between the treatments in the quantities of the mineral elements per unit dry-weight—except in the case of nitrogen (Table XXVI).

TABLE XXVI

COMPARISON OF *Pinus strobus* SEEDLINGS INOCULATED WITH GRANULATED CULTURE-MEDIA OF KNOWN MYCORRHIZA-FORMERS, AND UNINOCULATED SEEDLINGS. Results of Finn (1942).

	Inoculated seedlings	Uninoculated seedlings
Per cent short roots infected	87	10
Dry-wt. mg.	223·4	92·9
Root/shoot ratio	1·15	1·49
N mg. per seedling	2·05	1·17
P mg. per seedling	0·30	0·24
K mg. per seedling	1·52	0·98
N per cent root dry-wt.	0·98	0·79
N per cent shoot dry-wt.	0·78	0·71
P per cent root dry-wt.	0·13	0·16
P per cent shoot dry-wt.	0·14	0·15
K per cent root dry-wt.	0·64	0·58
K per cent shoot dry-wt.	0·71	0·68

In all these experiments where great differences were noted between mycorrhizal and non-mycorrhizal seedlings, the substrates used were deficient in one or more essential nutrients. Not only is the development of mycorrhizas less intense in soils that are well supplied with nutrients, but, also, it is often very difficult to obtain consistently infected samples in such circumstances. In the experimental beds of Mitchell *et al.* (1937), mycorrhizal development took place readily in the unfertilized beds even where no inoculum had been applied, but mycorrhizal roots were infrequent on all the root systems of the seedlings in the fertilized beds even when these had been inoculated. It has, therefore, proved difficult, and it is perhaps unimportant, to estimate the effects of mycorrhizal infection upon the growth and salt-absorption of young seedlings in highly nutrient soils. In very many cases, and perhaps in all, seedlings can be grown to a large size and to a healthy and vigorous condition by suitable treatments in the absence of mycorrhiza. As an example Table XXVII shows a comparison between non-mycorrhizal *Pinus strobus* seedlings grown for two seasons on heavily fertilized soil, with others of similar age that were unfertilized but fully mycorrhizal.

This kind of result has been obtained frequently, as for instance by

Hatch (1937) and Addoms (1937), and it has sometimes led to misunderstanding. Thus Addoms thought her results showed that mycorrhizal infection was not of great importance; such an assertion needs considerable qualification if it is to be made at all. Under natural forest conditions, the Loblolly Pine which she used in her experiments resembles other trees with ectotrophic mycorrhizas; it is normally infected, the mycorrhizal roots being its absorbing organs while its habitat is soil of low nutrient-supplying power.

Since plants which are normally mycorrhizal can be raised successfully as uninfected seedlings and can make very good growth in a short time

TABLE XXVII

Pinus strobus SEEDLINGS WITH AND WITHOUT MYCORRHIZAS. Results of Mitchell *et al.* (1937).

Treatment	Mycorrhizal development	Dry-wt. mg.	Nitrogen % dry-wt.	Nitrogen mg./seedling	Phosphorus % dry-wt.	Phosphorus mg./seedling	Potassium % dry-wt.	Potassium mg./seedling
None	abundant	337	1·60	5·39	0·21	0·72	0·63	2·12
Heavily fertilized	none	550	2·50	13·75	0·24	1·32	1·10	6·05

under heavily manured nursery conditions, a question of practical importance arises. Is it more important to the forester to grow large nursery stock although it is uninfected, or to ensure that the seedlings become mycorrhizal in the nursery at the expense of the degree of growth? It must at once be emphasized that this is a matter for experiment, the design of which must take into consideration the comparative physiology of the seedling transplants, the comparative ease of lifting, the susceptibility to damage during transit, and so on. In Great Britain and much of the rest of western Europe, in heavily afforested areas where a variety of trees developing ectotrophic mycorrhizas are habitually grown, the transplants will fairly readily become infected with mycorrhizal fungi when they are planted out in forest conditions, and the problem is purely a practical one. Where exotics, which are not habitually grown, are being raised, it is clearly good policy to raise infected seedlings by suitable inoculation and treatment—in order to ensure a source of inoculum in forest conditions.

In assessing the practical points of lifting, carriage and transplantation, the form of the root systems and the susceptibility to desiccation of mycorrhizal and non-mycorrhizal seedlings are worth consideration. It is evident that, in forest conditions, uninfected roots in the F and H layers are much more easily damaged by drought and frost than are mycorrhizas (Cromer, 1935; Harley, 1940), and this might be accepted as a pointer in the designing of tests. Theorizing on this subject must, however, be rigorously avoided. Björkman (1953), from his own experimental experience, is of the opinion that well-grown seedlings from nurseries in which manurial treatment is not too great to eliminate mycorrhiza formation, are generally the best. Whatever conclusions are reached in the examination of these important practical problems, they should not be confused with the results of physiological experiments which may help in their solution or else be quite irrelevant to them. A further discussion of these problems will be given at the end of the next chapter, when all the relevant data on the physiology of mycorrhizal organs has been considered.

Explanations of the Effects of Infection on Growth and Absorption

The increased rate of growth of seedlings which follows upon mycorrhizal infection in soils of low nutrient supplying power has been explained in diverse ways. Many of the early experiments on the establishment of exotics seemed to suggest that mycorrhizal infection resulted in healthy growth where treatment with mineral fertilizers did not. Kessell (1927) found this to be so, and McComb & Griffith (1946), working with *Pseudotsuga taxifolia* (*P. douglasii*), had similar experiences —as did Dale *et al.* (1955) with *Pinus banksiana*. Such results inevitably gave rise to hypotheses that mycorrhizal fungi produced some growth-promoting stimulus for their hosts, and such views are still held. In the areas of Western Australia where Kessell worked, it has now become apparent that nutrient deficiencies, of a more varied kind than nitrogen, phosphorus, and potassium deficiencies, are common (Stoate, 1950), and may be cured by suitable application of micronutrients. It is simpler, therefore, to erect the hypothesis that the increased growth of mycorrhizal seedlings, and their increased rate of absorption from deficient soils, arise directly either from an increase in the ability of their root systems to absorb nutrients or from an increase of availability of substances around their roots owing to fungal activity. In spite of this, the growth-promoting hypothesis, which must make the assumption that increased manurial absorption is an indirect effect of infection, has been subjected to some experimental testing. Lindquist (1939) obtained results purporting to show that the metabolic products of mycorrhizal fungi stimulated spruce growth, while those of harmless or indifferent fungi had no effect, and those of pathogenic fungi diminished growth. However, the results are not sufficiently clear-cut for acceptance; indeed, as will be seen, experimental examination of other hypotheses makes them appear more likely than Lindquist's.

M.C. Rayner's experiments upon Wareham Heath soils (Rayner & Neilson Jones, 1944) led her to suggest that mycorrhizal infection exerted a growth-promoting effect quite apart from the absorption of nutrients. 'The initial stimulus cannot therefore be due to improved nutrition depending upon mycorrhizal action in the ordinary sense.' She was indeed of the opinion that the organic composts which she applied to Wareham soil, and which brought about great stimulation of growth in the seedlings, were not active by virtue of their manurial content. The composts, she believed, affected the substrate in such a way as to make it suitable for the growth of mycorrhizal fungi. Subsequent researches showed that the soil contained toxins which inhibited the growth of mycorrhizal fungi and which were rendered innocuous or were destroyed after the application of the composts. A rationale for these observations was discovered by Brian *et al.* (1945), who demonstrated the presence in the soil of antibiotic-producing Penicillia which could antagonize the mycorrhizal fungi. Part of the effect of the composts could therefore be ascribed to their effect in changing the substrate for soil micro-organisms and so altering the soil population as to eliminate the antagonists. At the same time the manurial effects of the composts could not be assumed to be negligible. For instance, composts C_5, C_8, C_1, C_7 and C_4 had this decreasing order of phosphate content and the same decreasing order in stimulating the growth of pine. Although a great deal of valuable practical information was obtained by Dr. Rayner in her researches, many of her experiments have proved too complicated to be completely interpreted. A good deal of harm has moreover been done to the practical forester's understanding of mycorrhizal phenomena, by the mystical and obscurantist views which sprang up concerning the value and mode of action of organic composts upon seedling growth.

In any event a growth-promoting substance or influence produced by the fungus cannot alone explain the facts with any credibility. Mycorrhizal plants, as shown by the analyses given above, absorb nutrients from the soil more readily than do non-mycorrhizal plants, so that they contain more per unit weight. No one has yet suggested a model for this, based on growth-promoting substances, which can assist experimentation.

A further source of difficulty can and does arise from the use of bulky doses of humus as a source of mycorrhizal inoculum; for the effect of the organic matter and its constituent flora upon the availability of minerals may be very great. An example is given in Table XXVIII of observations on the growth of beech seedlings in two nurseries—one on a woodland site, the other on agricultural land. Here the addition of considerable quantities of beech humus, with or without the addition of compost (prepared to the recipe C_4 of Dr. Rayner), was made to the soil before sowing. As can be seen, the additions significantly stimulated the seedlings on the agricultural land which was initially low in humus and in its C/N ratio; but the humus and its flora diminished seedling growth in the woodland site, which was high in humus and in C/N ratio. The diminishing effect may be safely ascribed to the increased competition for nutrients between the microflora and the seedlings.

Returning to the simpler experiments, if we assume that increased absorption of materials into the host plant follows upon mycorrhizal infection and causes increased growth of the host, many points arising from the results fall into line. Mitchell (1934, 1939), Gast (1937) and others have shown that seedling growth and nutrient uptake in carefully controlled pot experiments with sand culture are closely related to nutrient availability. The acceleration of nutrient absorption by increase of nutrient supply is usually associated with a diminution of the root/shoot ratio. In the experiments described above, this was exactly the effect found with mycorrhizal infection. Mycorrhizal plants have higher internal nutrient levels and lower root/shoot ratios than non-mycorrhizal

TABLE XXVIII

EFFECT OF BULKY HUMUS INOCULA (I) AND OF RAYNER'S C_4 COMPOST (C) ON BEECH SEEDLINGS IN SOILS OF HIGH (H) AND LOW (L) HUMUS CONTENT. Data of A. R. Clapham & J. L. Harley (unpublished).

Measurement	Soil	None	Treatments I	C	CI	Significant changes $P > \cdot 05$
Height cm.	L	13·08	14·83	14·34	20·86	C, I and CI increase
	H	13·30	12·64	20·22	18·51	C increases, I decreases
Length root cm.	L	25·68	23·74	26·05	26·78	C increases
	H	23·75	22·32	25·88	23·08	I decreases
Dry-weight shoot mg.	L	484	786	826	1607	C, I and CI increase
	H	632	534	1428	1142	C increases, I decreases
Dry-weight root mg.	L	1389	1678	1982	2799	C and I increase
	H	1123	1002	2087	1789	C increases, I decreases

plants. Again, an increase in the number of short roots is observed when the nutrient supply is raised from starvation levels into the working range (Mitchell, 1939), and this is exactly what is observed after mycorrhizal infection.

Two main mechanisms have been postulated to explain the way in which nutrients become more available to mycorrhizal than to non-mycorrhizal plants. Burges (1936) put forward the view that the mycorrhizal fungi behaved in a way exactly analogous to other soil fungi, causing a breakdown of soil material and thus bringing nutrients into solution. In his view these fungi are essentially parasitic, and their morphological and histological effects on the host are irrelevant to their stimulatory activity. This last arises from the local increase of soluble material near the root surface which might far offset any damage that their parasitic activity might cause. Although this view was really based only on surmise, it cannot be dismissed without detailed consideration; certain lines of evidence may be interpreted as arguing in its favour. M. C. Rayner (in M. C. Rayner & Neilson Jones, 1944) observed a stimulation of seedling growth after inoculation of the soil (sometimes with compost addition) before mycorrhizal organs could be found on the seedlings.

Such an effect could be explained by the release of nutrients into the soil by the fungus of the inoculum. Some of the work of Levisohn (1953, 1956), in which plants not producing ectotrophic mycorrhiza were stimulated by ectotrophic mycorrhizal fungi, could be similarly explained. The experiments of Gerretsen, of Timonin, of Ark, and of Hoagland *et al.*, which were described in Chapter II, showed that rhizosphere associates of angiosperms could greatly affect the growth-rate of their hosts by changing the supply of nutrients to them. However, two lines of evidence throw doubt upon the importance of such processes in the physiology of ectotrophic mycorrhizas.

First, the observations on the behaviour of the mycelia of mycorrhizal fungi on complex substrates and more particularly in the face of competition with other micro-organisms suggest that many of them have a very limited activity in the soil. Of course, our present knowledge of the distribution of their mycelia is too scanty to be fully relied upon, but the pattern of their behaviour more closely resembles that of obligate root-inhabiting fungi than that of the active soil-inhabitants. This is not to say that mycorrhizal fungi may not be capable of considerably more diverse activity when attached to their hosts than when they are free-living; but such a view takes us away from what is known of them and into the realm of surmise.

Secondly, experimental analyses have been made of the soil in the neighbourhood of mycorrhizal and non-mycorrhizal seedlings in an attempt to determine whether an increase of soluble or available soil nutrients could be found (Mitchell *et al.*, 1937; Stone, 1950). Mitchell and his colleagues pointed out that, in experiments where pine seedlings were grown in nutrient solutions, relatively great increases in concentration in the working range of any nutrient were needed to bring about a relatively small increase in internal nutrient level. Hence, if mycorrhizal fungi exerted their effects by an external increase of available nutrients, this should be detectable by analysis of the soil from the rooting region. In their analyses no increases of ammonia, available phosphorus, or potassium were detected in the vicinity of mycorrhizal roots. The possibility that the avidity of the absorbing organs (mycorrhizas and roots) of the host for such nutrients might completely offset any increase and maintain a low concentration in the rooting region, is the main criticism of these results. In Stone's experiments with *Pinus radiata*, phosphate uptake by Sudan grass was used as an indicator of availability of phosphate in the soil. On two soils, Sudan grass absorbed less phosphate when grown with mycorrhizal seedlings than when grown with non-mycorrhizal seedlings. This indicates the unlikelihood of any increase of available soil phosphate resulting from the activity of mycorrhizal fungi. Again, soils which had borne mycorrhizal seedlings supplied less phosphate to Sudan grass than those that had borne non-mycorrhizal seedlings. Stone came to the reasonable conclusion that the passage of increased amounts of phosphate into mycorrhizal seedlings was through the fungal mycelium rather than directly into the root after release into the soil (*see* also p. 123).

By contrast, some of the experimental work of R. O. Rosendahl gave some evidence that, on substrates containing phosphorus as insoluble orthoclase, pure mycorrhizal fungi, in common with certain soil fungi, could increase the availability of phosphate to American elm, although this last does not form ectotrophic mycorrhizas. But like Levisohn's experiments noted above, this is a very special case. The evidence for the view that mycorrhizas exert an external solubilizing action is marshalled by Wilde (1954).

Lobanow (1960) reports the work of A. K. Eglite, and of Eglite & W. Jakobsone, on the capability of various mycorrhizal fungi to absorb nutrients from sparingly soluble compounds. They are described as being able to utilize phosphates from insoluble inorganic and organic phosphate, potassium from mica and silicate, and nitrogen from organic compounds such as ammonium salts of phytic acid, nucleic acids, and plant litter.

An hypothesis which has not yet been sufficiently explored is the possible mediation of surface-attached phosphatases in the processes of absorption of phosphate by roots and mycorrhizas (*see* Bowen & Theodoru, 1967). W. H. Woolhouse has recently shown them to occur in high activity on the surface of beech mycorrhizas. Since phytin and phosphates of inositols may constitute a large part of the insoluble phosphate present in the organic horizons of soils, it would be worth while to determine whether the surface phosphatases of mycorrhizas are able to release inorganic phosphates from them.* Even if mycorrhizal fungi have the kind of effect upon soil substrates which was postulated by Burges, i.e. of bringing nutrients into solution, his assumption that the morphological changes in the roots are of little significance is totally untenable with respect to ectotrophic mycorrhizas. (The endotrophic forms will be considered later.) The greater portion of the unsuberized root surface of trees which form ectotrophic mycorrhizas in their forest habitats is mycorrhizal. It is covered with a relatively thick layer of fungal tissue, which constitutes a sheath through which the greater part of the nutrients absorbed by the host must pass. Consequently the properties of mycorrhizal roots in respect of absorption, in terms of quantity and quality, are almost certain to be different from those of non-mycorrhizal root systems. Those who hold that the modification of the root into a mycorrhizal organ, and the actual structure of that organ, are essential considerations in the elucidation of the stimulating effects of mycorrhizal infection, are therefore undoubtedly correct, and indeed experimental justification of this view is now available. Whether or not it turns out that there is an external effect of mycorrhizal fungi on nutrient availability, an examination of the structure and absorptive function of mycorrhizal organs is an essential consideration.

* *See* Appendix §6.

Experiments comparing the uptake of salts by mycorrhizal and non-mycorrhizal seedlings have shown that there are greater quantities of nitrogen, potassium and phosphate in each mycorrhizal seedling. Frequently, too, as in the examples given in Tables XXII to XXVI, there is a greater amount, on a dry-weight basis, of some or all of these three nutrients. Hatch (1937), in his experiments on *Pinus strobus* inoculated with pure cultures of fungi, observed an increase in all these three nutrients which was higher proportionately than the increase of dry-weight in the mycorrhizal seedlings. He therefore put forward the view that mycorrhizal infection increased the uptake of nutrients in general, and he sought an explanation on the following basis. It was observable, but not measurable, that one of the morphological effects of infection was an increase in the area of the absorbing surface of the root system. This was due to an increase in the diameter of the roots, to prolongation of the life period of short roots, to stimulation of short-root branching, and to the provision of hyphae connecting the infected short roots with the soil. Hatch saw in this physical increase of area a possible explanation of the experimental results on mineral absorption. The absorptive capacity would be greater and hence the effect of infection would be to stimulate the absorption of all soil nutrients, so that those nutrients in shortest supply would be most affected. This presupposes two points which were not overtly stated by Hatch. The first is that mycorrhizal roots exert a selective influence upon the kind of salt which they absorb, similar to that exerted by normal roots; this point has now been investigated experimentally and it will be shown below to be justified. Secondly, it is assumed that the capacity of the sheath-surface to absorb nutrients is at least as great as that of uninfected roots, and moreover that the nutrients so absorbed are passed to the host at a rate not less than the absorption rate through the surfaces of non-mycorrhizal roots. Or alternatively, if the rate of supply of nutrients to the host is diminished by the sheath, the diminution is more than offset by the total increase in area of the absorbing surface.

McComb's results (Table XXII) show that phosphate absorption into mycorrhizal seedlings may be more markedly stimulated than the absorption of the other ions, and a similar result has been obtained by others. McComb & Griffith (1946), and Stone (1950), therefore incline to the view that the primary effect to be explained is an increase in phosphate absorption, which in turn is thought to result in a general increase in the metabolic rate and consequently a greater uptake of other elements. Certainly phosphate occupies a unique place in metabolic sequences, and this fact might provide a rationale for their views. But one must remember that phosphate is a radical which is often deficient in soil or else is present in unavailable form, so that in natural soils it may often be a limiting factor to growth. In any event, the problem is resolvable experimentally provided that the physiology of salt absorption by mycorrhizal organs is studied in sufficient detail for the place of phosphate compounds in their metabolism to be elucidated. Here it is sufficient to note that mycorrhizal infection results in increased salt absorption and in increased growth of seedlings; also that the absorbed salts must pass through the fungal sheath into the host. In the next section the nature of their route through the fungal tissue, and the mechanisms controlling their movement, will be studied, so that an assessment can be made of the importance of the physical increase in the absorbing surface of mycorrhizal root systems.

The Primary Site of Salt Absorption in Mycorrhizal Roots

The mycorrhizal rootlet, as described in Chapter III, consists essentially of four tissue systems: (1) an external hyphal system in the soil, of variable extent and connecting with (2) the fungal sheath, which is often of two layers; (3) hyphae running between the cells of the cortex of the host to form the Hartig net; and (4) the host cortical parenchyma and stelar tissue.

We can readily envisage the hyphae in the soil and the outside of the sheath as the potential absorbing surfaces. The Hartig net appears to provide a large surface of contact for exchange of material between fungus and host. The intervening sheath constitutes a membrane of living cells of considerable thickness, through which absorbed material must in some way pass.

In recent experiments, Melin & Nilsson (1950, 1952, 1953, 1953*a*, 1955, 1958), and Melin *et al.* (1958), have shown that the external hyphae of mycorrhizal roots of conifers can act in an analogous manner to root-hairs in respect of salt absorption. By an extremely elegant method, they fed isotopically-labelled materials to the fungal hyphae alone, and traced their movement into the host tissues. Pine seedlings were grown in aseptic cultures and, later, a culture dish inoculated with a suitable mycorrhizal fungus (*Boletus variegatus* and *Cortinarius glaucopus* were used) was placed upon the surface of the sand. The fungal hyphae grew over the lips of the dish and penetrated the sand to form mycorrhizas with the roots. A solution containing isotopic phosphorus, nitrogen, or other element, was now introduced into the culture dish. By measuring the radioactivity or the quantity of isotope in the various parts of the host plant, it was shown that material absorbed by the hyphae was passed to the host. By this means it was ascertained that phosphate, nitrogen from ammonium nitrate or from glutamic acid, and calcium, were absorbed in this way. Not only did the substances enter the root system but they were also translocated to the shoot.

As has been seen, the degree of development of external hyphae is very variable, and the proportion of material absorbed into mycorrhizas in this way must also be variable. Stone (1950) obtained comparisons of the rates of absorption of phosphorus by *Pinus radiata* with and without an extensive mycelium emanating from their mycorrhizas and projecting into the soil. In each case he found greater absorption by those samples having extensive external mycelia. The soils on which his plants were grown were of considerable phosphate content, but it appeared nevertheless that an extensive external mycelium increased both total phosphate absorbed and total phosphate translocated to the needles. These observations indicate that when an external mycelium is developed it does act, in the manner suggested by Hatch, as an absorbing surface. They also show that materials absorbed into the living fungal mycelium may be passed to the host and translocated throughout it.

Lobanow (1960) has drawn attention to the probable effect of the external mycelium in absorption, particularly of water. Using his own measurements and those of other Russian workers, he has collated estimates of the absorbing surfaces of root systems and the transpiring surface of the leaves of a number of plants. He points out that the ratio of the root surface to the leaf surface of strongly mycotrophic plants is small compared with that observed in weakly mycotrophic or non-mycotrophic plants. For instance, his own estimates of the ratio in one- and two-year conifer seedlings is less than unity, whereas that of winter rye is 140. Although some other estimates do not give such a very great difference (*see* Table XXIX), the conclusion is tenable that the external hyphae may make good the discrepancy in the mycotrophic plants.

TABLE XXIX

COEFFICIENTS OF WATER ABSORPTION (RATIOS OF ABSORBING SURFACE TO TRANSPIRING SURFACE) FOR A NUMBER OF WOODY PLANTS IN THEIR FIRST YEAR. *Root-hairs and hyphae have been omitted in estimating the coefficients.* (Data of Lobanow, 1960.)

Tree	Degree of mycotrophy	Coefficient
Spruce	M	0·9
Pine	M	8·31
Oak	M	5·60
Maple	+	64·0
Elm	—	72·5
Birch	M	181·5
Lime	+	198·0
Ash	—	237·9

M = strongly mycotrophic; + = somewhat mycotrophic.

The surface of the sheath itself has also been shown to be active in salt absorption. Kramer & Wilbur (1949), using excised roots of Loblolly Pine, obtained autoradiographs from them after exposure to ^{32}P (as phosphate) in solution; these showed clearly a greater accumulation in mycorrhizal roots than in uninfected roots. Similar results were obtained by Harley & McCready (1950), both with autoradiographs of beech roots after exposure to phosphate labelled with ^{32}P, and by counts with a Geiger counter. Some results of these counts are given in Table XXX.

TABLE XXX

UPTAKE OF ^{32}P FROM DILUTE POTASSIUM PHOSPHATE OF pH 5·5, BY INFECTED AND UNINFECTED EXCISED BEECH ROOTS. Values are relative. Data of Harley & McCready (1950).

Series	Fully infected	Uninfected	With local sheaths
1	5·18	0·88	4·76
2	6·68	0·75	1·20
3	1·97	0·42	0·62
4	2·72	0·61	—
5	1·69	0·72	2·13

Bowen & Theodoru (1967) have pointed out that absorption into uninfected roots is very rapid in the subapical region, where uptake of phosphate may be as fast as into mycorrhizas. They go on to stress that not only is this a small local area of high avidity, but that, compared with mycorrhizas, uninfected roots are impermanent and often short-lived.

In the experiments with excised beech roots it was observed that, wherever local infections involving the formation of small areas of fungal tissue occurred, locally high rates of absorption of ^{32}P were recorded both by autoradiograph and by counting (Table XXX, last column). By making autoradiographs of sections, it was found that, during short-term experiments lasting a few hours, the main primary site of absorption was in the fungal sheath.

A similar conclusion was reached through dissection of the fungal sheath from the host tissue of beech mycorrhizas. When excised beech mycorrhizas were allowed to absorb labelled phosphate from a solution of low concentration (less than 1 mM.) about 90 per cent was found to remain in the fungal tissue during the first few hours. Only when phosphate concentrations far above those expected in the soil were applied, did the sheath allow a greater proportion than 10 per cent to pass to the host (Harley & McCready, 1952*a*). Indeed in periods of time ranging from a few hours to a few days, it was found that the quantities of phosphate absorbed into the sheath and host tissue from 0·055 mM. phosphate or less, were in the proportion of about 9:1. That is, although absorption occurred and was linearly related to time, a constant and small proportion only, of about 10 per cent, passed into the host. When beech mycorrhizas still attached to the parent trees were fed with labelled phosphate solutions, conduction of the tracer up the mother roots into the tree was observed. Nevertheless, of the phosphate absorbed, the greater part was again retained in the fungal tissues (Table XXXI) during the period of uptake of phosphate.

The results obtained with excised roots have been criticized by Melin & Nilsson (1958) and by Lobanow (1960). They have stressed the point that in experiments with excised roots transpiration does not occur, and that therefore a distorted view of the effect of the fungus may be obtained. Melin & Nilsson (1958) performed experiments with *Pinus sylvestris* cultured aseptically with *Boletus variegatus*. They exposed the mycelium of the fungus to Na_2HPO_4 at 180 mg./l. in the manner described on page 119, and after 24 hours dissected the sheath from the host core of the mycorrhizas. Two sets of *Pinus* plants were used—intact plants and plants whose tops had been removed 24 hours before the experiment began. The

TABLE XXXI

ESTIMATE OF THE PERCENTAGE OF ABSORBED PHOSPHATE IN THE SHEATHS OF BEECH MYCORRHIZAS ATTACHED TO OR EXCISED FROM TREES. Bagley Wood 1950, pH 5·5, at woodland temperature (Harley & McCready, 1952).

Condition of roots	Attached				Excised			
Date	31 March	11 May	23 July		31 March	11 May	23 July	
Leaves	In bud	Expanding	Full leaf		In bud	Expanding	Full leaf	
Mg.P/litre presented	2·3	10·0	5·0	50·0	2·3	10·0	5·0	50·0
Per cent in sheath, range	74–96	78–94	81·0–93·5	85·5–93·0	85–96	79–93	83–88	91·0–92·5
mean	88	88	89·9	86·8	91	89·4	84·9	91·3

results of Melin & Nilsson, it should be noted, are given as comparative phosphate uptake per mg. dry-weight of core and sheath during 24 hours; but if they are calculated in the same way as those of Table XXXI, they undoubtedly show that a greater percentage is retained in the sheath of the decapitated plants (71–91 per cent) than in the intact seedlings (28–63 per cent). The reason for this discrepancy is not necessarily solely to be found in terms of transpiration, for the removal of the tops 24 hours before the experiment has other effects besides its elimination. Decapitation, especially in cultured seedlings, must greatly reduce carbohydrate supply and so may have a considerable effect on the uptake and movement of nutrients.

In any event it is clear that experiments conducted with excised roots must be interpreted in a clear understanding of the limitations of the experimental technique, and must be carried out over short time-periods with the aim of solving problems concerned with the initial processes of uptake. The experiments with whole mycorrhizal seedlings of oak (Lobanow, 1960), and with pine (Clode, 1956), show a greater relative accumulation of ^{32}P in the root system as compared with those of non-mycorrhizal seedlings after a period of uptake during which transpiration occurred; these will be considered later, against the background of work with excised roots.

These studies have demonstrated the greater powers of accumulation of phosphate by mycorrhizal roots than by non-mycorrhizal roots, and seem to suggest that the former are more able to compete with the other organisms of the soil for the available phosphate. There is, however, variation between different types of mycorrhizas in this regard. Using excised mycorrhizas of *Pinus radiata*, Bowen has shown big differences of uptake on a dry-weight basis to exist between types. The rates varied from equal or slightly less than uninfected roots, to about three times their rate.*

In this connexion the findings of Stone (1950), on the effect of mycorrhizal and non-mycorrhizal *Pinus radiata* on the absorption of phosphate by Sudan grass, are of interest. In the presence of mycorrhizal seedlings the Sudan grass clearly exhibited a reduced phosphate absorption and a much diminished growth-rate on both of the types of soil used. (Table XXXII).

TABLE XXXII

COMPARATIVE PHOSPHATE CONTENTS OF *Pinus radiata* AND SUDAN GRASS WHEN GROWN TOGETHER; MYCORRHIZAL AND NON-MYCORRHIZAL PINES COMPARED. Data of Stone (1950). (Dry-weights in gm.; phosphorus in mg.)

Soil	Mycorrhizal infection of the pine	Pine: Gain in P mg.	Pine: Gain in dry-wt. gm.	Sudan grass: Total P mg.	Sudan grass: Total dry-wt. gm.
Carrington silt loam	+	2·33 ± 0·07	3·71 ± 0·10	·066	0·112 ± 0·018
	—	0·06 ± 0·11	1·99 ± 0·25	1·73 ± 0·06	2·81 ± 0·15
O'Neill sandy soil	+	0·365	1·64	0·23	0·41
	—	0·085	1·27	0·68	1·11

Fewer results are available concerning the comparative absorption rates of other ions by mycorrhizal and non-mycorrhizal roots. Beech mycorrhizas have been shown to absorb alkali metals (J. M. Wilson, 1957) and ammonia very rapidly. The alkali metals in particular seem to be more mobile than phosphate and to pass through the fungal sheath more readily, so that no great disproportionate partition of them between the fungal and host tissues is observed. Radioactive rubidium, for instance, is distributed almost equally between sheath and host tissue soon after absorption. Absorbed ammonia seemed, on the other hand, in the experiments of Carrodus (1967), to be primarily accumulated in the sheath. Some 62 per cent was estimated to have accumulated there during an experiment of 4 hours' duration. Since the host tissue constitutes two-thirds of the total dry-weight of the mycorrhizal root-tip, the sheath clearly retains a proportionately greater quantity on this basis. J. M. Wilson has compared the uptake of potassium and rubidium by mycorrhizal and non-mycorrhizal beech roots. The rates of absorption by mycorrhizas exceed those by non-mycorrhizal roots by a factor of about 2.

All this work shows that Hatch's hypothesis must be slightly modified.

* *See* Appendix §6.

It is quite certain that the sheath cannot be considered to act in the same way as an uninfected root-surface. For although the absorptive surface may be much greater in extent in mycorrhizal root systems than in uninfected systems, the physical increase in area is associated with changes in the nature of the surface. The sheath is a living tissue with absorptive, assimilative, and accumulative properties of its own. Hence just as it is necessary to take into account retention of materials in the cortex of roots when considering their absorptive properties, so in mycorrhizal roots the accumulative properties of an additional living layer, the sheath, must be considered. The sheath does not seem so readily to afford a barrier to, or to accumulate excessively, the alkali-metal ions, but it does retain phosphate and ammonium ions, which are accumulated or assimilated in it. As will be seen later, the metabolism of the sheath also actively releases accumulated phosphate in conditions of phosphate deprivation.

The Mechanism of Salt Absorption

Although work with seedlings has been particularly valuable in demonstrating the efficacy of mycorrhizal infection in allowing greater rates of salt absorption than occur in the non-mycorrhizal state, a full appreciation of the importance of this can only be gained when something of the mechanism of absorption is understood. Already it has been seen that the fungal sheath appears to be the primary site of accumulation, particularly of the important element phosphorus; and, in this sense, the fungal sheath appears to behave as a barrier interposed between soil and host root.

The experimental work to be described in this section is concerned with an attempt to explain the mechanism of action of mycorrhizal roots by obtaining the answers to particular questions. We have seen, from the study of mycorrhizal root systems, that it is an inescapable conclusion that material from the soil passes through the fungal sheath into the host, and this has been experimentally demonstrated for nitrogen, phosphorus, and some other elements. We need to know, further, what external factors affect the rate of absorption of materials into mycorrhizal organs, and whether absorption rates are dependent on factors similar to those influencing absorption into non-mycorrhizal roots. A knowledge of the operation of these factors will, in addition, connect laboratory studies with ecological studies. Further, and to the same end, the effect of internal factors within the host-symbiont system must be considered. Besides these problems, the more difficult ones of translocation require consideration; in particular, the movement of absorbed materials through the sheath is of fundamental importance. We have already seen that this layer is of considerable thickness and corresponds to some 39 per cent of the total dry-weight of the root-tips in beech. It has very great powers of accumulation of some ions into its own hyphae, and yet the host appears to be better supplied with them than when uninfected.

It has therefore proved to be essential to study in detail the absorption physiology of ectotrophic mycorrhizas themselves, and not to assume that they behave in a similar manner to non-mycorrhizal roots.

Work in the last several decades has indicated quite decisively that the absorption of minerals by the roots of plants is dependent on metabolic reactions and, directly or indirectly, on respiratory processes. In almost all the writings upon mycorrhizal phenomena, physical processes, such as diffusion, adsorption and mass flow in liquid, are conceived to be either the sole processes involved or at least the rate-limiting processes in absorption. This outlook originates mainly from the application of the older ideas on salt absorption to mycorrhizal organs, and until recently very few experimental studies of the factors affecting mineral absorption by mycorrhizas had been made.

Mycorrhizal roots of *Fagus sylvatica* have proved to be very satisfactory material for physiological studies of salt absorption. They are relatively large in size, and the apices of racemose systems provide simple axes 0·5–1·0 cm. in length that are almost without branches. Such simple axes can be obtained in large quantities from the surface layers of soil in beech woodlands. Replicate samples of as few as twenty-five roots from a single site give results in respect of mineral absorption, respiration, etc., which agree within small limits, and the sheath tissue can, with little damage, be stripped off the host core under the dissecting microscope. In this way tissue samples can be obtained which again agree well in performance. Material of this kind has been used to study ion and carbohydrate absorption through the use of tracers or by conventional analytical methods; it has also been used to study related aspects of respiratory metabolism.

The details of the material of *Fagus* used in the experiments to be described are as follows. The sheath comprises some 20 to 30 per cent of the volume, about 39 per cent of the dry-weight, and is usually 30 to 40μ thick. The total volume of the mycorrhizas is about 0·7 ml. per 100 mg. of dry-weight, and they contain about 85 per cent of water. The results, however, like all those obtained with excised plant organs, must be interpreted with caution. Their value lies in the control of variability of the experimental material, and of the applied experimental conditions, as well as in the multiplicity of measurements and experimental procedures which can be obtained using excised mycorrhizas. In addition, low concentrations of nutrients and short periods of observation are possible in the experiments. Indeed, mycorrhizal roots sampled from the field and used within an hour or two of collection, are as uniform and useful for experiment as are most excised plant materials employed in the study of nutrient uptake. But results from them are of like value, and must be interpreted with the same understanding of the limitations in their bearing upon nutrient uptake by whole plants. They have given some insight into the effect of both external and internal factors on salt absorption, into the mechanism of ion accumulation, and into the selective action which mycorrhizas exert in respect of the kinds of ions they absorb. Some of the results have proved to be of immediate ecological interest; others, although of a more academic nature, have had a bearing upon the understanding of the process of salt absorption by plant tissue.

THE EFFECT OF EXTERNAL FACTORS ON SALT ABSORPTION

When excised beech mycorrhizas are immersed in solutions of nutrient salts, ions enter the tissues. The absorption rate from dilute solutions of the same order of concentration expected in the soil, is linear against time, but after prolonged periods of time the rate of absorption falls. With high external concentrations, a higher absorption rate is observed at first but it does not remain linear with respect to time. In what follows, emphasis will be given to the results obtained using low concentrations, for it is from these that the most reliable interpretation of how mycorrhizas may function in natural conditions can be obtained.

The uptake rate of all substances studied is particularly sensitive to temperature change over a wide range. Uptake rates are slow near 0°C. but increase greatly up to temperatures of 15°C. to 20° C. Over this range the temperature coefficients of absorption (Q_{10}) are high, being in the region of 2 or more—regardless of concentration. In many series of estimates of the Q_{10} of phosphate absorption over a wide range of concentrations, it was found that above 15° C. the values diminished, and that smaller effects of temperature are observed above 20° C. (*see* Table XXXIII). A similar result was obtained by Carrodus (1965), studying ammonium uptake. In his experiment a Q_{10} of 1·9 was obtained between 10° C. and 20° C., but between 20° and 25° C. the Q_{10} fell to 1·2.

TABLE XXXIII

ESTIMATED Q_{10} VALUES OF PHOSPHATE ABSORPTION BY BEECH MYCORRHIZAS. *Concentration* 0·03 mM. *Time of absorption* 2 *hours. Results of C. C. McCready* (1952).

Temperature range °C.	Q_{10}
5–10	2·27
10–15	2·36
15–20	1·42
20–25	1·34

The effect of temperature on the absorption of alkali-metals is similar but rather more complicated. These, as is usual in plant tissues, are very mobile ions. For instance, losses of potassium from the tissues into distilled water may be occasioned when the mycorrhizas are suspended in aerated water at a temperature above 20° C. In consequence of this, Q_{10}

After Hatch (1937). By permission Black Rock For. Bull.

(*a*) Short-root development in White Pine. A, long roots lacking short roots; B, average development of short roots in conditions of low nitrogen supply; C, mycorrhizal development in low-nitrogen soil.

PLATE 4

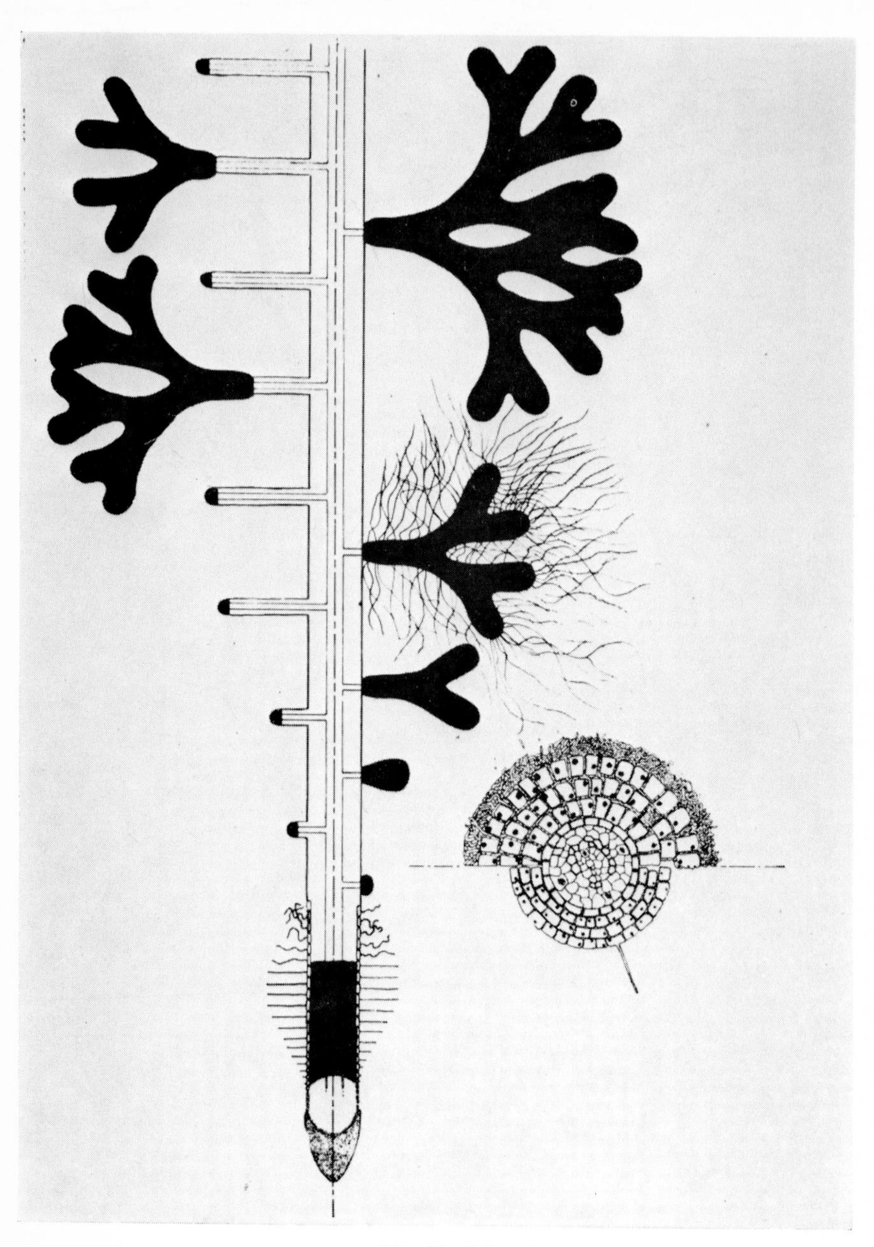

After Hatch (1937). By permission Black Rock For. Bull.

(*b*) Diagrammatic representation of the effects of mycorrhiza formation on the roots of pine.

PLATE 4

values of a very variable magnitude can be obtained by choosing experimental conditions. Below 15° C., however, mycorrhizal roots suspended in potassium or rubidium chloride solutions of low concentration rapidly absorb the alkali-metal ion, and Q_{10} values of 2 or more may be obtained. Above 15° C. the values obtained depend in particular upon the external salt concentration (Table XXXIV). There is, therefore, in contrast with phosphate absorption, an optimum temperature of about 15° to 18° C. for absorption of these ions.

TABLE XXXIV

THE EFFECT OF TEMPERATURE ON POTASSIUM UPTAKE BY BEECH MYCORRHIZAS IN TWO EXTERNAL CONCENTRATIONS OF KCl.

Note the loss of K at 25° C. in 0·1 mM. KCl. Uptake μg. equiv. per hour, per 100 mg. of dry tissue. Results of J. M. Wilson (1957).

	Temperature °C				
Concentration	5°	10°	15°	20°	25°
0·1 mM. KCl	+ 0·10	+ 0·16	+ 0·21	+ 0·12	− 0·13
0·6 mM. KCl	+ 0·19	+ 0·23	+ 0·32	+ 0·27	+ 0·09

In the experimental work on temperature we thus have a suggestion that mycorrhizas absorb alkali-metal ions most efficiently at temperatures below 20° C. This is in line with the temperature requirements of many of the mycorrhizal fungi of temperate climates, whose optimum temperatures for growth are not very high and indeed are usually little above this level. It is worth noting in passing that many of the trees which form ectotrophic mycorrhizas are temperate species for which the soil temperatures, even in the surface layers of woodland soil, may not rise very high.

The Q_{10} values of about 2 for salt absorption suggest that the processes involved include chemical metabolic reactions, rather than physical processes, as their rate-limiting steps. This suggestion is greatly strengthened by a study of the effect of oxygen supply upon salt absorption. Fig. 10 (p. 99) shows the effect of oxygen supply upon the respiration rate and the respiratory quotient (RQ) of beech mycorrhizas. The rate of oxygen absorption increases continuously from a low value at low oxygen concentrations up to 21 per cent oxygen. The respiratory quotient rises sharply from a value of 1·0 below 5 per cent oxygen; hence a significant rate of anaerobic carbon dioxide production occurs below this value. When phosphate absorption is studied over a similar range of oxygen concentrations, curves are obtained showing that phosphate absorption may be dependent upon oxygen supply. This relationship is emphasized in Fig. 12, where it is shown that provided a sufficiently concentrated phosphate solution (3·23 mM.) is supplied to the roots, phosphate and oxygen absorption are directly related. The absence of such a relationship when lower concentrations of phosphate are supplied could undoubtedly be explained by the limitation of phosphate absorption by other factors when O_2 uptake is high.

In the absence of oxygen, the rate of phosphate absorption is always low and, after a long period of anaerobiosis, there may occur a loss of phosphates from the tissues into the external solution. The loss is dependent on temperature and is the result of some metabolic activity under anaerobic conditions which releases phosphate.

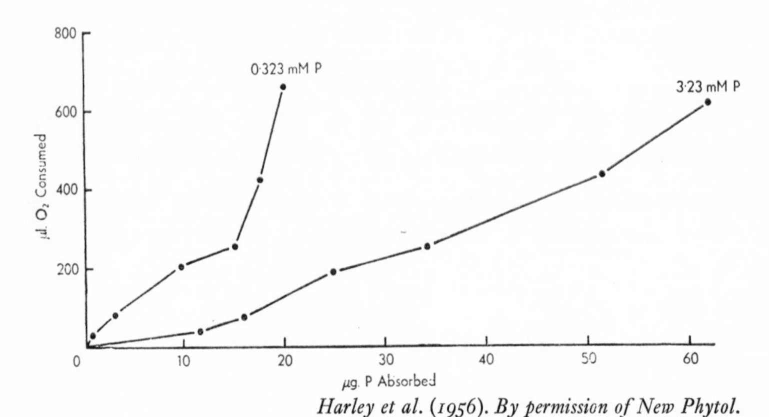

Harley et al. (1956). By permission of New Phytol.

FIG. 12.—Relationship between oxygen and phosphate absorption (per 100 mg. dry-weight of beech mycorrhizas) in 3·23 and 0·323 mM. KH_2PO_4.

The alkali metals show a somewhat similar dependence for uptake upon oxygen supply. The rate of absorption of potassium from a solution of 0·128 mM. KCl kept agitated by bubbles of different oxygen concentrations, is given in Fig. 13. As has been seen above, potassium is more mobile an ion than is phosphate, and it is more readily released from the tissues. When supplied with oxygen concentrations below 3 per cent by volume, potassium readily escapes from the mycorrhizal roots, but above this value potassium uptake increases with increased oxygen supply, reaching a high level at 21 per cent. Other alkali metals behave in a similar fashion. Ammonia uptake is diminished almost to extinction in the absence of oxygen, but is rapid at 21 per cent of oxygen.

The absorption of all these ions is so clearly decreased in all oxygen concentrations below that of air (21 per cent), that an important ecological point is immediately raised concerning the functioning of mycorrhiza as absorbing organs in the soil. They would not be expected to function at all below 3 per cent of oxygen, and undoubtedly if they act efficiently as salt-absorbing organs in the soil an oxygen concentration approximating to that in air would be expected in the horizon in which they exist. Brierley (1955) has found such conditions to obtain, as will be described later.

The linkage of salt absorption with metabolic processes in the tissues is further emphasized by the effect of poisons upon it. Substances which have upsetting effects upon respiratory metabolism and especially upon enzyme action—such as potassium cyanide, potassium azide, diethyl dithiocarbamate, 2,4-dinitrophenol, and sodium iodoacetate—have an inhibiting effect upon salt absorption by mycorrhizas. The form of the curve relating the rates of absorption against inhibitor concentration

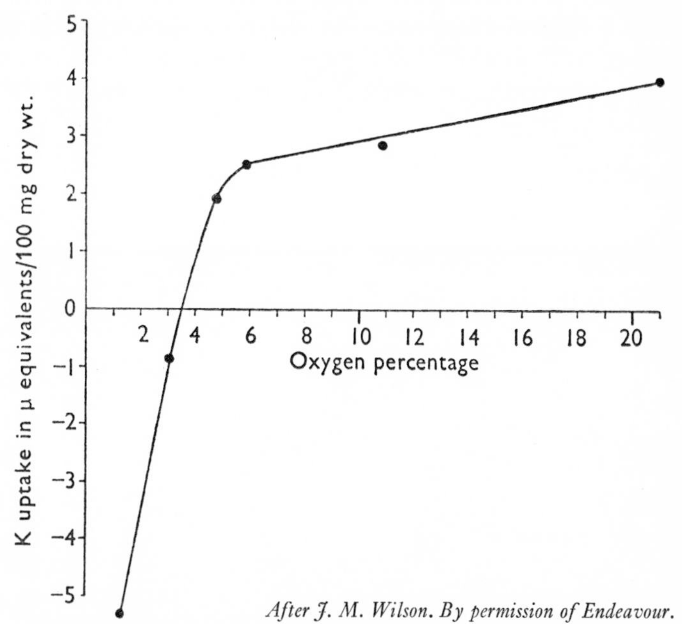

After J. M. Wilson. By permission of Endeavour.

FIG.—13. Graph indicating the effect of oxygen concentration on absorption of potassium from 0·128 mM.KCl during 17 hours by beech mycorrhizas.

in the case of sodium azide is shown in Fig. 14. The effects of other inhibitors administered under similar conditions (20° C., pH 5·5, 0·16 mM. KH_2PO_4) have a similar form, except that the range of concentrations over which they act may vary. Their effects may be compared by comparing the concentrations of inhibitor necessary to diminish absorption by one-half. In phosphate absorption these concentrations range from 5×10^{-5} molar for sodium azide which acts in very dilute solution, to about 10^{-3} molar for cyanide and iodoacetate.

These inhibitors, if applied in high concentrations or over extended time-periods, may also cause loss of salts of potassium and phosphate from the tissues. Potassium is, as expected, much more easily lost than phosphate.

Jennings (1964) has examined the effect of various concentrations of azide on the changes of phosphate fractions within mycorrhizas and on their loss into the external medium. Azide at 10^{-3}M. or higher concentrations causes a net decrease of insoluble and soluble, organic and inorganic phosphate within the tissues, and a rapid loss of inorganic phosphate from them.

Although the effects of inhibitors again demonstrate the dependence of salt absorption on metabolic processes, their effects are somewhat different on the respiration of beech mycorrhizas from those on other plant tissues, as has been described in Chapter 4. It is important to note, however, that in spite of suggestions to the contrary, beech mycorrhizas do not lack a cytochrome system. They are not a specialized example of an absorbing organ in which salt absorption is associated with an oxidase system that is insensitive to cyanide and carbon monoxide, as some have suggested.

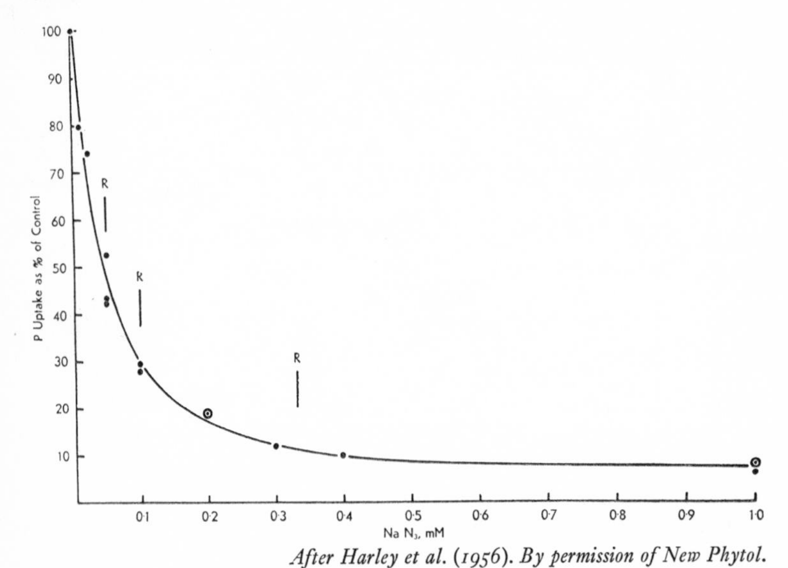

After Harley et al. (1956). By permission of New Phytol.

FIG. 14.—Graph indicating the effect of sodium azide in various concentrations on phosphate absorption from 0·161 mM. phosphate by beech mycorrhizas.

THE RELATIONSHIP OF SALT ABSORPTION TO RESPIRATION

The effects of metabolic inhibitors upon the salt absorption and respiration of mycorrhizal roots of beech indicate that the two sets of processes are linked, though they may be only indirectly linked. There are many hypotheses concerning the manner of linkage in other plant tissues. The electron transfer system may be itself a carrier of anions; or respiration may be involved in the synthesis of specific ion-carriers; or again, respiration may provide carbon radicals that are required for the assimilation of ions, energy to power accumulation against an electrochemical gradient, CO_2 for exchange for anions, and so on. We cannot concern ourselves here with detailed theories but it is of importance to consider further some aspects of the interdependence of salt uptake by mycorrhizas and respiration.

If mycorrhizal roots of beech are kept in aerated distilled water for a few days, their respiration-rate falls to a low value which is maintained for a long period. At any time during storage the application of salts to the roots results in an increase in respiration, which is dependent in magnitude upon the time of storage (Fig. 15) and also upon the nature and concentration of the salt. The increase in respiration-rate involves an increase both in carbon dioxide emission and oxygen consumption; at the same time the respiratory quotient (RQ) rises. The decrease in respiration-rate during storage is not the resultant primarily of carbohydrate deficit, for it occurs whilst the internal carbohydrate concentration remains high and also whilst carbohydrates are being absorbed by roots

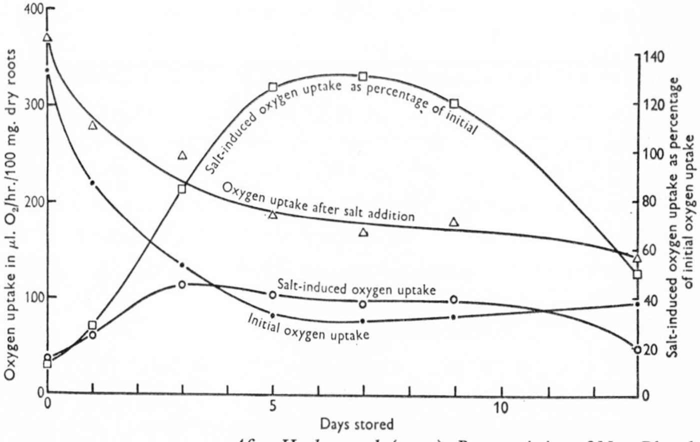

After Harley et al. (1954). By permission of New Phytol.

FIG. 15.—Graph indicating the progressive development of the respiratory response to phosphate by beech mycorrhizas when stored in distilled water at room temperature.

kept in sucrose or glucose solutions. Eventually, of course, long periods of storage in water result in carbohydrate deficiency, and then a respiratory response to salt is not observed.

A variety of substances in solution, besides nutritive salts, occasion a respiratory response in stored mycorrhizas. The following substances have been observed to give similar effects: salts of potassium, sodium, calcium, rubidium, and ammonium, with chloride, sulphate, nitrate, and phosphate; the sugars sucrose, fructose, and glucose; the inhibitors sodium azide, potassium cyanide, and 2,4-dinitrophenol. Amongst the salts, those of phosphate and ammonium are most effective, and the magnitude of the respiratory response to these far exceeds that of any other. Glucose has a greater effect than sucrose, and potassium has a greater effect than any of the other metals.

The inhibitors differ in their effects from ordinary salts. This is particularly true of sodium azide and 2,4-dinitrophenol, for they have a large stimulating effect on the respiration of freshly excised mycorrhizas. Potassium cyanide also may appear to stimulate, though to a lesser degree, the respiration of fresh mycorrhizas in some conditions. All these substances, however, show greatly increased stimulating effects during storage. These increases, especially with azide and 2,4-dinitrophenol, indicate that the dependence of respiratory rate upon the availability of substances capable of forming high-energy phosphate compounds in the tissues increases during storage ; they thus afford an explanation of the fall of respiratory rate of mycorrhizas stored in distilled water. During storage, energy-demanding processes such as synthesis or absorption, might be expected to diminish, so that as respiration proceeds, the availability of the immediate precursors of high-energy phosphate, which are generated in these processes, will fall; the respiratory rate then declines even though carbohydrate substrate is still available. The application of any factor, which stimulates the use of high-energy phosphate in the tissues, increases the supply of its precursor, and this is reflected in a rise in the rate of oxygen consumption. Substances which uncouple oxidative and phosphorylative processes and reduce a net synthesis of adenosine triphosphate,* such as azide and 2,4-dinitrophenol, will cause great stimulation. Substances absorbed and synthesized into organic compounds, e.g. ammonium and phosphate, will be particularly effective in this; and ions absorbed against a concentration gradient, e.g. potassium, will also cause some increase.

The effectiveness of the sugars, glucose and sucrose, in causing respiratory increase, agrees with the results of Rheinhold (1949) who studied carbohydrate absorption by disks of plant tissue, and indicates, again in agreement with many previous researches, that sugar absorption by plant tissues depends upon a phosphorylation of hexoses.

It has been found that, during storage, when equivalent samples all stored in identical conditions are compared, the magnitude of the respiratory response of mycorrhizal roots to any salt is dependent over wide limits on the concentration applied. This would be expected if the magnitude of the respiratory stimulation depended on the quantity of salt absorbed. Such an expectation has been tested in the case of phosphate, and it has been found that the extra oxygen uptake occasioned by the addition of phosphate as potassium phosphate, was proportional to the quantity of phosphate absorbed (Fig. 16).

The interrelation of respiratory metabolism and salt absorption of mycorrhizal roots of beech does not seem to be through the oxidase system but through the phosphorylation mechanism. The higher respiratory responses occasioned by ammonium and phosphate ions are explicable perhaps in the consumption of such substances as ATP in the synthesis of organic compounds. The effects of the concentration of salts, of inhibitors, and of carbohydrates, fall into place in this explanation.

These results may appear at first sight to give some point to the contention of E. L. Stone and others that phosphate plays a primary role in

* Adenosine triphosphate is conventionally abbreviated to ATP.

absorption by mycorrhizas. But as phosphate metabolism is of central importance in absorptive processes of other tissues, both plant and animal, it is clear that further research is needed before this contention can carry full conviction. The immediate value of these experiments lies in their incontrovertible demonstration that salt accumulation in mycorrhizal roots is a metabolic process, and although in detail the mechanism *seems* to be somewhat different from that believed to be

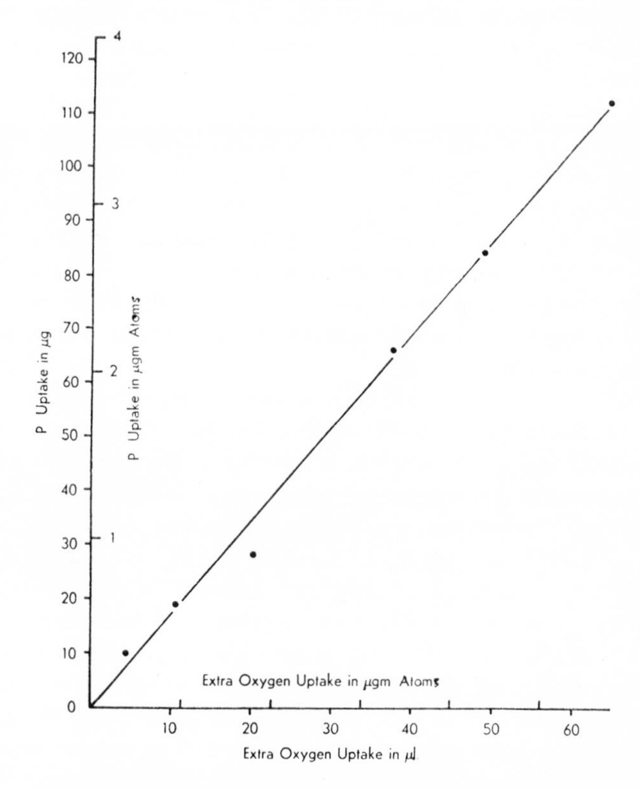

After Harley et al. (1956). By permission of New Phytol.

FIG. 16.—Direct relationship between oxygen and phosphate absorption by beech mycorrhizas. Material stored eight days in distilled water.

operative in the tissues of non-mycorrhizal absorbing organs, it has in a broad sense much in common with this latter mechanism. In a later paragraph, consideration will be given to the process of the translocation of phosphate through the fungal tissues of beech mycorrhizas into the host tissue, and it will be seen that one of the conditions for release into the host of phosphate accumulated in the sheath is exactly that in which respiratory responses to salts develop, namely storage in salt-free solutions.

Absorption of Nitrogen and Sulphur Compounds

One of the most commonly held hypotheses concerning the effect of mycorrhizas upon nutrient uptake and growth has been that they facilitate the absorption of nitrogen compounds from the soil. This hypothesis has taken two forms: first, that by intervention of the fungus, mycorrhizas are enabled to use organic nitrogen compounds and ammonium salts, and second that they have powers of nitrogen fixation analogous to the symbiotic system of legumes, alder, and other plants with bacterial (or actinomycete) nodules. It will be best to consider the question of nitrogen fixation first.

There have been recent experiments using modern methods such as the application of ^{15}N as a tracer. Bond & Scott (1955), who have examined many nitrogen-fixing systems, obtained negative results with naturally infected *Pinus sylvestris*. Stevenson (1959) by contrast reported positive fixation with *P. radiata*. Stevenson pointed out that ecological observation showed that there was an accretion of nitrogen in the ecosystem of developing conifer plantations. She observed that *P. radiata* was able to grow and increase the nitrogen content of the substrate in low nitrogen conditions, when the substrate was acidified to reduce the possible activity of the free-living nitrogen fixing organisms, *Azotobacter* and *Clostridium*, tests for which were negative. In other experiments *P. radiata* seedlings were inoculated with the mycorrhizal fungi *Amanita muscaria*, *Boletus luteus*, and *Rhizopogon rubescens*, and also with soil. The plants were grown in otherwise near-sterile conditions in Perlite, without addition of nitrogenous compounds. The control, uninoculated plants died early; the inoculated plants grew, although those associated with *Amanita* were much less thrifty than the others. These experiments were followed by others, using $^{15}N_2$ in conjunction with several species of plant. Both *Pinus radiata* and *Nothofagus fusca* showed evidence of incorporation of ^{15}N in their tissues. Stevenson pointed out that, in view of the inexplicable gap between the amounts of nitrogen added to the soil and the amount removed by the crop, and in view of her own experiments, nitrogen fixation was likely to have occurred. She ascribed this as possibly due to micro-organisms associated with the roots of the plants, perhaps including the fungi of mycorrhizas.

Richards & Voigt (1964) grew *P. radiata* on sterilized quartz sand inoculated with *Rhizopogon roseolus*. The intensity of mycorrhizal development was not great but the pine roots were colonized by mycelium, the cultures moreover were contaminated with bacteria. As in Stevenson's work, $^{15}N_2$ was incorporated into the plants in significant amounts.*

None of these experiments do provide convincing evidence that mycorrhizal roots or mycorrhizal fungi fix atmospheric nitrogen. In each case bacterial contamination was probably or certainly present. There

* *See* Appendix §6.

remains, however, the important possibility that nitrogen-fixing bacteria are encouraged in the rhizosphere of mycorrhizas. D. H. Lewis & Harley (1965) have pointed out that not only is mannitol the important carbon source in many of the media for nitrogen fixers, but also that it is commonly present in ectotrophic mycorrhizas. Moreover, Haasouna & Wareing (1964) recorded a marked stimulation of growth of *Ammophila arenaria* seedlings when they were inoculated with root-surface bacteria which were able to grow in a nitrogen-free medium only if they were also given exogenous mannitol as a carbon source. These results have much in common with the Russian work on nitrogen fixation in the rhizosphere (*see* Krasilnikov, 1961; Rubenchik, 1960), and so the problem is worthy of further study as an ecological possibility rather than a direct effect of mycorrhizal infection.

The absorption of nitrogen compounds from solution has been mentioned briefly before. Melin & Nilsson (1952, 1953) have shown that ^{15}N, derived from $^{15}NH_4NO_3$ and [^{15}N] glutamic acid absorbed by the mycelium of the fungal symbiont of pine mycorrhiza, may be translocated to the host and, through its tissues, to the leaves. The ability of beech mycorrhizas to absorb various nitrogen compounds has been examined by Carrodus (1966, 1967). He showed that ammonium, simple amino compounds, and amides, were absorbed and assimilated; but he could get no evidence of the absorption and utilization of nitrate. Ammonium salts were a readily utilized source of nitrogen, and the factors which affected their rate of absorption were those which affected metabolic activity and the synthesis of organic nitrogen compounds in the tissues. The amide glutamine and insoluble organic nitrogen compounds were formed from ammonia in the tissues, and the process was associated with an increased respiratory rate and required a carbohydrate supply. Hence at certain seasons, especially in spring, an exogenous carbohydrate supply might double the uptake of ammonia. Carrodus showed that during ammonia uptake the total non-volatile acids of the tricarboxylic acid cycle did not decrease significantly more than in mycorrhizas kept in the absence of ammonia, and he went on to confirm the earlier finding of Harley (1964) that, during uptake, dark CO_2 fixation was stimulated. The uptake of ammonia was also stimulated by the presence of excess bicarbonate up to a concentration of 7·5 mM., and by some of the acids of the tricarboxylic acid cycle if exogenously supplied. All the evidence confirmed a normal mode of ammonia assimilation by reductive amination of keto-acids to amino-acids, the drain on the TCA cycle acids being made good by CO_2 fixation.

As both glucose and fructose stimulated subsequent ammonia absorption if fed in advance to mycorrhizas, Carrodus expressed the view that it was probable that trehalose, glycogen, and mannitol, might all be potentially utilized in ammonia absorption. For he argued that D. H. Lewis & Harley (1965) had shown that glucose was converted mainly to glycogen and trehalose, whereas fructose was incorporated as mannitol in the tissues (*see* Chapter V).

In his second paper Carrodus (1967) re-examined the question of nitrate assimilation. It was surprising that nitrate did not seem to be utilized by mycorrhizas, since excess nitrate feeding had given satisfactory growth of uninfected host-plants and had inhibited mycorrhiza formation in previously reported experiments (e.g. Richards, 1965). Using

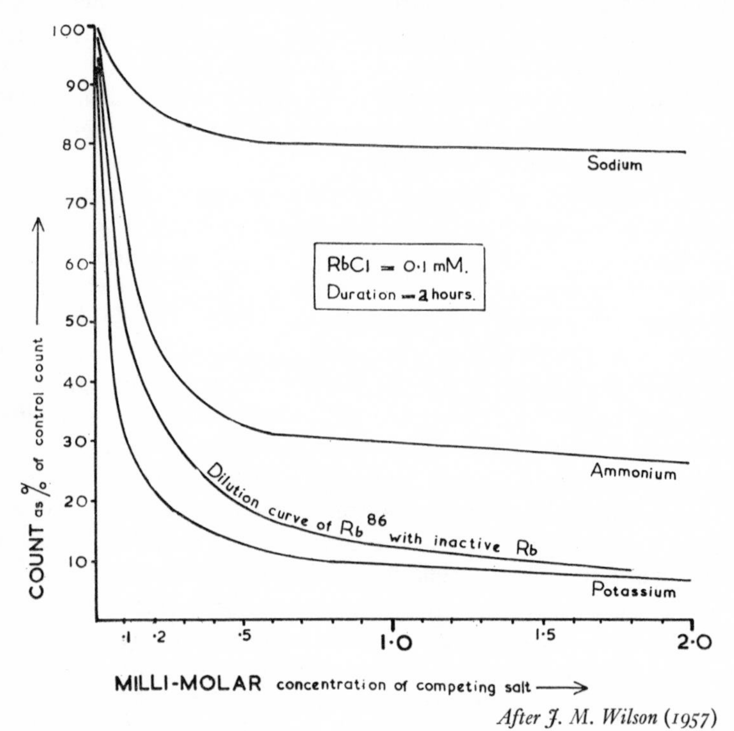

After J. M. Wilson (1957)

FIG. 17.—Graph indicating the effects of addition of various alkali chlorides on the uptake of rubidium (^{86}Rb) by beech mycorrhizas.

$NaH^{14}CO_3$, Carrodus showed that much less $^{14}CO_2$ was fixed by beech mycorrhizas in the presence of potassium nitrate than in the presence of ammonium compounds. Indeed the fixation in KNO_3 solutions was not different from that with KCl, and might be ascribed to absorption of the potassium ion in excess of the anions. It seems clear at least that nitrate is not readily utilized by excised beech mycorrhizas, and the problem needs further examination with other kinds of mycorrhizas.*

Morrison (1962*a*) examined the absorption of sulphate by *Pinus sylvestris*. He grew seedlings in low, medium, and high, sulphate media, and compared uptake by mycorrhizal and non-mycorrhizal plants. No evidence was obtained of a significant effect of mycorrhiza upon the

* *See* Appendix §6.

uptake or translocation of ^{35}S-labelled sulphate. There was no great accumulation of ^{35}S in the root systems of mycorrhizal plants such as he had observed of phosphorus in the uptake of ^{32}P.

The Selectivity of Mycorrhizal Roots for Cations

It was shown in Chapter V that the hypothesis of Hatch presupposes that mycorrhizal roots exert a selective influence, similar to that of non-mycorrhizal roots, on the ions that they absorb. This subject has been considered experimentally, in connection with cation absorption by beech mycorrhizas, by J. M. Wilson (1957) and Harley & Wilson (1959, 1963).

A little has already been said about the absorption of the alkali metals by beech mycorrhizas and especially has it been emphasized that their ions are more mobile than, for instance, those of phosphates. They are readily released from the tissues into the external solution at temperatures above 20° C., in the presence of low oxygen concentrations, or when traces of inhibitors are present. They tend to be exchangeable, one for another, between the tissues and the external solution.

TABLE XXXV

ABSORPTION OF K AND Rb FROM SINGLE-SALT SOLUTIONS AND FROM MIXTURES OF THEIR CHLORIDES.

Uptake expressed as μg. equiv. per 100 mg. of dry mycorrhizas per hour. Temperature 15° C. Data of J. M. Wilson (1957).

	Single salt 0·1 mM.	Mixed salt 0·1 mM. + 0·1 mM.	Single salt 0·2 mM.
K	1·50	0·82	1·78
Rb	1·61	0·64	2·01
Total Rb + K from mixed solution	—	1·46	—

The most striking result obtained so far in the study of the uptake of cations by mycorrhizal roots is the selectivity that the roots display in the absorption of cations from mixed solutions. The results have been obtained both by using analytical methods (flame photometry) and by the use of tracers. The rates of uptake of the metallic ions from pure solutions of their chlorides, kept at room temperature and aerated, are high, but from mixed solutions potassium is absorbed faster than are the others. Fig. 17 shows the results of an experiment in which sodium, potassium, rubidium, and ammonium, chlorides were added to a labelled rubidium chloride solution, their various effects on the uptake of rubidium by beech mycorrhizas being observed. Sodium, even at high concentration, has little effect upon rubidium uptake. Potassium chloride addition greatly diminishes rubidium uptake. The comparative effect of the addition of normal rubidium chloride to the labelled solution shows that potassium is somewhat more readily absorbed than rubidium (*see also*

Table XXXV). The addition of ammonium chloride shows that rubidium absorption is somewhat affected also by this metabolically rather different cation. Table XXXVI shows a comparison obtained by flame photometric analysis of potassium and sodium uptake both alone and in equimolar proportions in a mixed solution. In the mixed solution, sodium uptake by beech mycorrhizas is drastically reduced whilst potassium absorption is reduced by some 18 per cent only.

TABLE XXXVI

THE EFFECT OF SODIUM ON POTASSIUM ABSORPTION FROM CHLORIDE SOLUTIONS BY BEECH MYCORRHIZAS.

Absorption period 24 hours at 15° C. Results of J. M. Wilson (1957).

Solution	K absorption µg. equiv./100 mg. roots	K absorption per cent control	Na absorption µg. equiv./100 mg. roots	Na absorption per cent control
0·1 mM. KCl	1·68	100	—	—
0·1 mM. NaCl	—	—	0·41	100
0·1 mM. KCl + 0·1 mM.NaCl	1·37	82	0·06	14

In beech mycorrhizas the internal concentration of potassium has been found to be very great—70 µg. equivalents per 100 mg. of dry tissue. Therefore, on a volume basis the internal concentration is about 100 mg. equivalents per litre. In spite of their very high internal concentration, beech mycorrhizas in suitable conditions will absorb potassium from external solutions containing less than 0·1 mg. equivalents per litre. This in itself emphasizes the efficiency of ion accumulation of these organs.

The selective action of mycorrhizal roots upon the kinds of ions which are absorbed by them is in accord with the previous observations that the absorption rates are dependent upon metabolic reactions in the tissues. That mycorrhizal roots should exert this selectivity is, as we have seen, a necessary corollary to Hatch's hypothesis of the action of mycorrhizal infection, and this now has experimental justification.

Movement of Phosphates through the Fungal Sheath

The study of the behaviour of excised mycorrhizas has been undertaken in order to define more clearly certain aspects of their mechanism of salt absorption. It is an example of the necessary expedient of simplifying plant material in some phases of physiological inquiry. The experiments of Melin & Nilsson in particular, as well as some of those conducted with beech mycorrhiza, have shown that substances absorbed through mycorrhizas do gain entrance to the host plant and are translocated through its tissues. With excised mycorrhizas the observation that the greater part (some 90 per cent) of the phosphate absorbed from dilute solutions remains in the fungus, seems to indicate that, with this kind of material, treated in the experimental ways employed, the host tissue is limited or controlled in its absorption of phosphate by the fungus. Table XXXVII shows that there is really a greater avidity of the sheath than of the host for phosphate, and that the high rate of accumulation in the sheath is not merely a result of its position. When the tissues, fungal and host, are separately immersed in aerated phosphate solutions, both absorb phosphate rapidly, but on a dry-weight basis the sheath still absorbs 5–8 times as rapidly as the host.

TABLE XXXVII

RELATIVE PHOSPHORUS ABSORPTION (RADIOACTIVITY PER MILLIGRAM OF DRY-WEIGHT) BY MYCORRHIZAS DISSECTED BEFORE AND AFTER IMMERSION IN A BUFFER CONTAINING 0·16 mM. KH_2PO_4 LABELLED WITH ^{32}P.

From Harley & McCready (1952*a*).

	Sheath removed after immersion			Sheath removed before immersion		
	Total	Sheath	Host	Total	Sheath	Host
Experiment 1	518	1200	85	1200	2370	428
Ratio $\frac{\text{stripped}}{\text{intact}}$	—	—	—	2·32	1·98	5·04
Experiment 2	351	795	65·6	971	2153	265
	332	765	56·5	801	1671	228
Ratio $\frac{\text{stripped}}{\text{intact}}$	—	—	—	2·60	2·47	4·04

The operation of separating the two tissues greatly accelerates absorption by both kinds of tissue, the host's rate of absorption being increased about fourfold and that of the fungus being about doubled. We can explain the doubling in rate of absorption by the sheath by the fact that the area of presentation of phosphate solution to the tissue is about doubled after separation. No such explanation is available concerning absorption by the host tissue, for its surface area is unchanged. The conclusion is therefore inescapable that the sheath prevents the host from absorbing phosphates at their maximum possible rate from *dilute* solutions in the intact mycorrhiza. We may conclude that the higher absorption-rates exhibited by mycorrhizal as compared with non-mycorrhizal roots are a result particularly of absorption into the sheath, and that this tissue exerts a controlling influence upon phosphate absorption into the host when low concentrations are available.

The effect of the fungal sheath in diminishing the rate of absorption by the host is less marked when *high* concentrations of phosphate are applied externally; then the effect of improving the supply of phosphate to the host, by stripping off the sheath or by other means, is diminished. A comparison is given in Table XXXVIII of the effects of cutting mycorrhizas into thin slices, before exposure to potassium phosphate solutions of various concentrations. The slicing of roots increases phosphate absorption of both sheath and host in low concentrations but has a relatively lesser effect at high concentrations. It can therefore be concluded that, in some way, the phosphate moves more easily through the sheath when the mycorrhiza is immersed in a high phosphate concentration. At a concentration of 3·23 mM. phosphate, since slicing or stripping has little effect on absorption by the sheath, we may conclude that all its constituent hyphal cells are almost equally well supplied, presumably by diffusion through the interhyphal spaces. At a concentration ten times as strong as this (32·3 mM.), the host tissue also is relatively little affected by slicing, and the conclusion is that diffusion through the spaces of the sheath presents to the host, phosphate at a level not far below that applied externally.

TABLE XXXVIII

THE EFFECT OF SLICING MYCORRHIZAS UPON PHOSPHATE ABSORPTION FROM SOLUTIONS OF DIFFERENT PHOSPHATE CONCENTRATIONS.

Slices 0·5–1·5 mm. thick; time of experiment 7 hours. Results expressed as μg. P per 100 mg. of dry mycorrhizal tissue. From Harley & McCready (1952*a*).

P conc. mM.	0·004		0·032		0·32		3·23		32·3	
Treatment	Intact	Sliced	Intact	Sliced	Intact	Sliced	Intact	Sliced	Intact	Sliced
P uptake	4·67	7·85	17·7	35·6	63·0	124·5	190	239	407	460
P in sheath	4·35	6·20	16·1	27·5	56·0	94·6	146	141	258	267
P in host	0·32	1·65	1·5	8·1	7·0	29·9	44	98	149	193
Per cent P in sheath	93·0	79·0	91·5	77·0	89·0	76·0	77·0	59·0	64·5	58·0

Calculated ratios $\frac{\text{sliced}}{\text{intact}}$

	0·004	0·032	0·32	3·23	32·3
Total	1·68	2·03	1·98	1·26	1·13
Sheath	1·43	1·71	1·69	0·96	1·03
Host	5·05	5·50	4·27	2·23	1·30

Although these results may be fully interpreted in terms of diffusion of phosphate in solution through the interhyphal spaces, it is unlikely that this is the only route of passage. If it were so, the sheath would be acting rather like a very inefficient wick that is capable of great rates of absorption and accumulation—a wick that would be least inefficient in solutions of greatest concentration. By contrast, we know that mycorrhizal roots function in low external concentrations under natural conditions. Moreover, in Melin & Nilsson's experiments, movement into the host took place solely through the fungal hyphae and sheath.

It is readily demonstrated that phosphate primarily accumulated in the fungal sheath is not necessarily permanently retained there; under some experimental conditions it may move onwards into the host.

If mycorrhizal roots are allowed to absorb phosphate from a low external concentration labelled with ^{32}P, the exact distribution of the phosphate which is absorbed may be estimated by separating the sheath and host tissues. Replicate samples which have been kept, after the absorption period, in aerated phosphate-free buffer solution, may be dissected after various time-periods, and similar estimates made. During the period in the buffer solution, radioactive phosphate moves from the sheath to the host—first at a rapid rate and later more slowly. A study has been made of the effects of external factors on this process of translocation (Harley & Brierley, 1954, 1955). Its rate over 24 hours is dependent on temperature; it has a high temperature-coefficient and is therefore likely to be dependent in rate upon some chemical process going on in the tissues. The movement is also oxygen-sensitive and is reduced to negligible proportions when the phosphate-free buffer is agitated with 3 per cent or less of oxygen, but it proceeds rapidly in air.

Here again we have a clear indication that yet one more essential function of mycorrhizal roots of beech is dependent upon oxygen supply and operates successfully only with oxygen concentrations near that of air.

TABLE XXXIX

DISTRIBUTION OF THE PHOSPHATE ABSORBED IN RELATIVE UNITS (1) DURING A PERIOD IN LABELLED KH_2PO_4; (2) DURING A SUBSEQUENT PERIOD IN PHOSPHATE OR BUFFER; (3) DURING A THIRD PERIOD IN BUFFER AFTER TWO PERIODS IN PHOSPHATE. THE MOVEMENT OF LABELLED PHOSPHATE WHICH WAS ABSORBED INTO THE SHEATH DURING THE FIRST PERIOD IS SHOWN AS: P INTO HOST FROM SHEATH. From Harley & Brierley (1954).

	1st period 3 hours	2nd period 19¼ hours		3rd period 22¾ hours
	Phosphate 0·16 mM.	Phosphate 0·16 mM.	Buffer	Buffer
Total P absorption	36·7	77·9	0	0
P into host from external solution	2·5	15·3	0	0
P into host from sheath	—	1·3	4·4	3·6

This active, metabolically dependent translocating process has been shown to be suspended when the roots are kept at low temperatures or in low oxygen concentrations, and to resume activity when the roots are replaced in suitable conditions. Moreover, it becomes slow or ceases when the mycorrhizas are actively absorbing phosphate from the external solution. If mycorrhizas that have been allowed to absorb labelled phosphate are immersed in unlabelled phosphate solution, little movement of the labelled phosphate from sheath to host occurs although unlabelled phosphate is being rapidly absorbed and, in part, moved into the host.

The supply to the host during absorption is preferentially derived from that phosphate which is being absorbed, and is not derived from phosphates previously accumulated in the sheath. Hence when mycorrhizas which have been allowed to absorb labelled phosphate are placed in buffer solutions containing low unlabelled phosphate concentrations, some translocation from the sheath takes place. The higher the concentration of unlabelled phosphate, the smaller is the magnitude of translocation. It can be seen (Table XXXIX) that during the second absorption phase, phosphate passes from the external solution, into both sheath and host, but the movement of the accumulated labelled phosphate is sensitive to external phosphate supply.

Besides the pathway through the interhyphal spaces, where phosphate may move by diffusion, there therefore appears to be at least one more route in the living fungal hyphae along which phosphate moves by some metabolically activated means. The phosphate which moves is preferentially that which is being absorbed from the external solution but secondarily is that which has been previously actively accumulated into the sheath and is released in mobile form.

The Relative Importance of Diffusional and Metabolic Movement Through the Sheath

It appears at first sight that the diffusional pathway through the fungal sheath to the host is unlikely to be of great importance under natural conditions, because in the soil the available salt concentrations are likely to be relatively low. It is essential, nevertheless, to have some experimental justification of this—especially as, in experiments on salt absorption, it is often easier to work with fairly high concentrations.

We have already seen that the temperature coefficient of phosphate absorption by excised beech mycorrhizas is about 2, irrespective of the concentration of phosphate applied. The separated host and fungal tissues have similar Q_{10} values. It would therefore be expected that temperature would exert a negligible effect on the distribution in the tissue, of the phosphate absorbed by mycorrhizas. This is indeed found to be true where uptake is studied at low and high temperatures from very dilute solutions. The expected 90 per cent of the phosphate absorbed is found in the sheath at both extremes of temperature. Similarly, when very high phosphate concentrations (32 mM.) are applied to mycorrhizas, temperature does not affect or change the *proportion* retained in the sheath. By contrast, when roots are bathed in phosphate solutions of intermediate concentration (3·2 mM.), a temperature change from high to low results in a great decrease in the *proportion* retained in the sheath, and an increase in the proportion reaching the host. An experiment giving results of this kind is shown in Table XL.

The interpretation of this puzzling result is not far to seek. We have already seen that at high concentrations the host is supplied by diffusion through the interhyphal spaces almost as if it were bathed directly in the external solution. Hence the two tissue-systems, root and fungus, might reasonably be expected to react to temperature exactly as if they were free, and to exhibit similar Q_{10} values.

TABLE XL

THE EFFECT OF LOW TEMPERATURE AND DILUTE AZIDE ON ABSORPTION INTO BEECH MYCORRHIZAS AND INTO THEIR COMPONENT TISSUE-SYSTEMS. Data of Harley *et al.* (1958).

Conditions	Control 16·5° C.			2·5° C.			+ 0·2 mM. azide at 16·5° C.		
P conc. mM.	0·032	3·23	32·3	0·032	3·23	32·3	0·032	3·23	32·3
Total P absorbed μg.	6·2	62·0	185	1·0	19·0	52·0	0·6	15·3	67·0
Per cent of control	—	—	—	16·5	31·5	28·0	9·2	25·0	36·0
P absorbed in sheath μg.	5·5	52·0	96·0	0·94	12·2	29·0	0·34	5·3	33·5
Per cent of control	—	—	—	17·0	24·0	31·0	6·2	10·2	35·0
P absorbed in host μg.	0·7	10·0	89·0	0·10	6·9	23·0	0·23	10·1	35·5
Per cent of control	—	—	—	13·0	69·0	26·0	34·0	101·0	38·0

We know that the separate exposure of sheath tissue to low concentrations (0·032 mM.) doubles its absorption rate. This must arise because, at this concentration, the rate of uptake by the hyphae of the sheath of the intact mycorrhizas so completely drains the diffusion channels of phosphate that the inner face of the sheath receives no phosphate by this means. Indeed we could conclude that, effectively, only the outer part of the sheath was directly supplied with any phosphate at all. In this case it must follow that, at low concentrations, the diffusional pathway through the sheath is inoperative, and consequently that all the phosphate reaching the host comes by way of the living hyphal cells. This is in accord with the estimated Q_{10} value which is that of the sheath.

In intermediate concentrations (3·2 mM.) a considerable degree of diffusion through the interhyphal spaces occurs. The diffusion gradient at the higher temperatures is, however, significantly distorted by absorption by the sheath. The application of a low temperature reduces uptake by the sheath, decreases the distortion of the gradient, and results in a higher presentation concentration at the surface of the host. Low temperatures therefore have less effect on absorption by the host than on absorption by the sheath, and so the estimated Q_{10} of the host is spuriously low in these conditions.

This interpretation of the effects of temperature on movement to the host, may be tested by the application of other factors which diminish uptake of both sheath and host in roots kept in a similar range of phosphate concentrations.

The inhibitors sodium azide (2×10^{-4} M.) and sodium iodoacetate (10^{-3} M.) have been used for this purpose and the expected results have been obtained, as is shown for azide in Table XL. Each of the inhibitors greatly diminishes the phosphate uptake of both host and fungal tissue in 0·03 mM. or 32 mM. phosphate, but they have an insignificant or slightly stimulating effect on the host in 3·2 mM. phosphate.

We may safely conclude that the diffusional channel is the dominant route of phosphate movement through the sheath in all conditions when concentrations of 32 mM. are used. On the other hand the results of many experiments show that up to 0·3 mM. the diffusional channel plays a negligible part, and that it is not very important at normal temperatures even up to 1 mM. Above 1 mM. there is significant movement by both routes—through the living cells and also through the interhyphal spaces—though external factors may vary their relative importance.

The passage of phosphate through the fungal sheath may be represented by a diagram:

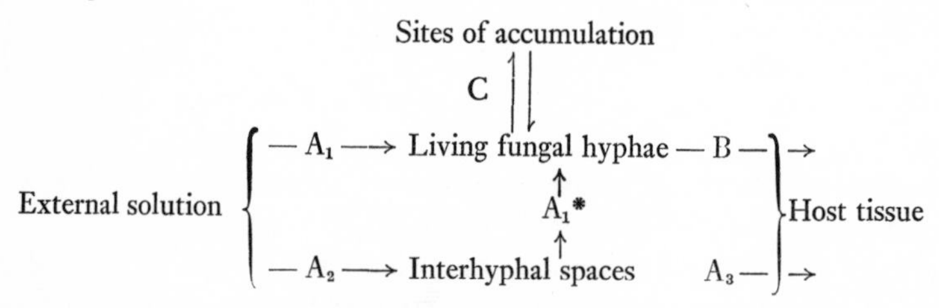

Phosphate may be actively absorbed directly from the external solution into the fungal hyphae which lie in contact with it (A_1). Alternatively, phosphate may diffuse into the interhyphal spaces (A_2). From these spaces, by an active process identical with A_1, phosphate may enter the living fungal cells (A_1*). Phosphate in the living hyphae is trapped by process C into organic phosphates, or in some organelle like the vacuole as inorganic phosphate, and is there firmly held as accumulated phosphate. Phosphate may pass to the host directly from the interhyphal spaces if the diffusion rate is great enough to present a significant concentration to the surface of the host (A_3). This will only occur when the external concentration is high and the absorptive processes of the sheath (A_1*) are fully saturated. In all other conditions the process B, an active release of phosphate from the fungus to the host, will be the dominant process supplying the tissues of the host.

When roots are bathed in a solution of phosphate, the process (B) by which phosphate passes to the host from the living cells of the hyphae, is satisfied wholly or in part by mobile phosphate in the fungal protoplasm; then little or no phosphate is derived from the accumulated phosphate by the reversal of process C. If, however, the mycorrhizas are kept in phosphate-free solution, phosphate from the accumulation sites moves to the host by a reversal of C, and at a rate dependent upon temperature and oxygen supply.

In their natural soil habitats, mycorrhizas are unlikely to be in contact with solutions of several millimols concentration. They develop, as we have seen, in soils relatively deficient in nutrients. It follows, therefore, that the diffusional pathway can be ignored ecologically, and the mechanism actively supplying the host with phosphate from the fungus can be assumed to function.

Biochemical Destination of Absorbed Phosphate

The distribution of phosphorus within the tissues of beech mycorrhizas is given in Table XLI. If the inorganic phosphate passing through the sheath during absorption equilibrated with the whole or any one of the main fractions during its passage to the core, experiments with radioactive tracers could give faulty estimates of the movement of phosphate into the core. There would be an underestimate of the rate of movement of phosphate to the core in the initial stages of any experiment. The extent of the underestimation can be used to determine something of the compartmentation of phosphate in the sheath tissues.

TABLE XLI

ESTIMATION OF PHOSPHATE CONTENT OF BEECH MYCORRHIZA AND THEIR CONSTITUENT TISSUES. μg. per 100 mg. dry mycorrhizas comprising 61 mg. core and 39 mg. sheath.

Phosphate fraction	Total	Sheath	Core
Total phosphorus	604	245	359
Total insoluble phosphorus	360	154	206
Total soluble phosphorus	244	91	153
Soluble organic phosphorus	113	47	66
Soluble inorganic phosphorus	131	44	87

The problem of determining the quantity of phosphorus that absorbed phosphate meets on its way to the host tissue was examined with the help of Dr. U. S. Haslam-Jones (Harley *et al.*, 1954). It was shown that the problems could be solved mathematically provided it was assumed—and the assumptions are justified by experiment—that phosphate uptake and movement to the host are linear with respect to time. During absorption of radioactive phosphate, if the sheath contains a quantity of phosphate, p, with which the phosphate passing to the core mixes, then the radioactivity of the core tissue, Q, would be related to time, t, as follows (ϵ, μ and λ being constants):

(i) where total uptake $\ll p$, $Q \simeq \frac{\epsilon\lambda\mu t^2}{2p}$ + terms in t^3

i.e. $Q \propto t^2$ approximately

(ii) where total uptake $\gg p$, $Q \simeq \epsilon(\mu t - p)$

i.e. $Q \propto t$ approximately

Hence during a period of absorption from a low external phosphate concentration, if a significant quantity, p, of phosphate is present in the sheath, with which the radioactive phosphate mixes on its way to the core, the radioactivity of the core tissue, Q, will increase initially approximately as t^2 and later as t. The extrapolation of the linear second phase to zero time in a plot of Q against time will cut the y axis at a value $-p$. Hence the magnitude of p can be estimated.

The results of an experiment on this model are shown in Fig. 18. The estimated value of p is small because the linear regression on the experimental points passes close to zero. The value of $p = 0{\cdot}017\ \mu g.$ per 100 mg. of dry mycorrhizas is very small compared with the total phosphate in the sheath or in any of the estimated fractions (Table XLI). It is also small compared with the total phosphate absorbed per 100 mg. dry-weight of mycorrhizas in experiments of short duration using low (10^{-4} to 10^{-5} M.) concentrations of phosphate in the external solution. It follows, therefore, that even if the experimental estimate of p were too low (as much as by a factor of 10 or 100), the magnitude of the mobile pool of phosphate in the sheath would still be too low to occasion a significant error, in estimates of accumulation in sheath and core, and (ii) the phosphate passing into the core by-passes the main phosphate pools in the sheath.

Jennings (1964) examined further the importance of this small pool of mobile phosphate in the sheath. He used both chemical analyses and radioactive methods. By chemical analyses he showed that, during uptake over short time-periods, only inorganic and easily hydrolyzable (7 min. in 2N HCl) phosphates increased in quantity. Insoluble organic and stable soluble organic phosphates did not show a net increase. By contrast, in experiments with radioactive phosphate, there was an incorporation into all the fractions, and this showed that they must all turn over metabolically. The form of the curves of incorporation of radioactive phosphate into the several fractions, showed that incorporation into the inorganic phosphate of the tissue preceded that into the others. Moreover, the high rate of initial incorporation into the inorganic fraction, that occurred in the first few minutes of an experiment, was followed by a slower phase. This indicated that the inorganic phosphate in the sheath was divided between two pools, one large and the other small. The small pool which rapidly equilibrated with absorbed phosphate was clearly that observed and measured by Harley and his colleagues. Jennings further showed that the size of this small pool might vary with conditions of uptake—but not so greatly as to cause error in the estimation of the movement of phosphate to the host by radioactive measurements.

In later experiments it was found that the phosphate compound which passes from the fungus to the core is inorganic orthophosphate (Harley & Loughman, 1963). Radioactive phosphate at a low concentration (10^{-5} M. or less) was fed to excised beech roots which were then rapidly

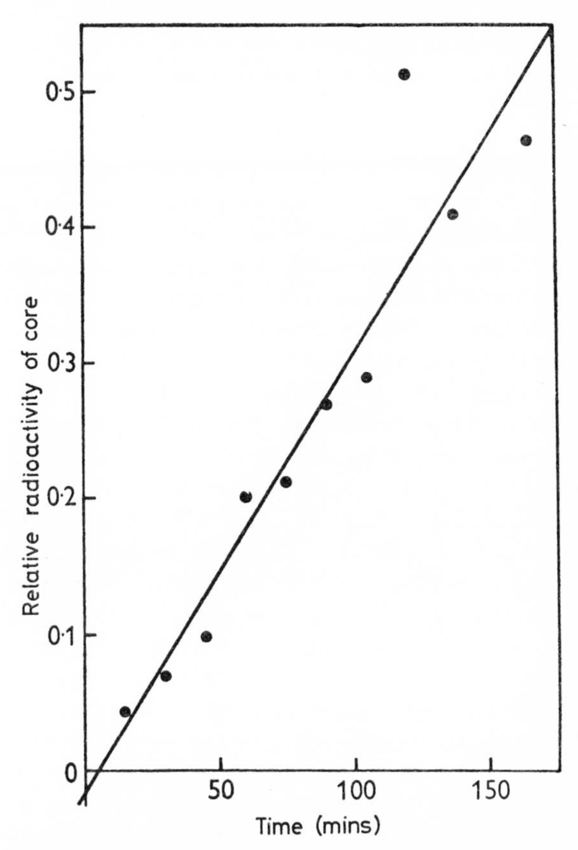

FIG. 18.—Graph showing the increase of radioactivity of core tissues of beech mycorrhizas kept in $1{\cdot}66 \times 10^{-5}$M. phosphate. The linear regression passes close to the origin.
After Harley et al. (1954). *By permission of New Phytol.*

dissected in the cold, extracted, and the extracts separated chromatographically. The main radioactive component in the core tissue during the whole of the experimental period was inorganic phosphate, the curve of percentage labelling in which extrapolated to 100 per cent at zero time.

It can be concluded that phosphate absorbed by mycorrhizas enters first into a small pool of inorganic phosphate which may be imagined to be located in the fungal cytoplasm. From there it is accumulated in the sheath cells as inorganic phosphate, and to a lesser extent as organic phosphate compounds. From the small inorganic pool, phosphate also passes to the core tissues. When absorption is prevented by the absence of an exogenous supply of phosphate, the accumulation of inorganic phosphate in the sheath is mobilized and tends to maintain both the small mobile pool and the movement from it into the core.

A comparison of these conclusions with those reached in work on phosphate uptake and movement in non-mycorrhizal plant material, shows that there is a great degree of similarity.* It remains to decide to what extent they are applicable to the functioning of intact mycorrhizal plants.

Translocation of Phosphate to the Shoots

It has been mentioned earlier that work with excised mycorrhizas has been sternly criticized by Melin & Nilsson (1958) and by Lobanow (1960). The main gravamen of the criticism is that the process of transpiration does not occur through excised absorbing organs, and that it plays a great part in the processes of uptake and movement in whole plants. This is not the place to review in detail the relationship of water and ion uptake. No one can deny that transpiration can effect ion uptake, but it is not usually correlated with it in rate in any simple manner when absorption occurs from low concentrations. It is therefore of interest to consider work that has been performed on the uptake of phosphate by whole mycorrhizal plants. Morrison, in a series of papers (1954, 1957, 1962) in which he fed radioactive phosphate to pine seedlings both in culture and in soil, has examined the question of the rate of phosphate translocation to the shoot. He first showed that the roots of non-mycorrhizal plants pass radioactive phosphate to their shoots at a rapid rate initially but later at a slower rate. Mycorrhizal plants under similar conditions pass radioactive phosphate to their shoots at a constant rate which, although slower than the initial rate of uninfected plants, is faster than their subsequent rate. Hence over an extended period, the uptake and movement of phosphate are both greater in infected plants than in non-mycorrhizal ones. Moreover, the phosphate status of uninfected plants more greatly affects both processes than it does in infected plants. As Morrison pointed out, there are good grounds for concluding that the infected pine seedlings in his experiments accumulated phospate in their root systems in the way predicted from the results of work with excised systems. Indeed he showed that, in plants transferred from radioactive phosphate to phosphate-free solution, translocation of radioactivity to the shoot continued in mycorrhizal plants at a steady rate for 21 days but rapidly fell in rate in non-mycorrhizal plants.

Clode (1956) measured the uptake of phosphate by whole seedlings of

* *See* Loughman & Martin, 1957; Loughman & Russell, 1957; R. S. Russell & Shorrocks, 1959; Crossett & Loughman, 1966; Loughman, 1966.

Pinus radiata. He showed that there was a great accumulation of phosphate in the roots of mycorrhizal plants, so that greater radioactivity was found there than in the shoots. Non-mycorrhizal plants not only absorbed less in total but also translocated more immediately to their shoots. Using autoradiography of microscope sections, Clode confirmed that very large accumulation of phosphate occurred in the fungal sheaths of the mycorrhizas—as would be expected from the work with excised roots.

These experiments appear to show the increased ability of the root systems of mycorrhizal plants to absorb phosphate from the soil. They also confirm that the sheath acts as a primary reservoir of accumulation from which phosphate passes steadily to the host. In uninfected plants, the smaller quantity of phosphate absorbed seems at once to get into all the tissues of the plant more readily. The actual mechanism whereby the increased rate of accumulation of phosphate in the sheath of mycorrhizal plants takes place is as yet unknown. It seems to add little to the matter to suggest with Morrison that mycorrhizas lower the free energy of soil phosphate more than do uninfected roots—unless the manner in which this may come about is also suggested, so that experimental verification may be undertaken.

VII

ECOLOGY OF ECTOTROPHIC MYCORRHIZAS

THE last four chapters have been concerned mainly with attempts to learn something of the structure and functioning of ectotrophic mycorrhizal organs. The procedure in these studies has been, first, to define the material under consideration by describing its detailed structure, mode of growth, and development under natural conditions. Then the material was pulled apart in order to learn something of the nature, physiology, and ecology, of the fungi concerned. Finally the organs themselves were investigated, mainly in the laboratory, in order to see how they *might* function. Only in the description of the behaviour of seedlings has a close approach yet been made, in this book, to consideration of the ecological significance of ectotrophic mycorrhizal organs. The present chapter sets out to rectify this deficiency and, as is common with ecological explanations, we approach in it the realm of speculation. In order to avoid the grosser errors common to some ecological writing, every attempt will be made to emphasize outstanding problems and make this a helpful exposure of ignorance rather than an outrageous pretence of understanding.

The tree species which we have been considering are usually dual organisms in the same sense as lichens are dual organisms. Their root systems are partly composed of fungal tissues, and their structure is such that at least the greater part of the minerals absorbed passes through the fungus. Although this is so, the infected condition is not obligate for the host, for which growth is possible in suitable conditions without the fungal symbiont. But so far as has been ascertained, uninfected hosts are rarely or never encountered in natural forest conditions.

MYCORRHIZA FORMATION

The intensity of infection depends upon conditions of the habitat, and the trees concerned are often forest dominants, in temperate or subarctic climates, upon soils on the acid side of neutrality. It is usual to find that they can flourish on a range of soils between the extreme podsols with a raw-humus horizon, and the brown forest soils that are characteristic of many temperate regions. Some, like the European beech, have a particularly wide ambit of tolerance of soil conditions, and in these in particular, variations in the intensity of mycorrhizal development may be observed. Two constellations of factors, those affecting availability of soil nutrients and those affecting photosynthesis, seem to be most important in affecting the degree of development of mycorrhiza. Intensity of infection has been shown to depend particularly upon the supplies of nitrogen, phosphorus, and potassium, in available forms. A deficiency of one or more of these, or a lack of balance in their availability, has been found to be a predisposing factor. Experimental work and the investigation of some species under woodland conditions, show that trees growing on soils which are well furnished with available nutrients are less intensely infected than their counterparts growing elsewhere. Light intensity, which affects carbohydrate synthesis in the leaves, has, it seems, a dual effect on mycorrhizal infection.

On any given experimental soil, mycorrhizal development has often been observed to increase with light intensity. And since it has been shown that the fungal symbiont absorbs carbon compounds from its host, some kind of relationship between intensity of infection and carbohydrate availability is undoubted. It is, however, now clear that many of the analytical methods used for the estimation of free carbohydrates in mycorrhizas are faulty. Hence the quantitative correlations of the intensity of infection with free carbohydrates are suspect, although there is a correlation with conditions which promote continued adequate carbohydrate synthesis.

In addition, light intensity may affect the kind of mycorrhiza formed. Some, such as those caused by *Cenococcum graniforme*, develop in lower light intensities than others, and are particularly prevalent on seedlings and saplings that are suppressed under a dense canopy, especially on nutrient-poor soils.

We do not know the importance of these factors in the natural regeneration of the trees. The interaction of light and root-competition in the forest is well known; Shirley (1945) has stressed its complexity. Here we have yet another complicating feature, for increased light intensity may not only increase relative root-growth and hence improve the power of the plant to overcome root-competition in this way, but it may also increase mycorrhizal development which again improves efficiency of absorption. This may be a worthwhile problem both to ecologists and to foresters interested in selective felling.

Although the mycorrhizal state is not obligate to the host plants, there is considerable evidence that it may be to the fungi or at least to some of them. It appears possible that some species, such as *Cenococcum graniforme*, *Boletus subtomentosus*, and *B. bovinus*, may be potentially free-living or have a larger host-range than the arborescent species with which they form mycorrhizas. On the other hand, most of those so far investigated have properties which align them physiologically with obligate root-inhabiting fungi in the sense of Garrett. It is, however, one of the outstanding defects in our knowledge of the Agaricaceae and Boletaceae, that little is known of the ecology of their mycelia. Amongst these, the mycelia of the known mycorrhiza-formers have no certain ecological site apart from the surfaces of their host-roots and the immediate environs of them. The assumption that this is their only habitat is based on ignorance rather than knowledge—upon probabilities estimated from laboratory investigation rather than upon actual field-work. There are some observers, such as Wilde (1954), who appear to make the assumption that mycorrhizal fungi persist in a vegetative state in the absence of hosts, because of the readiness with which young plants become infected when old woodland soils are afforested. Levisohn's observations (1955) on the distribution of rhizomorphs and hyphal strands of some mycorrhizal fungi in forest soil, as well as the work of recent Russian authors, give point to this view. Moreover, there have been reports that mycorrhizal fungi may persist in the rhizosphere of species with which they do not form mycorrhizas. Warcup (1959), for instance, reported that a presumed mycorrhizal Basidiomycete, repeatedly isolated from the mycorrhizas of *Pinus*, was also frequently obtained from adjacent dead or decaying grass-roots. On the other hand, the demonstration by Robertson (1954) of the efficacy of spore-borne infection, makes a general assumption on the available evidence untrustworthy. It also must be admitted that the balance of evidence upon the physiology of mycorrhiza-formers, which has been amassed especially by Professor Melin and his colleagues, strongly suggests that, with some few exceptions, these fungi have a restricted existence except in associations with their hosts growing in natural conditions. The mycorrhizal state, therefore, may well be the only normal state of both hosts and fungi under natural conditions—i.e. they may both be ecologically obligate symbionts.

The nature of the stimulus which selects the mycorrhizal fungi to become dominant on the root surface of susceptible hosts, has only been guessed. Internally high carbohydrate status is unspecific; it is a commonplace that obligate parasites become more readily established upon well-nourished hosts with adequate internal carbohydrate supplies. As far as can be seen, the stimulus promoting mycorrhizal infection is produced in conditions of adequate carbohydrate nutrition. The root secretions of Melin & Das (1954), which so greatly stimulate the growth of some mycorrhiza-formers, are not only produced by unsusceptible non-mycorrhiza-forming hosts but also have not been shown to be specific to mycorrhizal fungi. Nor do they have a morphogenic effect, but simply stimulate the growth-rate of mycelia. Interesting and important though these observations are, they do not go far to explain associated growth. They do, however, explain the vigorous development of mycelia in the rooting region of the host plant and, also, the rapid spread of infection along the root system.

Once infection has developed it is permanent, and short roots are infected from the mother-root. The spatially precocious maturation of the tissues of the short roots may well precede rather than follow infection. We are not yet sure whether the characteristic structure of the host in infected roots is, as a rule, caused by the fungus, or whether it is a state to which the short roots naturally tend. It is, however, fairly certain that the short axes are stimulated to proliferate after infection, so that the total area in contact with the soil of each short axis is increased. Moreover, the life period of each short axis is prolonged. In *Pinus*, estimates of the life period of the individual mycorrhizal systems developed by proliferation of one short axis, vary from several months to more than one year. Indeed, Lobanow reports (1960) that I. B. Shinkarenko observed them to have a life of 3–4 years.

In a series of observations recorded photographically on a set of marked long roots in beech woodland, Dr. F. A. L. Clowes and I followed the development of beech mycorrhizas from initiation to death. The main formation of mycorrhizas began in July and was at its peak in August (Plate 5). The short laterals of this crop were mostly dead by the following March—a life-span of some eight months. However, some root growth and mycorrhiza formation occurs throughout the year, and active mycorrhizas may always be found in fair quantity. Werlich & Lyr (1957) came to much the same conclusion by studying the proportion of living and dead mycorrhizas present on the long roots of pine and beech in various soil-horizons. They stated that the majority of short roots continued to be active for one year or less, and that very few lasted for two years.

Observations on spruce have led to the conclusion that its mycorrhizas are much longer lived. Orlov (1957) described an interesting method by which he fitted small frames holding cover-glasses over individual spruce mycorrhizal systems and so was enabled to follow and measure their growth with a binocular microscope. In his two papers (1957 and 1960) he showed that the main axes of the mycorrhizal systems grew at the average rate of about 0·01–0·03 mm. per day, and might on occasion grow as rapidly as 0·1 mm. per day. These systems persisted for four years, but as they aged their colour changed from light- to dark-brown. Eventually the sheath and primary cortex died. In the majority of cases, death did not occur for two years. Mikola & Laiho (1962) agree also that spruce mycorrhizas live more than one season, for they observed them to grow and branch in a second season at least. The variability of longevity of mycorrhizas has been discussed by Goss (1960). It is our experience with beech that, in mycorrhizas of large diameter, the apex of the host tissue may grow in the spring and rupture the fungal sheath. This form of growth and persistence of mycorrhizal apices has been widely described for many species and is, for instance, recorded and figured by Lobanow (1960) as well as by Masui (1926*a*) and Goss (1960).

The stimulus to prolonged persistence and increased branching of short roots appears to be fungal in origin. The work of Slankis (*see* Chapter IV) on the effects of auxin and of fungal metabolites has firmly indicated this. It seems possible that the tendency of short roots to cease growth and to differentiate completely, leaving only a small potential meristem, may be occasioned by many causes such as overdoses of auxin, effects of colchicine, or any ecological factor leading to reduced growth-rate as well as perhaps by fungal activity. The cause of their repeated branching and of their persistence seems to be of fungal origin. We must, however, appreciate the fact that only a start has been made in our attempts to understand the mechanism whereby the mycorrhizal root system develops its peculiar morphological and anatomical form.

The ageing of the short roots, so widely observed in conditions where infection does not or cannot take place, requires further study. It is possible that it is similar to the ageing observed in other plant organs—for instance in roots of groundsel and tomato in culture as described by Street & Roberts (1952) and Hannay & Butcher (1961). The latter authors showed that the process may be modified by carbohydrate supply and by growth factors. There is room for much experimental work on this subject, for it may be that ectotrophic mycorrhizal plants have an inbred tendency to ageing of their short roots which is offset by infection that is, therefore, necessary for the maintenance of an adequate absorbing system.

Mycorrhiza and Absorption

There is general agreement that relative increase of root-surface area is exhibited by seedlings which become mycorrhizal as compared with their uninfected counterparts. This is due to the increase in size and complexity of the short axes, and is further magnified by the hyphae which may radiate from the mycorrhizas into the substrate. It is equally certain that a simple increase of area does not alone explain the success of mycotrophs in difficult sites, nor the stimulation of seedlings which become infected. The fungal sheath, which is interposed between the host-root and the soil, calls the tune in the activities of absorption from the soil.

Work with excised mycorrhizas has been a stage in the experimental study of absorption—a necessary stage, but one with its own limitations. By experiments with this material, the mechanism of accumulation of salts in the fungal tissue and in the host-root can be investigated, but absorption into the host as a whole is prevented. The translocatory movement of substances away from their sites of accumulation in the host-root tissues is prevented, and this enforces an artificiality upon the whole method. It is, then, a study of the first stages of absorption and movement, and may be assumed to give valuable information only if the experiments are of short duration. We know from the work of Melin & Nilsson and of Harley & McCready (discussed in Chapter VI) that absorption of soil nutrients into, and translocation of them away from, mycorrhizas occurs; but we know nothing of the rates of the processes nor of factors affecting them. By simplifying the material, through the use of excised mycorrhizas, some of these difficulties are in part overcome.

The absorption of nutrients by mycorrhizal roots proceeds by a process which depends in rate upon factors affecting the rate of metabolism, such as temperature and oxygen supply. It is also affected by inhibitors and poisons—especially by those affecting energy-releasing processes in the tissues. In the cases tested, greater rates of absorption through each unit area are attained than with uninfected roots under similar conditions.

The dependence on an adequate supply of oxygen is an important feature. Beech mycorrhizas attain their maximum rates when supplied with an oxygen concentration near to that of air, and do not function below 3 per cent of oxygen. It is therefore essential to know whether, in their natural surroundings, mycorrhizas are amply supplied with oxygen. Brierley (1955) carried out a series of analyses of the oxygen concentration in the air spaces of the humus of beech forest in the actual vicinity of mycorrhizal roots. A large number of analyses, made at monthly intervals throughout the year, showed that the oxygen concentration was maintained above 19 per cent by volume during the whole period. Similar results were also obtained by Penningsfeld (1950) using different methods. He studied oxygen concentrations in the atmospheres of various soil horizons of woodland soils. The oxygen supply was always high in the surface and subsurface horizons, and also often in the deep-seated layers.

It is accordingly evident that the conditions of their habitat are such as to allow the mycorrhizas to accumulate salts rapidly, but this does not demonstrate that the host tissue is well supplied. The absorption of phosphate, ammonia, and alkali metals, is rapid at the oxygen level available, but the primary site of absorption is into the fungus. Phosphate, in particular, is mainly accumulated in the fungus, and during absorption the phosphate accumulated in the hyphae is fixed; of that which is absorbed, a small fraction, some 10 per cent, passes to the host. That which is accumulated in the hyphae is released and may pass to the host only when the external supply of the ion is very low. This translocation from fungus to host is again an oxygen-dependent reaction, and proceeds at its fastest rate when the oxygen supply approaches that in air. These features are correlated with the occurrence of mycorrhizal roots in the sub-surface humus layers of the forest floor where oxygen is plentiful; so we now have an idea of how mycorrhizas might work, and a demonstration that the ecological factors are suitable for them to do so in the case of beech mycorrhizas.

The experiments of Morrison (1957) and Clode (1956) with pine seedlings have gone far to verify the hypothesis of mycorrhizal functioning. Seedlings treated with radioactive phosphate were shown to accumulate phosphate in their mycorrhizal root systems and to translocate phosphate from these accumulation sites during periods of external deficit.

But this is not enough. Even though we have, from the work of Stone (1950) on pine seedlings and from observations on excised beech roots, evidence that mycorrhizas trap ions, especially phosphate, more actively than uninfected roots of the same and other species, we still have to consider the problem of the avidity of the fungus for nutrients.

Forest trees which bear ectotrophic mycorrhizas are typically shallow-rooting, and exploit intensively the upper horizons of the soil. In particular the absorbing roots are especially developed in the upper humus layers. Harley (1940), Redmond (1957), Werlich & Lyr (1957), Lobanow (1960), Mikola & Laiho (1962), and Lyr (1963), have all made this point. There is an almost horizontal exploitation by the spreading mother-roots, the mycorrhizal branches of which grow generally upwards into the newly-forming humus layers. The extent of localization of exploitation varies both with the species and with the type of soil. In the case of species with a wide ambit of soil preference, the localization of the root system in the sub-surface layers is most exaggerated in podsolic soils and in soils with surface humus accumulation. Werlich & Lyr (1957) estimated the distribution of absorbing roots and mycorrhizas with depth in the soil of beech woods and pine woods. Mycorrhizas were especially concentrated in the upper humic layers but were present in the deeper horizons also. Pine mycorrhizas went deeper than beech. Lobanow has also shown that pine mycorrhizas can be found at a considerable depth (1–1·5 m.).

The sub-surface accumulation of root systems results in intense root competition, so that the absence or sparsity of ground flora is in great part a resultant of it rather than of low light intensity. It moreover restricts the extent to which organic debris arising from leaf-fall is mixed with the mineral soil, because it enforces a resistance to the movement of soil denizens which help to bring about this mixing. Ectotrophic mycorrhizas are typically organs of absorption of nutrients from the humus layers and they are also one of the factors encouraging humus formation. It is reasonable to surmise, and indeed there is some supporting evidence, that the release of soluble material from decaying leaf-litter is not a steady process. Newly-fallen forest litter quickly loses its soluble minerals, which are carried in solution downwards through the soil horizons; much of this soluble mineral matter is removed as the solution passes through the biologically active surface layers. It is to be expected that the significance of the high accumulation rate by mycorrhizal organs lies in the exploitation of these flushes of released nutrients, and that mycorrhizas compete effectively with the other denizens of the surface layers. In other seasons, when the absorption rate is slow, phosphate and presumably other ions may pass by the active route of translocation to the host. And then again, the death of the mycorrhizal root returns the accumulated phosphate and other substances to the humus layer.

Mycorrhizas can be conceived as trapping agents—especially in soils of periodic release of nutrients. The work of Routien & Dawson (1943) has suggested, in addition, that absorption from colloidal surfaces into mycorrhizas is more rapid than into uninfected roots; and even if we do not agree with their ideas of a mechanism of exchange for CO_2, their results suggest that their ideas should be followed up.

Mycorrhiza and Disease

The demonstration of the efficiency of mycorrhizas as absorbing organs and of the avidity of their fungal sheaths in accumulation, may be thought to provide evidence of the selective advantage of the ectotrophic mycorrhizal habit. It is also of selective advantage that root systems, the short roots of which tend to precocious ageing, should have the capacity to form an association which results in greater functional longevity of their absorbing organs. The mycorrhizal root system has, in addition, other properties that may be ecologically significant. For instance, Cromer (1935) and Harley (1940) pointed out that infected roots are more frost- and drought-resistant than their uninfected counterparts. Goss (1960) has shown experimentally that mycorrhizal seedlings of *Pinus ponderosa* are more resistant to drought than are uninfected seedlings. Similar conclusions were reached in Russian work quoted by Lobanow (1960). Adverse conditions which lead to the collapse of the cortex and death of uninfected roots may not permanently affect mycorrhizal roots, so that they may recover physiological activity. Any differential effect of this sort may be viewed as an additional mechanism which increases the effective absorbing-area of the root system.

In a recent article, Zak (1964) has expressed the view that: 'Improved nutrition alone is insufficient explanation for the demonstrated benefit of mycorrhizae in many areas.' In support of this he pointed to results such as those of Briscoe (1959) and White (1941), in which the application of fertilizers alone failed to encourage the healthy growth of young trees and the added factor of inoculation with mycorrhizal fungi was necessary for successful growth. This might be because mycorrhiza formation protected the rootlets against parasitic invasion. Boullard (1960*a*) has supported this view in his discussion of biological factors relevant to the problems of reafforestation, and it is the contention of those who have studied the so-called pseudomycorrhizal infections that mycorrhizal fungi successfully compete with pseudomycorrhizal mycelia. There is a kind of 'capture effect' in that the mycorrhizal fungi so achieve dominance on the root-surface that fungi such as *Mycelium radicis atrovirens* cannot then exploit it. Mikola (1965) gives examples of how ectotrophic mycorrhizal fungi supersede ectendotropic fungi on the new rootlets when seedlings are transplanted from nursery soils to woodland soil.

Zak also made the point that the rhizosphere and sheath surface of mycorrhizas are inhabited by populations of organisms that differ from those of uninfected roots. He suggested, with reason, that this may endow the mycorrhiza with a 'different biological barrier to infection' from that of non-mycorrhizal roots.

These hypotheses clearly merit closer investigation, and a recent paper by Marx & Davey (1967) has shown that mycorrhizal seedlings of *Pinus echinata* were resistant to infection by spores of *Phytophthora cinnamomi* whereas uninfected seedlings were highly susceptible.*

Mycorrhiza and Forestry Practice

ECOLOGICAL DISTRIBUTION OF FUNGI

The results of all these investigations of the functioning of mycorrhiza have an important bearing on practical problems of forestry. They are important especially in the reafforestation of treeless areas, in the introduction of exotic trees, and in the raising of seedlings and transplants in nurseries.

It is usually true that mycorrhizal fungi are potentially present as spores or actually present as mycelia in old forest areas, and that any planted tree will encounter them. The lack of specificity of most species of mycorrhizal fungi to a single host is likely often to ensure that some sort of mycorrhizal development will ensue even if a new tree species is introduced. A good example of this is given by the work of Bowen (1963) on *Pinus radiata* in South Australia. This species of *Pinus* is an exotic but has been widely planted in this region. Bowen grew seedlings of it on 78 soils derived from various sources. Mycorrhizas were developed on 53 of these, but not on the remaining 25. Of the soils in which mycorrhizas developed, 25 came from areas with adult trees of *Pinus radiata* and the remaining 28 from areas bearing dense or scattered eucalypts. The soils on which mycorrhizas did not develop were either treeless or bore vegetation which included eucalypts. Two types of mycorrhizas were observed: one was a white form with well-developed sheath and Hartig net from *Pinus* soil, and the other was a brown form deficient in sheath development and was found in soils from eucalypt areas. It seems not impossible that the brown type was formed on *Pinus* by a fungus more positively adapted to eucalypt roots but capable of forming a sort of association with *Pinus* also. It is of prime interest that the brown form is recorded as having a much lower efficiency of phosphate uptake, and as bringing about a lesser stimulation of tree growth, than the white mycorrhizal forms.

Aseptic culture methods have also shown that a single host species may form mycorrhizas with several species of fungus belonging to more than one genus. Zak & Bryan (1963), and Zak & Marx (1964), found that a single host-tree could at one and the same time associate with as many as seven different symbionts, and a single root with two or three. Mever

* *See* Appendix §7.

(1966) quotes similar examples, and the extensive compilations of Trappe (1962) underline the magnitude of the problems, still almost unattacked, which concern this aspect of the ecology of mycorrhizal fungi.

Work on the ecological distribution of Basidiomycetes shows quite clearly that, in any area of forest, a large number of potential mycorrhiza-formers are usually present. Most of the species for which we have any experimental evidence, are not specific to any one host and so as a general rule no one species dominates a particular site to the exclusion of others. The interesting surveys of Boullard & Dominik (1960, 1966), conducted in beech forests in Poland and France, provide examples of how variable mycorrhizal types, which may be of different fungal origin, may be present in any locality. In each kind of beech community several types of mycorrhizas were found upon *Fagus sylvatica*, although certain kinds were common to all. There were significantly fewer kinds encountered on *Fagus* on dry calcareous sites, where mycorrhizas caused by *Cenococcum* were common. Gobl (1965) also recorded that in a comparison of a vigorous stand of spruce with a poor stand, there was a greater diversity of mycorrhizal types in the more vigorous stand. An inspection of the many records shows that, on fertile sites where tree growth is most active, a variety of light-coloured mycorrhizas of fully developed ectotropic structure are present (*see* Melin, 1927; Björkman, 1942; Bergemann, 1955; Mikola & Laiho, 1962), whereas on infertile, dry, shaded sites, or in heavily fertilized nurseries, ectendotropic mycorrhizas, pseudomycorrhizas, or *Cenococcum* mycorrhizas, may predominate.*

The difficulties of afforestation of *Calluna* moorland have stimulated research on the problems of mycorrhizal development there. Handley (1963) has shown that the reduction or absence of typical ectotropic mycorrhizas in trees planted on *Calluna*-mor soils may be explained by the presence there of substances inhibiting fungal growth. He showed that in cultural experiments, aqueous extracts of *Calluna* humus inhibited the growth of many mycorrhizal fungi in conditions where the supply of nutrients and the acidity of the medium were favourable to their growth. Some strains of *Boletus scaber* were, however, capable of resisting the inhibitor. This fungus was one of two (the other was *Amanita muscaria*) observed by Dimbleby (1953) to form fruit-bodies in locations where birch regeneration was successful on *Calluna* moorland. Levisohn (1959, 1960) has also recorded the existence of strains, within the species *Boletus scaber*, which differ in respect of their cultural characteristics. One strain from Wareham Heath in Dorset was recorded as reducing the growth of *Betula* in culture, although *Boletus scaber* is generally considered to be one of the most important mycorrhizal fungi for that genus.

In contrast to sites within naturally forested zones, the treeless areas (such as the prairies of North America or the steppes of Russia) are devoid of naturally occurring mycorrhizal fungi. Many cases have been reported from such places of trees failing to become established if they were not

* *See* Appendix §3.

mycorrhizal before planting, or if they were not inoculated with suitable fungi. Mikola (1953) has shown that if seedlings of *Pinus strobus* or *P. banksiana* were planted in prairie soil taken at different distances from shelter belts, the rate of mycorrhizal development varied. Even after five months, the seedlings had not become mycorrhizal if they were grown in soil collected more than 6·5 metres from shelter-belt trees. The effects of mycorrhizal development following inoculation of such soils has been mentioned above, and striking examples are given by Lobanow (1960) as the result of work on Russian steppe soils.

ARTIFICIAL INOCULATION

The failure of schemes of afforestation and the establishment of exotic tree species has often been ascribed, sometimes correctly, to the absence of suitable mycorrhizal fungi or of favourable conditions for mycorrhizal development. Success has often accompanied the inoculation of the stock with mycorrhiza-formers. It must, however, be emphasized that mycorrhizal development is not the only factor determining successful growth. It would seem superfluous to stress this point were it not for the fact that mycorrhizal development has sometimes been viewed as the cure of all ills, incompetent cultivation and husbandry being excused by blaming mycorrhizal inadequacy. Nevertheless the mycorrhizal condition is normal for many forest dominants, and it is clearly logical to ensure that transplants become infected.

It is important, however, to get the matter into proper perspective. It is possible to grow mycotrophs in uninfected condition, for they are not obligately associated with their fungi except in an ecological sense. Heavy applications of nutrients to tree-nurseries will often bring about this result, but it is not an economic proposition. Not only is it impracticable to maintain high levels of soil nutrients during the whole period of growth of forests and plantations, but also the land available and utilized for forest establishment is most frequently 'marginal' land that is unsuited to intensive preparation, cultivation, and fertilization. The aim of nurseries should, therefore, be to raise transplants of maximum vigour and size which are adequately equipped with mycorrhizal organs. As Björkman (1956, 1961) has pointed out, fertilization to excess in forest nurseries, which results in an impairment of the development of mycorrhiza, results in a reduction of the vigour of plants when they are transferred to normal forest soil. This result is due partly to the absence of the mycorrhizas, which are able to absorb nutrients from the soil, and partly to the effects of treatment on the form of the root system, and on the suitability of the seedling for lifting, carriage, and re-establishment. The uninfected plant, with its extensive root system, is more susceptible to damage by drought, less easy to replant, and less able to establish itself, than the more compact, more drought-resistant, infected plant. Björkman has shown that the establishment of mycorrhizal infection, coupled with a transplanting procedure within the nursery, results in a compact and sturdy root system of greatest efficiency for transplanting to the forest. He has further experimented with methods of cutting the root systems of seedlings *in situ*, in order to obviate the expensive procedure of transplanting within the nursery.

Following the demonstration that mycorrhizal fungi may compete successfully with weak pathogens of a pseudomycorrhizal type and possibly of more virulent sorts which may abound in forest nurseries especially on agricultural land (Levisohn, 1954; Thornton, 1956; Zak, 1964; Marx & Davey, 1967), further point has been given to the advisability of ensuring early mycorrhizal development. The important problems involved are first, how best to bring about mycorrhizal inoculation, and second, to what extent it is advisable or practicable to select the kind of mycorrhizal fungi to be used for this purpose.

The actual process of inoculating sites or plants may be performed in various ways, and these need consideration in the light of field and laboratory researches. The use of soil or humus from forest or other selected sites where mycorrhizal development readily occurs, is open to criticisms—some of which have already been mentioned. The process usually involves the introduction of unknown mycorrhizal fungi along with a soil population which may contain pathogens—as well as, perhaps, humus of high carbon-to-nitrogen ratio. The inoculum is bulky, and its preparation involves the cost and dangers of collection and transportation. It has been used, however, with great effect in many different parts of the world in connexion especially with the establishment of exotics. The use of pure sources of known mycorrhizal fungi is possible, and its practicability is the subject of current research. The work of many students of mycorrhiza on the different effects of different mycorrhizal fungi on the growth of the hosts, has an important impact on this problem; but it has not gone far enough for full consideration. Young (1940) observed that the different species of mycorrhizal fungi which he used had different effects on the growth of *Pinus caribaea*. Moser (1956) observed that *Larix* grew faster when infected with *Boletus* (*Suillus*) *viscidus* than with *Boletus elegans*. Levisohn (1957*a*, 1959, 1960) has reported not only that *Boletus scaber* and *B. bovinus* vary in their effects on birch and Scots pine in different soils, but that there is also a strain variation within the single species, *B. scaber*, so that different strains may exert different physiological effects. These facts, taken together with the knowledge that different kinds of mycorrhizas on one host have different efficiencies of absorption of nutrients, and that different mycorrhizal strains have different growth requirements in temperature and acidity, expose a great area in which new researches of practical importance are urgently needed.

The methods available for the preparation and use of pure sources of mycorrhizal fungi appear to be three in number. The demonstration by Robertson (1954) that spores constitute a possible inoculum, points the way to a method which would provide sources of fungi easily; besides being pure, they could be cheaply stored and transported. Research is, however, needed on methods of collection and storage, before this kind of inoculum can be widely used. The observed spread of mycorrhiza from an infected seedling to its neighbours in a nursery-bed, has been used occasionally as an inoculation procedure—by interplanting lines with infected plants. This method, harnessed to a system for ensuring inoculation of selected plants with pure sources of fungi, would seem to be a possible means of providing inocula for wide distribution. Indeed the method of culture of plants in polythene sachets, described by Boullard (1960), would seem to lend itself conveniently to this process of provision and distribution of inocula.

Dr. M. Moser is carrying on most useful researches on inoculation with fungal mycelia (1956, 1958, 1958*a*, 1958*b*, 1959*a*, 1963). He has outlined methods for the culture, preservation, and use of mycorrhizal fungi, and has examined practical means of applying the inoculum in various forms to plants in the nursery. The complexity of problems which have yet to be adequately studied may be comprehended if one considers the variations, genetic and phenetic, of the fungi and of their hosts against the background of environmental variation. As Moser has shown, the fungi vary in their dependence on temperature, humidity, pH, and nutrient supply, and there is a genetic strain-variation within species with respect to their physiological demands and mycorrhizal potential. Amongst the hosts there is, as Linnemann (1959, 1960, 1964) has pointed out, also variation of extent and form of mycorrhizas with habitat and genetic origin.

The value of these methods in practical forestry will not only depend upon their scientific perfection, but also upon their relationship to other practices, and upon economic considerations. The use of herbicides and pesticides has, for instance, an important effect on soil micro-organisms, and the sensitivity of mycorrhizal fungi and of the infection process is already under study (*see* Laiho & Mikola, 1964, for references). The results of these researches may well help to define the methods and timing of the application of biocides and fertilizers, and the timing of sowing of nursery plots.

One further point may be worth mentioning concerning mycorrhiza and forestry practice. Even if it is true that mycorrhizal development will permit trees to grow satisfactorily on soils that are more nutrient-deficient than those which uninfected trees can colonize, it does not provide something for nothing. Extraction of timber removes soil-derived nutrients from the site (*see* Rennie, 1955), and these in the long run must be made good. If mycorrhizas improve the growth of trees on deficient sites, then the rate of removal of nutrients from such sites in timber extraction is likely to be increased.

PART III

ENDOTROPHIC MYCORRHIZAS

VIII

ENDOTROPHIC MYCORRHIZAS—INTRODUCTION

It is more difficult to discuss coherently the results of work on endotrophic mycorrhizas than those of work on ectotrophic mycorrhizas. As has been shown in an earlier chapter, the kinds of association between roots and fungi which have been described as non-pathological, are very diverse. It is of first importance to obtain a natural classification of these, so that groupings may be made about which generalizations are possible. In many of the endotrophic infections, a considerable degree of intracellular penetration by the hyphae is present, and in such cases some kinds of haustorial structures are formed. To these, the host-cells react by digesting the endophytes. Such features, together with morphological characteristics of the endophytic hyphae themselves, have been used in classifying the various types of endotrophic mycorrhiza.

The most obvious character which divides the kinds of endotrophic mycorrhiza into two sorts is the septation of the hyphae. The hyphae may be sepate, like those of Ascomycetes, Fungi Imperfecti, and Basidiomycetes; or they may, by contrast, be aseptate, like those of Phycomycetes. The two types of hyphal construction are not specifically associated with any particular plant phylum; for instance, root systems or absorbing organs of angiosperms, gymnosperms, pteridophytes, and bryophytes, may all be inhabited in a regular fashion by aseptate hyphae, and yet other species of the same phylum may be inhabited by septate hyphae. As will be seen later, the aseptate hyphae are on the whole more uniform in structure, and produce in most cases more characteristic organs, than do the others; and they have been seen not only in living plants but also in fossils preserved from the Devonian period onwards. This kind of infection may, therefore, be separated for present purposes as comprising a single or a closely-allied series of natural kinds of endotrophic mycorrhiza.

The septate hyphae are not so uniform, but are as widely dispersed in the plant kingdom. Many have been shown to belong to Basidiomycetes, and they may bear clamp-connexions; others have been placed in the 'imperfect' genus *Rhizoctonia*, some of the species of which are sterile forms of Basidiomycetes. Others, by virtue of their simplicity and sterility, cannot be at all surely placed taxonomically. Moreover, the diversity of the composite structures, the mycorrhizas, which these fungi having septate hyphae form with their hosts, is very great. Structures not unlike the ectotrophic mycorrhizas of forest trees are found in *Arbutus* and *Monotropa*, where an external hyphal sheath is present. Most Ericaceae have a considerable external hyphal plexus loosely developed around the root, which is penetrated intracellularly. The Orchidaceae are usually profusely colonized within their cells by hyphae which form fewer and simpler connexions with mycelia in the soil. *Lobelia dentata* is described as having intercellular penetration only.

These few examples show that it is impossible to generalize very confidently or very profitably about endotrophic mycorrhizas. In a light-hearted and rash moment I have previously made three general points about endotrophic mycorrhiza which all types have in common. First, a host; secondly, a fungus penetrating the host; and thirdly, our ignorance of their physiological function. One may, however, go a little farther than this. It seems more common to find some form of fungal infection in an absorbing organ of a plant grown in natural conditions, than to find that it is free from fungal hyphae. If no detrimental effect of infection is obvious, such as the breakdown of cells and cell-walls or other clear signs of parasitism, the infection may well be negligible in effect, or alternatively it may be of some selective advantage to the host in particular ecological conditions.

The distinction here is difficult to make. Killing diseases caused by unspecialized parasites are only very prevalent under natural ecological (as opposed to agricultural and horticultural) conditions in the seedling phases and in the senescent phases of plant growth. In these two phases the plant is liable to be killed by a variety of adverse factors. In the seedling stage, the summation of all the chances of death up to the attainment of the state of reproduction is offset by the rate of seed production in those individuals that reach the mature state. The addition of yet another relatively small chance of death, by parasitic attack, may be of no great selective disadvantage. Again, the senescent organism, which has passed its most active reproductive state, may be killed without great loss of reproductivity to the species. But there is considerable selective advantage in those strains whose mature reproductive members are not subject or susceptible to killing diseases. It is found, indeed, that the frequent parasites of adults are specialized parasites capable of a prolonged symbiotic period of life, and which permit considerable physiological and reproductive vigour in their hosts. In the extreme we find cases of symptomless infection such as that of *Lolium*.

One must not assume from this that the most specialized of the endophytic and epiphytic organisms, the permanent denizens of the tissues, are stimulating in any way or are of selective advantage to their hosts, but rather that they are of little disadvantage. We would not assume that the fungi of the skin of man, prevalent though they are and although they may infect 50 per cent of even the cleanest of the adult population, and which increase in prevalence in the most athletic and physically developed members of the species, have a selective advantage. It is easier to assume them to be of a physiological disadvantage so small, that they add little to the total chance of diminished vigour and reproduction which is the summation of all the inhibitory factors in the habitat. The frequency with which a particular type of infection is present in a population in its natural habitat is no measure of selective advantage, but may indicate the relative absence of disadvantage.

If an infection of the organs of a plant by fungi increases the survival rate of infected individuals, its influence may originate in various ways which may be grouped under two headings. The fungi may influence the germination of seeds or of spores and the early development of the young plant, or they may stimulate the growth and activity of the adult. In the first case, if the influence is solely on the young phase, infection in the adult may be a relic of a condition physiologically effective in the seedling. For the infection to be effective in the adult phase, the fungus must in some way provide food material, i.e. material of which the body is built, or growth-factors or enzymes, i.e. substances essential to the metabolism of food substances, to a system that is limited in its efficiency of operation by the lack or paucity of these substances.

In most kinds of endotrophic mycorrhizas, where intracellular penetration occurs, some form of digestion of the hyphae by the host-cells has been described. This usually involves the clumping together of the infecting hyphae within the cell, a loss of contents from the hyphae, and the formation of a mass of undigested cell-wall material. This has been held to be either a simple defence mechanism on the part of the host, or a mechanism by which substances are absorbed by the host from the fungus. There is no doubt that digestion may have both of these results. It is, however, essential to realize that movement of material from fungus to host has been proved to occur in ectotrophic mycorrhiza where no digestion has been described. It also occurs in various wilt-diseases, when fungal exudates or secretions cause wilting. Digestion results in a release of only those substances actually present in the hyphae during digestion, whereas there can be a continued release of material from fungus to host if the hyphae remain living. There is no difficulty in visualizing interchange through the adjacent walls of host and fungus, as the process need be little different from that between adjacent parenchyma cells. In both instances we have evidence of a retention of enzymes in the cell-walls (e.g. Rothstein, 1954; Harley & Smith, 1956; Jennings, 1956, Harley & Jennings, 1958), and these enzymes may be operative in the interchange. It is therefore important not to make too much of digestion as evidence of a balanced nutritive relationship: it may be primarily a defence reaction which results in some degree of absorption of substances into the host.

The relative infrequency of digestive patterns in the cells of hosts attacked by specialized parasites of a damaging kind, such as Erysiphales and Uredinales, has led these to be contrasted with infections in which digestion is a well-developed feature. The very possibility of an absorption of material by the host by means of digestion, has resulted in a separation by biologists of the two kinds of infection as disease and symbiosis, respectively—with concomitant neglect of the defensive function of digestion. Burges (1936), however, viewed digestion as a purely defensive reaction, and believed that mycorrhizal fungi in penetrating the tissues were purely parasitic, and that any stimulatory effect which they might have on their hosts was a result of their activity in the external substrate. However much this may be an extreme view, it is, and has been, a satisfactory corrective to those who tend to neglect the defensive function of digestion in favour of a purely nutritive function.

The degree of penetration of the hyphae into the host tissue in endotrophic mycorrhiza is very variable. In extreme instances no intracellular penetration occurs. This is the case in *Lobelia dentata*, where Fraser (1931) has described how the intercellular mycelium collapses in the mature tissues, the hyphae losing their contents. Sometimes, as in *Monotropa*, simple hyphae penetrate the cells, where they may become encapsuled in cellulose walls. In most of the plants that are infected by aseptate mycorrhizal fungi, branched 'arbuscules' are found in the cells, which grow within the protoplasm towards the nucleus. In the Orchidaceae the endophytes are mainly intracellular, forming coils of hyphae within the cells.

The hyphal branches which penetrate the cells as simple projections or as arbuscules are called haustoria. The word haustorium is here, as elsewhere, a name given to a structure penetrating a cell or tissue and looking as if it might have a special absorbing function. Thus in spite of its etymology, the word haustorium is not applied only to structures of known absorptive function. It is usually reasonable to assume that the formation of haustoria may increase the area of contact between host and endophyte, and so an exchange of material between them may be facilitated. But haustoria are not naked protoplasm; usually they are enclosed in a fungal membrane, and they may also be encased by a wall laid down by the host (*see* Fraymouth, 1956). The contact between the two organisms is not necessarily better, but it may be of greater extent when haustoria are formed. Haustoria do not explain the exchange of material, but merely provide an increased surface over which it might take place.

The exchange of material between host and fungus involves its passage across the cellulose and chitin cell-walls of the two organisms. These walls are often conceived as inert membranes, but recent study has shown them to possess enzymes attached to their surfaces. It would appear, too, that R. D. Preston (1952) does not consider the cell-walls of the plant, which he has studied by biophysical methods, to be inert in a biological sense. Their composition, and the methods and rates by which they grow and become modified, indicate a high degree of localization of enzyme systems in them. Similarly Rothstein (1954) has shown, from his own work and from consideration of the work of many others, that a great variety of enzyme systems are localized in hyphal walls and on the cell-walls of plants, and this has been repeatedly confirmed in the experiments of many workers.

The study of fungi in pure culture has also provided information about absorption of various substances by, and their loss from, hyphae. Organic substances, ranging from simple acids and alcohols to complex organic molecules including precursors of enzymes and vitamins, may be released into the substrate, and the rate of their release depends both upon the conditions of nutrition of the hyphae and upon the physical factors of the environment. These substances pass to and fro across the so-called inert membranes.

These aspects of fungal physiology—the absorption and loss of material, and the action of surface and extracellular enzymes—have often been ignored in mycorrhizal studies because patterns of digestion of hyphae were visible in the cells. Yet it seems likely, especially in view of the fact that in many kinds of mycorrhizas digestion is absent, that movement of material across intact cell-walls is of greater importance in the nutrition of the symbionts than has usually been believed.

In considering ectotrophic mycorrhiza, emphasis was laid on two striking points which are relevant to the present discussion. The fungi are characteristically selective in nutrition and deficient in those activities of hydrolysis and breakdown of complex organic compounds which are characteristic of aggressive parasites. Furthermore, most of the work on the process of infection has emphasized the conditions of the host-plant which encourage or allow it to occur, rather than the consideration of the fungus as an aggressor. Indeed, defence reactions of the host-plant against fungal exploitation have been hardly mentioned; the only examples were tannin-impregnated cell-layers in the cortex of some ectotrophic mycorrhizas such as those of *Fagus*, *Castanea*, and *Eucalyptus*. By contrast, many of the fungi of endotrophic mycorrhizas are not only capable of active breakdown of resistant substrates, but they also may have an independent existence apart from their host with which they form mycorrhizas. Some, indeed, are aggressive destructive parasites. *Armillaria mellea* and *Rhizoctonia solani* are two such examples, strains of which are symbiotic with diverse Orchidaceae. It also seems that species of the parasitic genus *Pythium* may form symbiotic unions with a number of green plants. In the study of endotrophic mycorrhizas, therefore, some emphasis has been given to the study of resistance to invasion. Digestion is only one aspect of this. Another aspect is chemical resistance, which was first studied in orchids by Noël Bernard in the early part of the century. This subject has lately become of great interest because of its relationship to the functioning of phytoalexins in plant pathology.

It must not be assumed, however, that there is a complete or striking contrast between the physiological potentials of the fungi of ectotrophic and endotrophic mycorrhizas. In some senses a series can be arranged, showing a gradation between the extreme forms of ectotrophs which seem to be obligately dependent on their hosts, and the free-living litter-decomposing Basidiomycetes. *Boletus subtomentosus* and *Lactarius deliciosus* seem to lie between these two extremes. Similarly, the fungi of ectendotrophic mycorrhizas and pseudomycorrhizas seem to have powers of tissue invasion which bring them closer to endotrophs than are most other ectotrophic fungi. We will, indeed, not only meet examples such as the fungus of *Monotropa* which seems to form both ectotrophic mycorrhizas with pine and spruce and a sort of ectendotrophic mycorrhiza with *Monotropa*, but also a range of endotrophs of very variable potential for utilizing carbohydrates, and exhibiting very variable dependence on accessory nutritional factors.

In the succeeding chapters of this part of the book, various kinds of endotrophic mycorrhizas will be described, beginning with those which have most in common with ectotrophic forms. As far as possible the greatest emphasis will be given to experimental work and as little as possible to speculations that are not experimentally based. No attempt will be made to mention all the work that has been done on endotrophic mycorrhizas; that would be an impossibly tedious and lengthy task.

IX

MYCORRHIZA IN THE ERICALES

MOST of the families of the Ericales, and especially the Ericaceae themselves, have been reported to include a majority of mycorrhizal genera. The mycorrhizal infections which have been described in these plants generally conform to one type, although there is considerable variation in detail. In all cases there is mycelial penetration into the cells and tissues by septate hyphae, and there is also a considerable development of an external mycelium on the root surface and in the soil nearby. In extreme cases, such as *Arbutus*, this external mycelium may take the form of a fairly closely-knit fungal sheath, so that the whole structure is not unlike the ectotrophic mycorrhizas of forest trees.

For present purposes we must divide the kinds of mycorrhizas found in this group into two. The Arbuteae, the Pyrolaceae, and the Monotropaceae, have relatively stout underground organs which, as in *Arbutus*, may be differentiated into long and short roots in a manner similar to those of forest trees; the short roots are enclosed in a thick mycelial pseudoparenchyma. The rest of the Ericaceae, and the Epacridaceae, contrast with these in having a fine root system in which the cortex of the rootlets is narrow. Here the rootlets are loosely clothed in a weft of mycelium, and there is usually a considerable degree of penetration of the hyphae into the cells. These two types of mycorrhizas may be called the arbutoid and ericoid types, respectively.

In spite of the differences in the structure of their root systems, these two types of mycorrhizas are undoubtedly very alike and closely related in a taxonomic and evolutionary sense. Henderson (1919) made a comparative study of the Pyrolaceae and Monotropaceae. He pointed out that a series, possibly showing increased dependence on saprophytic nutrition, could be arranged from Ericaceae through Pyrolaceae to Monotropaceae in the following way. The most vigorous shoots in many genera of the Ericaceae, such as those of *Rhododendron*, bear flower-buds in such a manner that the shrubs grow sympodially; this habit is often more obvious in other Ericaceae. *Vaccinium* and *Andromeda*, for instance, may produce vigorous whitish runners below the surface of the ground which break through the soil to form new aerial shoots. They bear scale-leaves, in the axils of which buds arise to form roots and runners. The mode of growth of many species of *Chimaphila* and *Pyrola* is similar, though the distinction between runners and roots is less clear in other species of *Pyrola. P. aphylla* produces many adventitious shoot-buds upon its roots. The related genus *Moneses* has horizontal root-like runners that are devoid of scale-leaves but from which adventitious shoot-buds arise. *Pterospora*, *Sarcodes*, and *Monotropa*, have underground fleshy root systems that are much-branched and closely compacted. These form endogenous buds which give rise to the chlorophyll-less flowering scapes. This morphological series is, according to Henderson, approximately correlated with an increasing degree of fungal infestation of the root system and an increasing degree of development of a closely-knit sheath of fungal tissue upon the root surface. The series also shows a decreasingly woody habit, with a reduction of the relative preponderance of xylem, increasing phloem development, a reduction in the extent of development of the embryo in the ripe seed, and, in the Monotropaceae, the completely saprophytic habit.

Similar series of increasing saprophytism may be arranged, as we shall see in other groups, e.g. in the Orchidaceae. The possibility of arranging species in this way has given point to the hypothesis that mycorrhizal infection may be essential to the saprophytic habit in Angiosperms. It is a reasonable hypothesis which has been deservedly investigated further, but the fact that series of the kind given may be *arranged*, does not constitute conclusive evidence in favour of the hypothesis; physiological evidence of a direct experimental kind is necessary to give it full credence.

The investigation of the physiological significance of mycorrhizal infection of the Ericales has been bedevilled with controversy of a very confusing kind, but it is becoming more and more certain that the physiological properties of their mycorrhizal organs resemble those of forest trees. Much has been said of the possibility that the endophyte of Ericaceae can fix atmospheric nitrogen; however, it now appears that this is not very likely. In any event, fixation of atmospheric nitrogen could not by itself explain saprophytism and the absorption of carbonaceous material from the substrate. In the ensuing pages a discussion of nitrogen fixation and of earlier views on mycorrhizas in the Ericaceae will be given, because these are frequently repeated in a most uncritical manner in textbooks. The opinion will be expressed that much of the work involved, and many of the cognate statements, may be based upon misinterpretation.

Ericoid Mycorrhizas

The roots of many Ericaceae are of a very uniform structure. They are very fine, the ultimate branches consisting of a narrow central stele and few, sometimes only one row of, cortical cells. The rootlets are enclosed in a loose weft of septate mycelium which makes contact with and penetrates the cortex in many places, commonly entering the cells. The intracellular mycelial masses are often compact, and the intracellular hyphae may be of somewhat larger diameter than those lying outside. Although the hyphae seem to pass readily from cell to cell, they do not normally penetrate the stele, but infection is most widespread in the enlarged cortical cells of the ultimate branches of the root system. Older branches, while still retaining the investing surface hyphae, are less infected in their cortices, and the extent of the infection varies with the individual plant and also shows some seasonal variation. Usually, however, there are restricted areas in which the penetrating hyphae colonize all the cells, these areas being separated by uninfected ones.

The hyphae within the cells may be digested by the activity of the host. The stages in the process, as described by M. C. Rayner (1927), are: the increase in size of the host nucleus, the collapse of the mycelium about the nucleus (followed by the disappearance of sharp hyphal outlines), and, finally, the complete loss of filamentous structure with the formation of a clump of fungal material. When digestion is complete the nucleus returns to its normal form. This process occurs repeatedly throughout the growing-season.

Burgeff (1961) has described three forms of infection. The young epidermal cells behind the apical meristem are penetrated by fine hyphae, which cause the nuclei of the host-cells to become deeply stained and amorphous before the hyphae disintegrate. Elsewhere a strong hypha may penetrate the cell-wall, breaking through a cellulose cap deposited by the cell around its apex, and branch out to form an arbusculoid structure within the cell. A third, less common type of infection which occurs in *Calluna*, results from a direct attack on the apical meristem, which is destroyed, so stimulating a lateral root to be initiated. This process may occur many times, causing garland-like complexes.

The sequence of development of the young roots of *Calluna* was described by M. C. Rayner (1925). In the early spring the roots grow fast and are lightly infected; digestion of the endophyte occurs very rapidly, especially near the meristematic region. As the season advances, more complete infection occurs. The freedom from infection of the apex and its subjacent tissues depends, no doubt, on the relative rates of cell division and growth and digestion of hyphae, compared with the rate of colonization of new cells by hyphae. Gordon (1937) has described the initiation of infection in seedlings of *Rhododendron*. Seeds were sown at the end of May and the seedlings remained uninfected throughout June and July, but infection began in August and became increasingly intense until October.

The seasonal variation in the intensity of infection seems to be general. Hasslebaum (1931) has described it in *Vaccinium oxycoccus** and *Empetrum nigrum*, M. C. Rayner (1929) in *V. macrocarpum* and *V. oxycoccus*, and Councilman (1923) in *Epigaea repens*. Hence any claims that uninfected

*The names of Ericaceae etc., given in this chapter are those used by the authors quoted.

plants occur in nature must be viewed with suspicion, for a relative lack of infection is of seasonal occurrence.

In all these respects the development of ericoid mycorrhizas is similar to the development of mycorrhizas in forest trees. More recently, Morrison (1957*a*) has further emphasized this similarity in his work on *Pernettya*. The intensity of infection of the roots of *P. macrostigma* was reduced by mineral fertilization of the soil. Ammonium nitrate and sodium diphosphate, added in reasonable amounts either singly or together, reduced the intensity of infection. Great imbalance of soil nutrients, obtained by the excessive addition of phosphate, increased mycorrhizal infection and caused stunted growth. Calcium phosphate application also reduced the intensity of infection. Burgeff (1961) has similarly shown an inverse relationship between nitrogen level and infection of plants. Plants with low nitrogen levels produced relatively large, profusely-branched root systems with heavy infection but poor shoot-growth. These results are similar in many ways to those obtained with forest trees, and show that the intensity of infection decreases in soils which are well supplied with nutrients and that it depends in some measure upon the internal nutrient status of the host.

Apart from the formation of intracellular hyphal complexes which are eventually digested, M. C. Rayner (1927) also described a type of infection by fine hyphae. She figured these as passing from cell to cell, and she stated that *Calluna* was always infected in this manner, even when no normal mycorrhizas were formed. She also remarked that 'the development of the endophyte in the mycorrhiza cells of *Calluna* is markedly inhibited by certain conditions in the rooting medium and roots exposed to such conditions may appear uninfected'. Among the conditions leading to relative immunity to infection of the roots, were the differential effect of low temperature on root and fungal growth; other unspecified internal factors were also believed to be important. This type of 'attenuated' infection was described by Rayner to be present throughout the tissues of the root and the shoot, so that the whole plant, even the seed-coat, contained fungal hyphae. Rayner stressed the fact that it was extremely difficult to show the presence of the mycelium in the shoot tissue, and wrote 'it can be accomplished only with the aid of a delicate specialized technique'. This must indeed be so if such a mycelium commonly occurs, for the majority of more recent observers, as well as many earlier ones, have failed to find any attenuated hyphae in the shoot, even after the most exhaustive search.

The mycelium in the aerial shoot was said to be relatively easily traced in the mesophyll of the young leaves. The intracellular complexes, when formed, were subject to digestion as in the root, but the greater part of the fungal hyphae were to be found in the middle lamella. M. C. Rayner also observed that the tissues of the flower, the ovary wall, and the seeds, were all colonized by hyphae of the same fungus which infected the roots: 'Ripe seeds removed from unopened fruits carry fine mycelium, sometimes profuse, more often scanty and difficult to put in evidence.'

On germination, the seedlings of *Calluna* were described as becoming immediately infected by the mycelium on the seed-coat, so that there was a cyclic infection which passed from generation to generation in this organism.

Similar descriptions of infection of root and shoot tissue have been given for several other species of a few genera of Ericaceae: *Vaccinium oxycoccus* and *V. macrocarpon* (M. C. Rayner, 1929), *V. macrocarpum* (Addoms & Mounce, 1931, 1932), *Epigaea repens* (Barrows, 1936, 1941), and *Vaccinium* spp. and *Ledum* (F. J. Lewis, 1924). The seed of *Vaccinium* was said by M. C. Rayner to be more extensively infected, even into the endosperm, than that of any other genus, and an extremely extensive list of species of this and other genera infected in their ovaries is given in her papers. These species include *Vaccinium vitis-idaea*, *V. myrtillus*, *V. pennsylvanicum*, *V. ovatum*, *V. vacillans*, *V. corymbosum*, *Ledum palustre*, *Rhododendron ponticum*, *R. indicum*, *R. sinense*, *Leiophyllum buxifolium*, *Kalmia angustifolia*, *Pentapterygium serpens*, and *Erica carnea.*

In spite of these descriptions of cyclic infection, it now seems certain that the condition described is neither usual nor normal in healthy ericaceous plants. As we shall see, the hypothesis that infection of the stem and fruit-tissue occurred, preceded the observation of fungal hyphae in the stem. Hence there existed, inevitably, the danger that unusual stem-infection could be taken as normal. Many observers have made the most careful examination of very many species of Ericaceae in order to determine whether infection of the stem-tissue occurred, with negative results. Christoph (1921) concluded, from anatomical observation and from the uninfected growth of cuttings, that *Calluna* stems were fungus-free. Doak (1928) found hyphae to be absent from the stems of *Vaccinium corymbosum* and *V. pennsylvanicum*, while Freisleben (1933) obtained negative results with species of *Vaccinium* and other genera; the same was observed by Gordon (1937) with *Rhododendron*, Hasslebaum (1931) with *Empetrum* and *Vaccinium oxycoccus*, Bain (1937) with *V. macrocarpum*, *V. canadense*, *Chamaedaphne calyculata*, and *Ledum groenlandicum*, and Brook (1952) with *Pernettya macrostigma.*

SEED GERMINATION

Those who believed that the stems of Ericaceae were infected by the same fungus as that which formed mycorrhizas with the roots, also assumed that the growth of uninfected seedlings was impossible unless the seed-coats were effectively freed from the fungus.

As early as 1900, E. Stahl had germinated seeds of *Vaccinium* in sterilized soil and obtained plants free from fungal infection. In 1907, however,

C. Ternetz was unable to raise *Calluna* seedlings free from fungal infection, and suggested that the seed might become infected whilst still in the fruit—possibly by spores. It was this suggestion which was followed by M. C. Rayner in her work on *Calluna* and *Vaccinium*. As has been stated, the ripe seeds when removed from the capsules bore hyphae upon their coats, and these were thought to belong to the mycorrhizal fungus. When these seeds were surface-sterilized and set to germinate under aseptic conditions, the early stages of growth were normal. Soon, however, the shoot became arrested in growth, and root formation became inhibited. These features were attributed to the absence of the necessary fungus. Unsterilized seeds under the same conditions of germination were said to become infected, not with typical mycorrhizal development, but to form an 'attenuated' infection of the type described in the stem. In 1925 Rayner wrote of this phenomenon, 'fungal infection and the stimulus to development associated with it on the one hand, and the formation of root mycorrhiza on the other hand, must be regarded as distinct phenomena'. In assessing all the work of M. C. Rayner on *Calluna*, this particular point of view must be remembered; for in her experiments and morphological observations she was more commonly looking for infection by hyphae than for the formation of that particular morphological and histological pattern in the roots which is typical of natural mycorrhizas. Indeed, as far as I am aware she did not describe typical mycorrhizas as being produced in experimental material.

The great majority of those who have germinated the seeds of Ericaceae do not agree that they become infected in any way from their seed-coats. Christoph (1921) grew seedlings and cuttings of *Calluna* which developed normally in sterile soil for two years although they were thought to be free from mycorrhizal infection. Melin (1921) reported that seeds of *Azalea* germinated and developed normally, and that the seedlings remained uninfected. Doak (1928) showed that the development of seedlings of *Vaccinium corymbosum* and *V. pennsylvanicum* did not depend on mycorrhizal infection but that they developed in aseptic conditions. Knudson (1929) germinated sterilized seeds of *Calluna* and grew them for two months in agar. In a later paper (1933) he described a repetition of his experiments, using various seed-sterilization methods, and showed by anatomical investigation that they were fungus-free although they developed normally. In the course of his work, Knudson discovered that the growth of *Calluna* seedlings was frequently inhibited by some organic constituents of agar media. Potato dextrose agar and peptone agar, for instance, inhibited root formation, while other media allowed normal development. It seems that this effect of the medium upon root growth may explain some of the difficulties encountered by M. C. Rayner and others in early seedling growth.

Freisleben (1936) made an extensive study of seedling development of many Ericaceae. He discovered that the seeds of thirteen genera, twenty-one species in all, developed very poorly, with growth inhibition of the seedlings, on sterilized peat and on other media containing much organic matter. Bain (1937) obtained similar results with *Vaccinium macrocarpum*, the seeds of which were readily germinated aseptically in sterile water agar; but germination was irregular and growth abnormal on all media to which much organic matter, especially peptone, had been added. He obtained, in the course of four months, well-developed seedlings with extensive root systems on a variety of agar media and on a cork-and-sand mixture to which nutrients had been added. Molliard (1937) obtained satisfactory germination and growth of sterilized *Calluna* seed on cotton-wool moistened with sterile water, soil extract, or nutrients; and Gordon (1937) had no difficulty in obtaining uninfected *Rhododendron* seedlings by similar means. Brook (1952) has produced similar results with *Pernettya macrostigma*, McLennan (1935) with *Epacris impressa*, and K. G. Singh (1964) has germinated sterile *Calluna* seeds and grown the plants to the flowering stage.

This impressive list of mutually confirmatory results, speaks strongly against the *absolute* necessity of fungal infection in cultural conditions for seed germination to occur, and for growth and root development to take place. Nor can it be accepted that it is normal for systemic infection of the shoot system, the inflorescence, and the seeds to occur. Most of the observers listed above failed to observe hyphae in these organs. Of course, it is very difficult to prove this negative point in the face of the statement that the hyphae exist in such an attenuated state as to require delicate specialized techniques to make them visible. However, it is an important consideration that nobody, not even those who believe that the seeds become infected in the ovary, seems to have shown that seedlings developing upon sterilized agar with or without seed sterilization produce true mycorrhizas. In none of these experiments has the histological picture of cellular infection by wide hyphae and hyphal knots been seen. The claim that conditions are not right for this does not ring true in view of the very extensive investigations which have been made on seedling growth in varied conditions.

The question may fairly be asked as to whether the 'hyphae' described in the shoot system were truly fungal, or whether they were artifacts of some sort. The colonization of healthy tissues is not unusual in many kinds of plants; already in Chapter II, various cases were described. Particularly interesting examples are found in species of *Lolium*, where the shoot tissues are regularly colonized by septate mycelia. At one time it was believed that these belonged to the same fungus as formed mycorrhizas with the roots of *Lolium*, but later it was discovered that that in the roots was aseptate and hence like those of the roots of other Gramineae. Strains of species of *Lolium* may be found in which shoot infection is absent, and it seems possible that a kind of sporadic shoot infection occurs in some Ericaceae. On the other hand, the sizes of the 'attenuated' hyphae which have been described are so small as to make it possible that they were artifacts of microscopy, though considering the experience of the observers, this does not seem to be likely.

THE FUNGI OF ERICOID MYCORRHIZAS

One of the earliest attempts to isolate the fungi from the roots of Ericaceae was made by Ternetz (1907), who was interested in nitrogen fixation by fungi. She was particularly impressed by the fact that available nitrogen in combination seemed to be deficient in the peaty soils upon which the Ericaceae grew, and suggested that their mycorrhizal fungi might fix atmospheric nitrogen. She isolated five strains or species of *Phoma* from the roots of *Vaccinium oxycoccus*, *Andromeda polifolia*, *Vaccinium vitis-idaea*, *Erica tetralix*, and *E. carnea*. These formed pycnidia in culture and were named *Phoma radicis*; each strain was given a second epithet relating it to the plant from which it had been isolated. Ternetz was unable to obtain proof that these were mycorrhizal fungi, because of difficulties in obtaining sterile seedlings of the host plants.

M. C. Rayner (1915) attempted to isolate the mycorrhizal fungus from the roots of *Calluna*, but without success. She obtained, by the method employed, only common soil moulds. As she had already become convinced at this time that *Calluna* was systemically infected, she assumed that isolation of mycelia from the seed-coat would be an easier way of obtaining the mycorrhizal fungus. In this way she isolated a strain of *Phoma* and, in consideration of Ternetz's results, assumed it to be the root-infecting fungus. In her subsequent work, Rayner assumed that systemic infection always occurred, and also that Ternetz's fungi were correctly assumed to be mycorrhizal fungi. Hence she appears to have made no more attempts to isolate fungi from the roots of any ericoid plant although, according to M. C. Rayner & Smith (1929), *Phoma* was isolated several times from the seed of *Calluna*. In 1940, in response to criticisms contained in a paper by Bain (1937), M. C. Rayner & Levisohn reported the isolation of a strain of *Phoma radicis oxycocci* from surface-sterilized roots of *Vaccinium oxycoccus*. This appears to be the only published account of isolation of *Phoma radicis* from the roots of any ericoid plant, apart from the original isolations made by Ternetz.

The assumption that *Phoma radicis* may form mycorrhizas with Ericaceae depends essentially on the following evidence only:

(*a*) Isolation from roots by Ternetz (1907) and by M. C. Rayner & Levisohn (1940).

(*b*) Isolation from seeds of *Calluna*.

(*c*) The hyphae that were reported to be present in the tissues of the stem and fruit were described as resembling those in the roots, and also those of *Phoma*.

(*d*) *Phoma*, on back-inoculation, was shown to infect young seedlings, causing systemic infection by 'attenuated' hyphae.

(*e*) Infection of young seedlings by *Phoma* was found to stimulate their root development.

Points (*c*) and (*d*) lose a good deal of their weight by virtue of the fact that so many careful observers have failed to see systemic infection in so many ericaceous species. Moreover, *Phoma* has never been shown to form the histological pattern found in natural ericoid mycorrhizas.

In 1928, Doak, working on *Vaccinium*, reported the isolation of a sterile mycelium, from the roots of *V. corymbosum* and *V. pennsylvanicum*, which would form typical mycorrhizas when back-inoculated. This appears to be the first report of normal mycorrhizas being formed in culture; later, similar but more extensive results were obtained by others. Freisleben (1933) isolated a sterile mycelium form *V. myrtillus* and, with it, synthesized mycorrhizas with sterile seedlings in a peat-and-sand culture. These plants produced mycorrhizas of a normal kind, but no mycelium could be seen in the shoot system. In the next year Freisleben reported the isolation from *Vaccinium* roots of further mycelia which were named by him after their species of origin. Thus *Mycelium radicis myrtilli* (2 strains, α and β), *M. r. vitis-idaea*, and *M. r. uliginosi*, all formed true mycorrhizas in culture with all the species of *Vaccinium* that were tested, not only with their own original host.

Again, in 1936, Freisleben tested the ability of *M. r. myrtilli* strain α to form mycorrhizas with many ericaceous plants. He succeeded in synthesizing mycorrhizas with fifteen species* in addition to the *Vaccinium* species previously tested. He failed to obtain successful synthesis with *Rhododendron catawbiense*, *Chamaedaphne* (*Lyonia*) *calyculata*, *Leucothoë catesbaei*, *Pernettya mucronata*, *Erica arborea*, *Arbutus unedo*, and *Arctostaphylos uva-ursi*.

Bain (1937) used a dissection technique to obtain groups of mycorrhizal cells from the roots of several species. When these were mounted in a nutrient culture medium, he was able to follow the growth of hyphae from the intracellular hyphal knots. Hyphae grew onto the agar after about 40 hours and were kept under observation for several days, after which they were transferred to tube cultures. In this way, endophytes were actually seen to grow out from typically infected cells of the mycorrhiza before being isolated in pure culture. The endophytes of *Vaccinium macrocarpum*, *V. canadense*, *Chamaedaphne calyculata*, and *Ledum groenlandicum*, when so obtained, proved to be composed of septate hyphae which grew very slowly in culture and produced no reproductive bodies.

* *Ledum palustre*, *Rhododendron maximum*, *R. flavum*, *R. sinense*, *Menziesia pilosa*, *M. ferruginea*, *Leiophyllum buxifolium*, *Kalmia latifolia*, *K. angustifolia*, *Daboecia cantabrica*, *Erica tetralix*, *E. vagans*, *Andromeda* (*Pieris*) *japonica*, *Pernettya phyllyrifolia*, *and Calluna vulgaris*.

They were distinguishable by colour, growth-rate, and colony characteristics. All the isolates were able to infect *V. macrocarpum* seedlings, producing typical mycorrhizas in aseptic culture. No infection of the shoot tissue, and no 'attenuated' hyphae, were observed in these seedlings either by Bain or by Addoms* to whom they were submitted.

Similar endophytic fungi have now been isolated from numerous species of Ericales by Burgeff (1961), McNabb (1961), K. G. Singh (1964), and D. J. Read (unpublished). Plate 6 compares cultures of Freisleben's isolates of *Mycelium radicis myrtilli* α & β and Read's isolates from *Calluna* and *Erica tetralix*, which have been conclusively demonstrated to be mycorrhiza-formers, with *Phoma radicis callunae* and *Mycelium radicis atrovirens*, which have not been satisfactorily shown to be true mycorrhizal endolytes.

Because of the lack of reproductive structures, the exact taxonomic position of all these fungi except *Phoma* is uncertain. Ecologically they may be grouped with the many sterile mycelia that are commonly isolated from the root systems of heathland plants and customarily described as dark sterile forms. This population seems to include both mycorrhiza-formers and fungi which may be able to infect the root-cells in an incomplete fashion. Indeed there is a close parallel to the population of root-surface mycelia of mycorrhizal and pseudomycorrhizal fungi encountered on tree roots. Even the mycorrhiza-formers are not specific to one host but may be common to many species. McNabb (1961) suggests that many form mycorrhizas outside the Ericales.

The sterile mycelia of the roots of Ericales are difficult to isolate into pure culture because their growth-rate is slow and they become overgrown by other fungi in artificial media. The most successful techniques of isolation have proved to be the dissection and maceration methods of Stover & Waite (1953), or the serial washing method of Harley & Waid (1955). These methods have been compared by K. G. Singh (1965), who showed that a large proportion (*c.* 70 per cent) of the fungi isolated by both methods were 'dark sterile forms'. Many of the other methods of isolation which have been used, select the fast-growing fungi (to which group *Phoma radicis callunae* belongs). It is of great significance, therefore, that this species has been so rarely isolated from roots.

The status of *Phoma radicis* as an endophyte of Ericales is thus called into question by these results and one begins to wonder whether the explanation is not simply that *Phoma* spp. occasionally colonize the healthy stem-tissues of Ericaceae as they have been reported to do in other plants (R. W. Rayner, 1948).

If we do not accept the view that the Ericaceae are systemically infected by mycorrhizal fungi, and there is strong evidence that we should not, the initiation of infection in young seedlings is of interest. Bain (1937)

* Addoms & Mounce (1931, 1932) had described systemic infection in cultivated cranberry.

observed that *Vaccinium macrocarpum* seedlings became infected about a month after germination on peat, and developed typical hyphal knots and coils in the cells. It must therefore follow that the endophyte was present in some form in the peat. This was also the experience of Gordon (1937) with *Rhododendron*, and of Brook (1952) and Morrison (1957*a*) with *Pernettya*. The impression gained by Brook with *Pernettya* was that the endophyte was widely present in some form in natural soils, but that it spread very slowly from a source of inoculum. The use of excised mycorrhizal roots as a source of infection failed to give rise to mycorrhizas except in a few cases—a result not unlike those obtained with tree mycorrhizas, and which may indicate a low competitive saprophytic ability in the endophyte. The work of Morrison, which has already been noted above, has further served to emphasize this comparison.

The cultural characteristics of proved mycorrhizal fungi of Ericales have not so far been described in detail.* The large majority of isolates are dark, especially some shades of grey. On the Ridgway Colour Scale the centre is often olive-grey, the aerial mycelium a pale grey, and the edge white. On the reverse side the cultures are frequently olivaceous-black. Young colonies may be vinaceous fawn or grey-violet, becoming grey with age. There is of course some variation of colour on different media. The growth-rate of these fungi is less than a millimetre a day, and the colony usually ceases growth before reaching the edge of a standard Petri-dish (*see* Plate 6). For instance, Bain gives growth-rates of 0·1–0·4 mm. per day, and McNabb's isolation from *Pernettya macrostigma* grew at 0·1 mm per day at 25° C. K. G. Singh records the highest growth-rate of 1·0–1·5 cm. in seven days at 25° C.

There is a great need for further information about the cultural behaviour and physiology of these fungi. Burgeff (1961) showed that catechol tannins, found in the roots of many members of the Ericaceae, can be utilized by the endophyte both in the living root and in culture. Since the fungi can apparently utilize a wide range of simple sugars in culture, the importance to them of the tannins is obscure.

THE EFFECT OF FUNGI ON GROWTH

The early growth of the seedlings of ericoid plants has been found to be very sensitive to the constitution of the substrate. But M. C. Rayner (1915) observed that the stagnation phase in the growth of the seedlings, particularly in the growth of their roots, could be prevented when the cultures were infected with *Phoma radicis*. This was accepted as proof that *Phoma* was the mycorrhizal fungus.

This conclusion has been questioned on account of more recent work on the effect of fungi on the germination and early growth of ericoid plants. The media on which the stagnation phase is most strongly evident are those containing a high content of organic matter, particularly of

* This work is now being undertaken by D. J. Read at Sheffield University, England.

organic nitrogen compounds. Freisleben (1933) showed that his strain of *M. r. myrtilli* α had the same effect on *Vaccinium myrtillus* growing on peat-and-sand mixtures as *Phoma* had on *Calluna* in the experiments of Rayner. Later, Freisleben (1934) showed that each of his isolated endophytes of *Vaccinium* also had a growth-promoting effect on the seedlings of all the *Vaccinium* species which he tested. In addition, many saprophytic soil fungi had a similar effect. A culture of *Phoma radicis callunae* obtained from Rayner was also tested, and this stimulated the root growth of *V. vitis-idaea* but parasitized *V. oxycoccus* seedlings. Later again (1936), Freisleben made a very extensive further investigation of the effect of various fungi upon the early growth of the whole list of Ericaceae given in the footnote on p. 179. *Mycelium radicis myrtilli* α caused a stimulation of root-growth of variable magnitude with all the species tested except *Arbutus unedo* and *Arctostaphylos uva-ursi*, which showed no stagnation phase. The magnitude of the stimulation was quite unrelated to the extent of mycorrhizal formation. Dead mycelia and mycelial extracts had a slightly stimulating effect at times, but none which was of equal magnitude to that of living inoculations. Freisleben concluded that the retardation of root-growth was an effect of the medium, perhaps due to a toxin, which could be neutralized by the activity of growing mycelia of many kinds. He concluded that the fungi associated with Ericaceae, and those which cause mycorrhiza formation in particular, might be ecologically important in the early stages of seedling growth on peat, but might be relatively harmless parasites in later life.

Bain's results with *Vaccinium* (1937) agree on the whole with those of Freisleben concerning the effect of the endophytic fungi on early growth in peat soil. But the assumption of the lack of effect of the endophyte or its parasitic activities in later life have not been experimentally considered until recently. Brook (1952) obtained indications that, during eleven months, *Pernettya* seedlings growing on 'fertilized' leached clay were stimulated in growth by mycorrhizal infection. His numerical results were, as he pointed out, somewhat fragmentary, but in five out of seven treatments the mycorrhizal plants exceeded the non-mycorrhizal plants in dry-weight. Morrison (1957*a*) has followed up Brook's results. He grew mycorrhizal and non-mycorrhizal *Pernettya* on a range of soils, and in most cases found a slightly greater growth-rate in the uninfected than in the infected seedlings. On the most nutrient-deficient soil, however, uninfected *Pernettya* seedlings failed to grow whilst the infected, mycorrhizal, seedlings increased in size. These results are no doubt the beginnings of a serious study along the right lines. Morrison's experiments suffer from the defect that the non-mycorrhizal seedlings were grown on autoclaved soil, so that the comparison is somewhat vitiated by the different treatment. The possibility that ericoid mycorrhizas may function in a manner similar to tree mycorrhizas has been credible ever since the probable elimination of *Phoma* as a mycorrhiza-former, and the results of Brook and Morrison strengthen this view a little.

Although it is not now possible to believe entirely confidently that *Phoma radicis* is a mycorrhizal fungus of Ericaceae, its effect upon their growth has been widely studied. The results of this study must be mentioned here because they have received such wide publicity and consequently must be put into perspective.

Ternetz (1907) examined the ability of *Phoma radicis* to utilize atmospheric nitrogen and obtained evidence of a slow, highly efficient (in terms of carbohydrate utilization) fixation. Her results were examined by Duggar & Davis (1916) to determine whether they could be relied upon. Duggar & Davis concluded that, even allowing for the errors of the Kjeldahl method as used by Ternetz, her results showed nitrogen fixation. The results of Ternetz were also accepted by M. C. Rayner (cf. 1915–27), especially as seedlings infected with *Phoma* developed in agar or silica gel to which only water had been added. W. Neilson Jones & Smith (1928) examined the possibility that *Calluna* seedlings inoculated with *Phoma* could fix atmospheric nitrogen. They grew the plants on a nitrogen-free medium and passed a gas stream containing nitrogen over one set, and a gas stream lacking nitrogen over the other. The former showed slight fixation. But they also found that growth was greatly increased by the provision of combined nitrogen. Their work on the whole did not provide a convincing proof that nitrogen was fixed in anything like sufficient quantities to support the growth of both *Calluna* and the fungus.

In recent years further experimental studies have thrown additional doubt upon the potentiality of *Phoma* to fix nitrogen. Using ^{15}N as a label for gaseous nitrogen, P. M. Wilson (1952) failed to obtain fixation with *Phoma casuarinae*. Bond & Scott (1955), who have successfully used the isotope technique to study symbiotic nitrogen-fixation of a large number of kinds of root nodules, tested the ability of *Calluna* obtained from natural habitats to fix nitrogen. Their results were negative. They did not, it is true, prove the presence of *Phoma* in the tissues, because this would be inevitable if one accepted the whole hypothesis of systemic infection by *Phoma*.

In spite of the defects in the evidence that *Phoma* is a constant endophyte of Ericaceae, that it can fix adequate amounts of nitrogen, and that infected *Calluna* plants can fix nitrogen, the hypothesis put forward by Rayner concerning the significance of mycorrhizas in the Ericaceae is still widely considered to have been proved. It is high time that the other work bearing on the problem was properly appreciated.

SUMMARY

No apology is offered for expending so many words upon this rather inconclusive account of researches on the Ericaceae. The evidence for systemic infection is much less convincing than that for its absence. Other fungi than *Phoma* have been proved to be mycorrhiza-formers with a large number of Ericaceae, and there is no sound evidence for nitrogen fixation on anything like the scale required for normal growth of both the host and the fungus.

If this be so, what do we *know* of the significance of mycorrhizas in Ericaceae?

The stimulation of germination and early growth of seeds and seedlings is not a property of proved mycorrhiza-formers only. Nor is it yet shown to be peculiar to any particular fungus or group of fungi. We are therefore left with a sound knowledge of the structure of the mycorrhizal roots, but very little else. The gross similarity of these with the roots of other plants that have been ectotrophically infected, might lead us to consider the possibility that they have a similar absorbing function where the outgoing hyphae exploit a greater volume of soil than would the roots alone. The works of Brook and of Morrison on *Pernettya* point the way for further work. We have, in their results, the first good indications that similar external factors to those which affect the development of tree mycorrhizas may effect the development of mycorrhizas on *Pernettya*. And in Morrison's attempt at comparing infected and uninfected plants on a range of soils, we have the first indication that the mycorrhizas may be more efficient salt-absorbing organs on nutrient-deficient soil than are uninfected roots.

In considering the properties of the fungi of tree mycorrhizas, we have seen that in most cases they have a restricted power of carbohydrate utilization. There is, however, a fair amount of variability in this. Some types, such as *Boletus subtomentosus*, can bring about considerable litter-destruction. The suggestion has been made that the observed variability in carbohydrate utilization explains the occasional variations in structure—the ectendotrophic types of mycorrhiza being formed by fungi with considerable powers of breakdown of carbohydrates. I believe that the carbohydrate physiology of the endophytes of ericoid plants should be investigated, and that the results will show that these fungi as a group are related to those of ectotrophic mycorrhizas, but that they possibly have greater powers of utilization of complex humic substances.

It has been suggested by Stalder & Schultz (1957), and by Handley (1963), that the established endophyte may indirectly contribute to the growth of ericaceous plants by exerting an antagonistic effect on other soil micro-organisms. Stalder & Schultz investigated a disease of nursery-grown plants of *Erica gracilis* caused by *Olpidium*. The disease was restricted to richly fertilized plants in which mycorrhizal associations were lacking. High nitrogen levels were particularly effective in reducing mycorrhizal infection and increasing susceptibility to attack. These workers have suggested the possibility that wild plants of *Erica* are free from attack because the mycorrhizal association of their roots is antagonistic to the pathogen.

Handley has suggested a possible antagonism between the endophyte of *Calluna* and the ectotrophic mycorrhizal associates of coniferous trees in heathland soils. An antibiotic substance produced by the endophyte was considered to be responsible for the failure of 'heather sensitive' plants such as spruce to produce normal mycorrhizal root systems.

ARBUTOID MYCORRHIZAS

As was stated earlier, *Arbutus* and *Arctostaphylos*, together with the Pyrolaceae and Monotropaceae, produce mycorrhizas that are even more like those of the ectotrophic organs of forest trees than are those that have already been considered in this chapter; and it is also of interest that D. J. Read has observed fungal sheaths on roots of *Vaccinium myrtillus* in certain restricted localities. The roots in *Arbutus* are differentiated into long and short branches, and upon the short branches a tight hyphal sheath may be formed. The long roots, according to Rivett (1924), are not abundantly invaded but, as in *Pinus*, the hyphae penetrate the walls and intracellular spaces of the cortex in its outer cell-layers. A limiting zone may be formed in the central cortex by the suberization of the cell-walls, and through this the hyphae do not pass. A similar layer may be seen sometimes in *Fagus* roots and, more often, in those of *Castanea*. In most long roots there is some intracellular penetration into the outer cortical cells: short, blunt hyphae penetrate the walls and grow towards the nucleus.

The short roots are produced endogenously from the long roots, and become infected as they pass through the cortex. They are covered, as in other Ericaceae, with a mucilaginous coat which is first colonized by the fungal hyphae. As the external weft of hyphae thickens to several cells in depth, the rootlet swells and a tubercle covered by a dense pseudoparenchyma is formed. Soon the fungus penetrates the cortex of the tubercle and forms coils in the cells. These become clumped together, digested, and disintegrated. After digestion is complete the cell may become recolonized by new hyphae and the process may be repeated. In these short tuberculate rootlets, as in the long roots, a suberized layer is formed which appears to limit penetration, by the hyphae, to the outer cortex.

The cells of the apical regions of both long and short roots remain uninfected, although the apices of the short roots are completely overarched by the fungal sheath. Penetration of the cells by the hyphae occurs proximally to the actively dividing meristem.

The sequence of development of the root system of *Arbutus* was also described by Rivett. The long roots are white at the apex but gradually form a dark corky layer behind, which, as it is shed, sloughs off the cells colonized by fungi. A month or more after root-growth has begun, strands of fungal mycelium may be seen on the root surface. Branches from these strands form the initial infecting hyphae which penetrate the cortex. After five to six months, short roots are formed but the earliest produced do not become infected. Subsequently-developed short roots are rapidly colonized by hyphae, which soon form a sheath that may become seven hyphae thick. The rootlet forms a white tubercle as it swells and the sheath becomes very dense, giving rise to stiff straight setae which radiate into the soil.

The swelling of the rootlet precedes hyphal invasion of the cortical tissue and appeared to Rivett to be initiated by the fungus during its ectotrophic phase. Tuberization becomes more rapid as the hyphae penetrate the cells, but the tubercles continue to grow and branch in an apparently dichotomous manner. The tubercles persist, and grow throughout the summer and autumn; in some cases they last through the winter, to be abscissed in spring.

The germination of the seeds of *Arbutus* and *Arctostaphylos* was studied by Freisleben (1936). They germinated readily on all the media tested, and were not susceptible to the toxic effects of some of the highly organic media—unlike seedlings of some other Ericacae. They were neither stimulated by *Mycelium radicis myrtilli* α, nor did they form mycorrhizas with that fungus. The readiness with which the seeds of *Arbutus* germinate has been confirmed by Sealy (1949) and Sealy & Webb (1950), who studied the ecology of *A. unedo* in Ireland. They recorded a 50 per cent viability, and noted that the best substrates for germination were humic materials—especially *Sphagnum* peat. Under natural ecological conditions, *Arbutus unedo* differs from many of the plants of the Ericaceae, discussed above, by the absence of vegetative reproduction (except for sprouting from stem-bases) and by its consequent dependence on seed germination as the sole means of spreading.

We know nothing of value about the physiology of *Arbutus* mycorrhiza. There have been reports of systemic infection and of mycorrhizal fungi on the seed-coat, but these reports are of very doubtful validity. The interest of this type of mycorrhiza lies in its similarity, in many ways, to the ectotrophic mycorrhizas of forest trees. The long roots resemble in behaviour and in infection those of *Pinus*, and the short tuberculate mycorrhizas have much in common with all ectotrophic organs. The presence of considerable intracellular penetration and hyphal digestion makes the *Arbutus* mycorrhiza a connecting link in structure between the ectotrophic forms and the endotrophic forms such as those of other Ericaceae. To me, this example gives added point to the suggestion that, functionally, the mycorrhizas of Ericaceae may resemble those of forest trees, and indicates a promising line of experimental work.

PYROLACEAE AND *Monotropa*

We may view the partial or complete saprophytes of the Pyrolaceae and Monotropaceae as a culmination of the series which we have been considering. Thus the forest trees are typically associated with fungi which have restricted powers of digesting complex carbohydrates, while the green Ericales may represent a middle condition, and the saprophytic members of the Pyrolaceae and Monotropaceae the other extreme where the fungus actively absorbs organic substances from the humus or from other host-plants. We shall see that we have not as yet enough information about the fungi concerned to propound this hypothesis at all fully, but sufficient data are available to emphasize the need for undertaking experimental work on the fungi.

It is of interest to note that *Monotropa* was very early recognized as having its roots invested with fungal hyphae. In the middle of the last century there was considerable controversy as to the mode of its nutrition, whether it was parasitic or whether it nourished itself from decomposing humus. M. C. Rayner (1927) has given a full account of this discussion, and has pointed out that F. Kamienski noted the similarity between what are now called mycorrhizas of *Monotropa* and those of beech. Kamienski concluded that the mycelium of both was of the same fungus, and that *Monotropa* was dependent upon it for nourishment and might also be essentially parasitic on the roots of beech and pine.

The likeness between the mycorrhizas of *Monotropa* and the related Pyrolaceae on one hand, and those of the forest trees on the other, is considerable. *Monotropa* roots are clothed with a fungal sheath, hyphae from which ramify in the surrounding humus. Francke (1934) showed that the development of this sheath is greatest in humic soils and less in sandy, humus-poor soils. Hyphal branches from the sheath penetrate the tissues to form a network, similar to a Hartig net, in the cell-walls of the outer cortex. As growth proceeds in the summer, these hyphae and the sheath hyphae form haustoria which are usually simple in form and which penetrate into the lumina of the outermost cells of the host. There is usually only one haustorium in each cell, and its apex seems to grow towards the nucleus. These haustoria may become encapsulated, as has been described for the penetrating hyphae of the fungi of ericoid plants by Freisleben (1936) and Brook (1952), by a cellulose wall which seems to be formed by the host-cell. The haustorium, as it grows, does not usually become completely enclosed but has its tip remaining free. Later, when the end of the haustorium is in the region of the nucleus, digestion of the free tip may occur, and the disorganized haustorium become completely sealed off by the host. These histological changes take place during the period when *Monotropa* forms its flowering scape.

Francke attempted to isolate the mycorrhizal fungus from *Monotropa* and obtained a sterile septate mycelium which grew reasonably easily in culture. He believed, on grounds of comparative hyphal structure, that this was a species of *Boletus*. Francke's first attempts to germinate *Monotropa* seeds were unsuccessful. However, when the seeds had been well soaked in water, asymbiotic germination occurred and a small cellular body was formed which might, by analogy with orchids, be called a protocorm. The seeds, of course, are exceedingly small, and the protocorm after development consisted of only a few dozen cells of which the apical group constituted a growing-point. No further development was obtained in the asymbiotic condition on a variety of nutritive media. If the protocorms were inoculated with the fungus which had been isolated from the roots, they became infected and a hyphal sheath developed, with some penetration into the tissues. Growth proceeded and cell division and expansion occurred. Francke did not succeed in obtaining further development into a plant more like the adult, but we must realize that, in complete saprophytes, growth is very slow, and indeed later work has shown that a second host may be necessary to the fungus (*see* below). It appears that seed germination of *Monotropa* up to embryo differentiation is possible in the absence of the endophyte, and no doubt depends primarily on the small food stores of the seed. The cessation of growth may be due in some media to the absence of nutrients in suitable form, but it may also be due to the inability of the embryo alone to synthesize essential metabolites.

It will be seen in the next chapter that the mycorrhiza of *Monotropa* shows striking similarities with that of some of the Orchidaceae, so that it appears to form a kind of connecting link with the types of mycorrhizas occurring in saprophytes of that group. The need for prolonged soaking and washing of the seeds in water, the limited potential for asymbiotic growth, and the stimulation by infection, are all found in the Orchidaceae.

The physiological significance of infection in *Monotropa* has been under investigation for a longer period than that of most mycorrhizal associations. Early on, Kamienski (1881) pointed out that as its underground organs were clothed in a weft of fungal tissue, all materials passing into it must pass through the fungal mycelium. He took the view that the same mycelium which infected *Monotropa* formed close connections with the neighbouring roots of beech and pine. In 1960, Björkman published an account of an experimental reinvestigation of the physiology of *Monotropa*. He pointed that Francke had classed its mycorrhizal fungus as a species of *Boletus*, a genus well known to form mycorrhizas with trees, He went on to determine whether the ultimate source of the carbon supplies of *Monotropa* could be the photosynthesis of the associated trees. Björkman showed, following earlier casual observations by W. J. Hooker, that if *Monotropa* plants growing in woodland were carefully isolated from tree-roots by sheets of metal, their growth in subsequent years was very weak, whereas unisolated control plants grew as vigorously as ever. Hence it appeared that the severing of connection with tree-roots cut off some essential supply.

Björkman (1960) isolated a white fungus from *Monotropa* roots which was similar to that obtained by Francke, and found it to be capable of forming mycorrhizas with pine in culture. The association with spruce was atypical in its lack of a Hartig net. The fungus had a much greater growth-rate in culture than the mycelia from known mycorrhizas with which it was compared, but both were greatly stimulated in growth by extracts of *Monotropa* tissues—over and above stimulation by vitamin B1. ^{14}C-labelled glucose and ^{32}P-labelled orthophosphate were injected into the tissues of mature spruce and pine trees having species of *Monotropa* growing in their immediate vicinity. After five days the radioactivity of *Monotropa* tissues (as counts per minute per gram of dry-weight) was estimated in young plants, old plants, and in control specimens at a distance. Small but significant radioactivity was accumulated in *Monotropa* within 1–2 m. of the injected trees. As expected, movement of both ^{14}C and ^{32}P was greater into young than into mature plants. In the experiment with ^{32}P, the other plants besides *Monotropa* (*Vaccinium vitis-idaea*, *V. myrtillus*, and *Calluna vulgaris*), growing near the injected tree, showed no significant accumulation of radioactivity.

These results not only show that *Monotropa* is to a great degree an epiparasite depending for carbon on the tree host of its mycorrhizal fungi, but they also explain the growth of *Monotropa* on relatively humus-deficient sand-dunes where it may be epiparasitic. There are examples of epiparasitism among the saprophytic Orchidaceae, such as *Gastrodia elata*, and the habit is possibly more widespread than has hitherto been assumed. Those saprophytes (e.g. *Corallorhiza trifida*) which have sand-dunes as one of their habitats, would certainly repay investigation in this regard as would other Monotropaceae.

The green Pyrolaceae are similar in many ways to *Monotropa*. Their mycorrhizas have been investigated particularly by Christoph (1921), Lück (1940, 1941), and Lihnell (1942*a*), but we still know too little of real value about them because of the difficulties of experimenting with them. A sheath of variable thickness and density is formed on the root surface, and hyphae penetrate from it into the epidermal and subepidermal tissues. Some haustoria penetrate the cells, which may at times become filled with and killed by the fungus. According to Christoph, the sheath of *Pyrola secunda* is most fully developed, and the cells are most completely penetrated, under conditions where growth is least healthy. Lück (1940) observed similar features in *P. chlorantha*. In this respect they are comparable with *Pernettya macrostigma*, which was observed by Morrison (1957*a*) to be most heavily infected in poor soils, where growth was very meagre.

Pyrola species vary in the ease with which their seeds germinate. Here again, many seem to need soaking in water—perhaps to elute some substance from the testa before germination can begin. Lihnell germinated seeds of *P. rotundifolia* and *P. secunda* aseptically to form a colourless axis, the 'procaulome', which had a simple root-like anatomical structure and which branched and grew in culture for three-and-a-half years without developing further, although its total length in *P. rotundifolia* was 30 cm. (including branches). *P. uniflora* did not germinate aseptically.

Lück (1940) and Lihnell (1942) both isolated fungi from the roots of adult plants. Lück obtained a clamp-bearing Basidiomycete which did not appear to increase seed-germination and seedling growth in *Pyrola chlorantha*. Lihnell isolated five mycelia of which two were obtained from *P. rotundifolia*. These fungi had the effect of stimulating the germination of *P. uniflora* to some degree, but none of them formed any close associations with the procaulomes or encouraged them to develop further.

We shall be able to see these results in suitable perspective after considering in the next chapters the mycorrhizas of the partial and complete saprophytes of the Orchidaceae.

X

MYCORRHIZA IN THE ORCHIDACEAE—GENERAL FEATURES

It may, at first sight, seem surprising that during the present century, almost as much laboratory effort has been applied to the investigation of physiological problems of seed germination and growth of the Orchidaceae, as to the elucidation of the mycorrhizal problems of forest trees. The Orchidaceae appear to exert a considerable fascination over botanists and horticulturists because of their great variability in life-form and habit, in reproduction and distribution, and in the complexities and beauties of their floral structure. More than this, they are economically important on account of the luxury trade in their blooms and also on account of a demand for a few of their products, such as vanilla.

It is impossible here to do full justice to this diverse family of plants or to the results obtained by those who have investigated problems concerning their ecology and nutrition. No one who has not had the opportunity of wide travel, especially in the tropics, can adequately appreciate the problems. We are fortunate that certain aspects of their study, notably those of physiology and ecology, have been dominated by two central figures—Noël Bernard, whose pioneer work, particularly in the first decade of this century, set the scene for later investigations, and Hans Burgeff, who from 1909 onwards has published several books and many research papers in which he has considered all the most important aspects of mycotrophy of orchids.

Regardless of their morphological and anatomical diversity, the members of the Orchidaceae have many outstanding physiological characters in common, the most important of which arise from a close association with fungal mycelia. All Orchidaceae, at some phase in their lives, are infected with fungi under natural conditions, so that their absorbing organs maintain fungal mycelia within their cortical cells. There is, here again, some diversity in the kinds of organs which are infected and in the distribution of the fungus within those organs; but as will be seen, the orchids as a group have in common a kind of mycorrhiza of their own. Similar kinds are formed also by a few other saprophytes, by partially saprophytic and holophytic angiosperms, and by bryophytes. Their fungi, too, in many ways constitute a definable group. They are either Basidiomycetes (often with clamp-bearing hyphae), imperfect Basidiomycetes with clamp-connexions but whose fruiting-bodies have not been found, or imperfect fungi of the genus *Rhizoctonia*, some of which are already known to be of basidiomycetous affinity. These fungi contrast with the Basidiomycetes which form mycorrhizas with forest trees, especially in regard to their carbon nutrition: for all of them seem to utilize a greater range of sources of carbon, and many can destroy lignin and cellulose. They continue, as it were, the series already noted in the fungi of ectotrophic mycorrhizas. Of these, as we have seen, most are able to utilize simple sugars only but a few can utilize cellulose. Of the orchid fungi, most can use complex carbohydrates.

Correlated with the more vigorous carbon nutrition of their fungi, the orchids show an increased dependence upon a supply of carbonaceous material in organic form. Some orchids are, in the adult phase, capable of enough photosynthesis to be self-supporting, whilst others are partly saprophytic, and yet others, even including some of the largest lianes, geophytes, and epiphytes, are completely saprophytic. In spite, however, of this variability in carbon nutrition as adults, all orchids are saprophytic for some period of their seedling life.

We have already seen that the intensity of fungal infection in adult plants of Ericaceae and related families is apparently correlated to some extent with the saprophytic habit. This is also to be observed in the Orchidaceae, but of them we know more. More investigation has been made of their germination, growth, and physiology, and also of the nature and physiology of their fungi. It will be simplest here to describe first the seed germination, seedling growth, and adult structure, of the orchids in order to give a general impression of the problems involved in a study of their mycorrhizas. Specific problems will then be considered in more detail.

Seed Germination and Seedling Growth

The minute seeds of Orchidaceae are produced in large numbers in each capsule. The fertilization of the several thousand ovules involved may be the result of a single act—the placing of pollen masses from the pollinium upon the receptive face of the stigma. The fertilized ovum develops to form a small group of cells, the embryo, which is little differentiated but has the form of an oval or fusiform body of which the distal end is a potential growing-point. The number of cells comprising the embryo may be as large as about 100 or as few as 8. The basal extremity is attached to a suspensor. These embryos typically lack cotyledons but in a few cases they may be more completely differentiated. They are enclosed at maturity in a loose, thickened testa which may vary greatly in form and ornamentation. The range of weight of the mature seeds is given by Burgeff (1936) as between 0·3 and 14 μg. per seed. The number of seeds in each capsule may vary from a few thousands to several millions. Burgeff estimated the average number in each of the capsules of *Anguloa ruckeri* to be 3,932,948 seeds. Typically, many capsules may be produced by any one plant in a season.

In assessing the experimental work on seed germination and seedling establishment under natural ecological conditions, these seed characters must be appreciated. The probability of any one seed becoming established is very low, being, over the life of a perennial orchid, many millions to one against. The small size of the embryo and the relative absence of food reserves in the seed, demand that the greater part of the food material utilized, up to the period when photosynthesis begins, must come from the substrate. As we shall see, this period of dependence on external carbon supplies varies in green autotrophic orchids and, of course, is a permanent feature of saprophytes.

Seed germination of orchids has been widely studied—both by setting them to germinate in the greenhouse or laboratory, and by observing seedling stages in the field. Of course, details of the development of many of them are as yet unknown, but enough have been studied to allow generalizations to be made. Seedling development may be divided into several phases, which may not be exactly delimited in practice but which are fairly readily discernible.

(1) The embryo absorbs water, swells, and ruptures the testa.

(2) The food material of the seed is utilized and some growth occurs.

(3) Organic matter is absorbed from the substrate and organs are differentiated.

(4) Chlorophyll formation occurs in leaves or other photosynthetic organs and the embryo becomes gradually autotrophic.

(5) The plant matures and produces an inflorescence.

Although these phases are variable in extent and definition, there is always a period of saprophytic absorption. In complete saprophytes the plant passes from stage (3) to stage (5) and remains permanently saprophytic. In most green orchids, stage (4) occupies a fairly prolonged period during which growth is relatively slow and a larger plant-body is gradually built up. As a larger photosynthesizing surface is formed, growth becomes more active until a flowering spike is ultimately laid down. The length of time for the whole course of events to take place is very variable; even in relatively small green orchids it usually comprises several years and seems to vary according to external conditions.

Ziegenspeck (1936) gives many detailed examples with figures of stages of development. The first foliage-leaves may appear in the second year in many orchids, e.g. *Dactylorchis* and many species of *Orchis* such as *O. morio*, but they may not appear until about the eleventh year in *Spiranthes spiralis* or even the fourteenth year in *Orchis ustulata*. The last two examples show the possible variation within a genus, for *Orchis morio* and *Spiranthes aestivalis* produce leaves in their first few years whereas *Orchis ustulata* and *Spiranthes spiralis* have a very prolonged subterranean saprophytic existence.

Wells (1967) made a very valuable study of a population of *Spiranthes spiralis* in which the individual plants were recorded so that their behaviour could be followed over three years. The results showed that the individuals were able to persist over one or more years without forming foliage leaves above the ground. When they reappeared above-ground, some produced flowering shoots in the year of reappearance. This clearly suggests that this species, at any rate, is able to revert to the saprophytic habit and is quite likely to be partially saprophytic all the time. Summerhayes (1951) gives the examples of *Cephalanthera rubra*, *Goodyera repens*, and *Cypripedium calceolus*, to illustrate dependence of adult orchids upon infection. In very heavily-shaded woods these species may continue to exist in green vegetative form without flowering, or even as purely underground saprophytes.

In some circumstances the swelling of the embryo may be followed by some cell-division, after which there may occur a pause in development. This has been the cause of some differences of opinion as to the results of germination experiments. If the criteria accepted for successful germination are only those of the first phase of the sequence given above, different opinions may be formed of the factors necessary for germination from those which would be formed if a longer period is studied. For this first phase is only the phase of utilization of food reserves, and quite readily occurs at suitable temperatures if there is an adequate water supply. Significant growth of the seedling involves absorption of food, and, consequently, a different constellation of factors may affect it. In this account the use of the term germination will imply that growth has occurred over and above that which is involved in the first phase.

In unsterilized natural substrates, the swelling of the seed is followed by the penetration of hyphae of the fungal endophyte into the embryonic tissues. The sequence was well described by Bernard in his pioneer work on the mycorrhizas of orchids. Plate 7 is from his paper of 1909 and gives the sequence for *Phalaenopsis amabilis*. The fungus makes its entry through the suspensor end of the embryo and immediately a great stimulation of growth occurs. Bernard (1909) stated that *Phalaenopsis amabilis* made more growth in several days after infection than it did in several months without the fungus. The central region of the embryo is completely colonized, and as this occurs, differentiation begins. The first stage is the formation of epidermal hairs and the division of the apical meristem, so that a protocorm is eventually formed. This structure is radial or asymmetric, according to the orchid species. It develops a central conducting-strand, an apical series of leaves, and roots from the axis near the leaf-bases. The fungus becomes restricted to the cortical region, being connected to the substrate by many hyphae which pass particularly through the epidermal hairs. The fungus does not, however, remain uniformly healthy within the tissues. In the central region of the embryo, the hyphal coils lying in the cells become disorganized. The hyphae swell, lose their regular form, and are digested—leaving a mass of unorganized fungal matter in the cell. With the division of cells in the young meristem, the hyphae rapidly penetrate the new cells which are formed, and the process of colonization and digestion follows hard upon growth. As the embryo differentiates, the fungus-infected area becomes restricted so that it at last occupies that part of the protocorm which is in contact with the soil. Young roots, as they are formed, become infected from the substrate, and in their cells colonization and digestion occur in regions behind their apical meristems.

It is in these early stages of germination, during the infection of the plantlet by fungal hyphae, that a considerable mortality of seedlings may occur. Not infrequently the fungus may become wholly parasitic and incapable of control by the digestive action of the cells. Bernard stressed this danger to the seedling and stated that, although he set many thousands of seeds to germinate, only a few hundreds grew successfully in the presence of the fungi. In other cases digestion may be so rapid that the infection is eliminated, and then, according to Bernard, no further development takes place. In the majority of experiments where there is careful control of conditions, and where a selection of fungal strains is made, uniform germination in the presence of fungi is obtained. This matter will be reconsidered below, in the discussion of specificity of orchid fungi.

The duration of the primary axis of the seedling, its form, the extent of infection in it, and the length of its period of saprophytic life, are all highly variable in the various orchids. Accounts of the process of germination of many kinds of orchids are to be found in the works of Bernard, Irmisch, Burgeff, and others. The classification given by Burgeff in his book on seed germination of orchids (1936) may be briefly summarized to illustrate the most important variants. He divided the methods of germination into some thirteen types, which he classified into two main groups—those yielding seedlings whose axes are radial, and those yielding seedlings whose axes are soon dorsiventral.

Amongst the radial forms are the two interesting genera *Bletilla* and *Sobralia*, of which Bernard (1909) and Burgeff (1936) have described species, *B. hyacintha* and *S. macrantha*, the seeds of which will germinate upon natural substrates without being infected by fungi. These seeds contain embryos with well-developed cotyledons. In the absence of fungal invasion, *Bletilla* produced in five and a half months a thin leafy plant some 12 millimetres long, but lacking roots. When the seedlings were infected by a suitable endophyte, a stout plant 16 mm. long having expanded leaves, thickened axes, and many roots, was produced in four and a half months. Germination and early growth were possible without infection, and the plant passed rapidly through the heterotrophic to the autotrophic state.

The radial seedlings of the epiphytes *Cattleya*, *Laelia*, and *Odontoglossum*, although not capable of germinating aseptically in natural substrates, also pass rapidly to the photosynthetic stage. Some of these, although their normal saprophytic stage is short, are capable of growing in the dark on organic substrates in the infected condition.

The dorsiventral seedlings show a very much greater variety of developmental sequences and also more diverse adult forms than do the radial ones. In this group also, there are seedlings with a short saprophytic period, such as those of some species of *Angraecum*, *Vanda*, and *Phalaenopsis*, and others with a long saprophytic phase, such as those of species of *Cypripedium*, *Cymbidium*, *Listera*, and *Orchis*. The extreme examples of this latter group are found in the complete saprophytes, some of which belong to otherwise holophytic genera (such as *Cymbidium macrorhizum*), while others belong to wholly saprophytic genera, e.g. *Neottia* (Plate 8), *Corallorhiza*, *Galeola*, etc.

The great size which some completely saprophytic orchids can attain is vividly described by Burgeff (1936, p. 139). The climbing, liane-like *Galeola hydra* is a chlorophylless saprophyte which produces long branches a centimetre or two thick that may be some 16 metres long. These bear scale-leaves and develop roots which bind the plant to its support. The blossoms are produced in large quantities, and the fruits ripen to give rise to large numbers of winged seeds, which, according to the description, may be so plentiful as to cover the ground like sawdust in a carpenter's shop. One single plant is capable of covering a large dead forest tree with yellow flowers. In the same genus, the terrestrial species *G. septentrionalis* has been investigated by Hamada (1940). It has an underground rhizome which gives rise to roots—perhaps twenty long roots up to about 20 metres long—bearing short side-branches each 5 to 15 cm. long. Many erect flower-shoots, perhaps a metre high, produce the yellow or reddish-yellow flowers and fruits. A somewhat less imposing but sufficiently remarkable example is *Gastrodia orobanchoides*, which springs from a rootless underground tuber about 5 cm. by 2·5 cm. in dimensions, and produces a flowering shoot some 60 cm. high. Like other members of the genus *Gastrodia*, it forms, after flowering, new lateral tubers which lie dormant for a period below the ground. The related species *Gastrodia elata* has been investigated by Kusano (1911) who described its relations with *Armillaria mellea*. The fungus first parasitizes the dormant tuber, and later is kept under control by digestion as the growth of the flower-shoot proceeds; *Gastrodia elata* is thus epiparasitic on the hosts of *Armillaria*.

Infection in the Adult Plant

The entry of fungal hyphae into the young seedling at germination seems to be a necessary condition for the initiation of further growth in orchids. Seedling growth can take place in special experimental conditions without the presence of the fungi, but under natural conditions on a natural substrate, infection or at least the presence of fungi seems to be essential. The young protocorm is the first organ to be infected, but the fungus does not spread directly from it to the young roots. These when formed are almost always infected anew from the soil. In the Ophrydeae, where two kinds of roots—tubers and absorbing roots—are formed, only the absorbing roots are infected. The axis of the young seedling, the protocorm, is heavily infected by the endophyte—except near its growing-point. It gives rise to an apical series of leaf rudiments, from the base of the lowest of which one or more absorbing roots grow out. The first tuber eventually arises at the base of the second leaf, near the growing-point, and is essentially a composite organ—a bud with an adventitious root. It is carried down into the soil by a dropper. At the end of the growing-season the protocorm and its roots and leaves die away—leaving the tuber, an uninfected structure, to produce in the next year a rosette of foliage leaves and adventitious roots. The tuber never becomes infected, but the roots are newly infected, direct from the soil, as they are formed.

Amongst the Ophrydeae, certain forms have a primary seedling axis which is small and short-lived (e.g. *Orchis morio*); others have a seedling axis which may persist for many years before the first tuber is formed, e.g. *O. ustulata*. This variability in the size and longevity of the primary axis is found throughout all orchid groups. An extreme example of a permanent axis which persists from the earliest seedling stages until the adult, is found in such rootless saprophytes as *Corallorhiza* where the whole underground organ of the plant consists of the branched primary axis. An excellent summary of many of the variations is to be found in Burgeff's *Samenkeimung der Orchideen* (1936).

The infected organs of adult orchids are the ones through which material might be expected to be absorbed from the soil. Those roots which become modified into storage tubers, and also those which become green and photosynthetic, remain uninfected. Infection does not spread into the stems and leaves, nor into the stalks of inflorescences. The rootless forms whose branched rhizomes function as absorbing organs, harbour fungi in their rhizomes where these are in contact with the substrate. The lianes which are attached to their supports by climbing roots, may show a differential distribution of infection even within their roots. These, in the Vanilleae, are often differentiated into long spreading roots and short roots which make contact with or penetrate the support. The short roots are, typically, completely infected and short-lived, whereas the long roots may be almost uninfected. This condition is also found in some terrestrial saprophytes, such as *Galeola septentrionalis*, and in species of *Didymoplexis* which have a root structure somewhat reminiscent of the forest trees and *Arbutus*.

The infection of the embryo, which takes place during germination, does not usually initiate a permanent state of infection in the adult. A seasonal or periodical reinfection of the new absorbing organs is usual.

Many of the European terrestrial orchids may show a seasonal cycle in fungal activity. Thus the new roots formed in the spring may become infected by fungi which invade the cortex, forming, in the cells, coils of hyphae that at first appear healthy and active. In some or in many of the cells, the fungus then becomes disorganized, and eventually a structureless mass is left. This sequence of infection and digestion may take place more or less at random in the mature cortex, or it may be restricted to a roughly-defined digestion layer. According to Burges (1939*a*), the new roots of *Orchis incarnata* become infected in April and show signs of digestive activity in May. He obtained evidence also of repeated invasion of, and digestion in, individual cells throughout the season. In addition to this activity in the spring, other workers have described a digestive activity in autumn and winter. Vermeulen (1946) described two periods of digestion in *Dactylorchis*.

The relationship of periods of active digestion to periods of vegetative or fruiting vigour of the host, has led to the view that a considerable gain of material may accrue to orchids as a result of digestion. Ziegenspeck (1936) made analyses of the tubers of *Dactylorchis* in October and in the spring, when the plant bore no leaves. The roots in autumn were full of active fungus and digestion took place extensively. Ziegenspeck estimated the change in dry-weight, polysaccharide, and nitrogen, of the tubers during the winter, and his results seemed to show an increase of some 42 per cent in nitrogen and 25 per cent in polysaccharide. His sampling method is open to some criticism, but it can be accepted that he observed a gain, the magnitude of which he believed to be under-estimated.

The absence of invading hyphae in the tubers of terrestrial orchids suggested that they possess either a mechanism of resistance to, or a substance that is toxic to, mycorrhizal fungi. In 1911 Bernard, in a post-humous paper (Bernard, 1911*a*), followed up this idea, for he had already conceived the notion that the resistance of some orchid seeds to certain strains of fungi during germination might be due to the development of a mechanism of immunity within their cells. He showed that a substance toxic to mycorrhizal endophytes was released on agar from the tubers of *Loroglossum*. It did not, however, kill *Rhizoctonia solani*, which has since been proved to be a mycorrhizal fungus of *Orchis* (*Dactylorchis*) *purpurella* (Downie, 1957). These observations were confirmed by Nobécourt (1923) and later by Magrou (1924). It was believed by Bernard and Nobécourt that the toxic principle was formed as a result of fungal attack, but Magrou did not agree. As will be seen, both these views have some substance. The matter was not followed up in any detail for many years, but in 1939 Burges, using a micro-dissection technique, extracted droplets from the cells of tubers and also from cells in which fungi were undergoing digestion. He showed that the contents of digestion-cells caused hyphal disintegration in culture.

More recently E. Gäumann and his colleagues have re-examined the whole question in considerable detail and have isolated and characterized some of the fungistatic substances. Gäumann & Jaag (1945) showed that fragments of the tubers of *Orchis militaris* inhibited mycorrhizal fungi in agar culture in the same manner as had been observed by Bernard. Later experiments showed that, although living tissues gave a zone of inhibition of *Rhizoctonia repens* around them, tissues killed by low temperature, by heat, or by poisons, were unable to secrete the fungistatic substance (Gäumann *et al.*, 1950). This suggested that it was formed only in living tissues by the activity of the fungus upon them—a point which was confirmed when the nature of the substance had been determined. By extraction of a large quantity of *Orchis militaris* tubers which had been kept in contact with *Rhizoctonia repens* for several days at 24° C., a benzene-soluble fungistatic compound of molecular weight 258 was extracted (Gäumann & Kern, 1959). This was called orchinol and was found to be a dehydroxyphenanthrine derivative which could be separated and identified chromatographically in reasonably small quantities.

By exposing cylinders of tuber tissue to cultures of fungi, Gäumann & Hohl (1960) were able both to confirm that living cells were involved in orchinol production, and also to estimate the time taken for its synthesis. The cylinders were cut into slices 2 mm. thick and the orchinol content of each was estimated. It took two days for the orchinol to appear in the slice in contact with the fungus, and a maximum quantity of 920 μg. per gram of tissue was present after eight days. The slice 10–12 mm. from the point of contact of the tissue with the fungus, showed a positive reaction after eight days. Orchinol was found to have a wide spectrum of activity on the fungi, and it was by no means specific against mycorrhizal fungi. Gäumann *et al.* (1960) observed that it had an inhibiting action at concentrations of 10^{-2} to 10^{-5} M. against 17 orchid mycorrhizal fungi and also against 19 common fungi from the soil—including soil-borne parasites. The ability of various species of fungus to initiate orchinol formation was examined, using the same range of mycorrhizal and soil fungi, on *Orchis militaris* tubers. All the mycorrhizal fungi tested could bring about the defence reaction, but 9 out of 24 soil fungi were unable to do so. In addition, three bacteria which caused orchinol production were isolated from the soil. In a test of the tubers of 24 European orchids, 16 were found to form orchinol in the presence of *Rhizoctonia repens*. The remaining 8 produced other toxic substances when infected by their own mycorrhizal fungus as judged from the zone of inhibition around tuber slices.

One of the orchids which failed to form orchinol was *Loroglossum hircinum*, a species used by Bernard and by Nobécourt. Gäumann *et al.* (1960) isolated *Rhizoctonia versicolor*, a new species, from its roots. This stimulated the production of an inhibiting substance called hircinol, which was believed to be chemically related to orchinol (*see* Nuesch, 1963), and it had similar properties with regard to its specific effect on fungi.

Although roots of orchids may also produce small amounts of fungistatic substances, they are not so efficaceous in this respect as are tubers. The roots, therefore, become infected by fungi with slight resistance to orchinol, whereas the tubers are heavily protected from infection not only by mycorrhizal fungi but also by many other fungi. Nevertheless, certain species of fungi were found by Gäumann and his colleagues to attack orchid tubers rapidly. For instance, one strain of *Rhizoctonia solani* could rapidly destroy orchinol and hence was unaffected by it, and very rapidly growing fungi, such as *Fusarium solani*, attacked tuber tissue so fast that there was not time for orchinol synthesis.

It remains to be seen whether the differential effect of these fungistatic substances can in certain conditions be such that mycorrhizal fungi can exist in the roots whereas others are inhibited. If this were so, some explanation of the ability of special fungi only to set up a symbiosis with orchid roots, would have been achieved. In a wider field, the inducible chemical defence reactions of orchids are comparable with phytoalexins which are produced by other plants. These have been reviewed in detail by Cruickshank (1963, 1965), and are compounds formed as a result of interaction between fungus and host, which inhibit non-specifically the growth of micro-organisms. Their routes of biosynthesis are incompletely known, as is the case with orchinol and hircinol, and they are also formed in relatively high concentration in the tissues.

Both the degree of infection and the mode of digestion of the fungus vary greatly in different Orchidaceae. The saprophytes are, typically, heavily infected. Many of these, such as *Neottia* and *Corallorhiza*, have a form of digestion which is very similar to that already described; but there may be a greater differentiation of the tissues, so that more definite 'host' and 'digestion' layers can be discerned. The large saprophytes, especially those with differentiated roots, may show digestive patterns only in the short roots. The species of *Didymoplexis* described by Burgeff produce short swollen lateral roots which may be branched, and also tuberous structures like those of *Arbutus*. In these swollen roots and tuberous structures the fungus appears healthy in the sub-apical regions, but undergoes digestion nearer the base, and they may show alternating bands of host tissue and digestion tissue along the length of the organ. In these plants the apices of the long, uninfected roots are normal, with a well-developed cap and meristem, whereas the cells in the short roots tend to differentiate completely. Like the short roots of other plants, these last are very short-lived affairs—perhaps organs of one season only.

Hamada (1940) has described infection of *Galeola septentrionalis* by its endophyte *Armillaria mellea*. Colonization of the roots occurs in the spring and summer—especially when the roots of *Galeola* penetrate a substrate, such as twigs, on which the *Armillaria* is growing. Either single hyphae or rhizomorphs may enter the root of the orchid. If a rhizomorph enters, hyphae attack the tissues, forming an active rhizomorph in the outer cortex. This causes extensive death of the external cells, but in the inner layers of the cortex the fungus is digested vigorously. Infections spread down the long roots by these rhizomorphs. Short roots may be infected from the parent long root, or directly from the mycelium in the soil. The digestive pattern found in this species is similar to that observed in most green orchids.

In *Gastrodia elata* and in some other saprophytes, the fungal coils do not become digested as a whole. There is, in the infected organs, a well defined layer or series of layers in which the fungus appears healthy. If the hyphae penetrate from this region into the deeper tissues, their apices become swollen and distorted and may break up into small digestion bodies. H. Burgeff called this digestive pattern 'ptyophagy' —as contrasted with the more usual 'tolypophagy', in which hyphal coils are digested.

At the other extreme are many of the green terrestrial orchids which have been reported to be virtually fungus-free as adults, or at any rate either to be free for a great part of the year or to have many uninfected roots. Burgeff has described many series of orchids within genera or larger groups which he has arranged in order of increasing saprophytism. The increased dependence on saprophytically absorbed carbon compounds may be shown by decrease of pigmentation or decrease of proportionate leaf-area. Hence some intensely green orchids may have such a small leaf-area in proportion to their total size, as to suggest that they are partially saprophytic. Others show obvious lack of chlorophyll pigments or the complete absence of them. Correlated with these series is an increased infection by fungi. The green orchids with wide leaf expanse which show little fungal infection as adults, are not necessarily those which have a short saprophytic life as seedlings. *Listera*, for instance, has a long saprophytic phase and the related genus *Neottia* is a complete saprophyte as an adult (Plate 8). But *Listera ovata* may be virtually fungus-free in its adult phases. Similarly, *Cypripedium calceolus* has a prolonged developmental phase and may take many years to reach the blooming stage. Ziegenspeck (1936) suggests sixteen years for this, whereas Irmisch (1853) gave half this time. Nevertheless there is a very gradual change-over from the saprophytic existence to the holophytic adult stage when little infection is present. *Cephalanthera oregana*, *Oncidium* spp., *Epipactis* spp., and *Orchis globosa* have all been reported to be virtually fungus-free as adults (Ramsbottom, 1922; Knudson, 1927; Niewieczerzalowna, 1932).

It is characteristic of all these orchid species which are virtually fungus-free as adults that they possess many well-developed roots, a good vascular supply in their roots and stems, and green leaves of considerable photosynthetic area. They have the appearance of normal autotrophic plants. In some cases the degree of saprophytism shown by adult individuals seems to vary within a species. Albino forms of green species of *Cephalanthera* and *Epipactis* have been described by Beau (1920) and Renner (1938), and are occasionally encountered in the field. Many species have been shown by Montfort (1940) and Montfort & Kusters (1941), to vary in their development of chlorophyll and leaf area and are believed to vary in their degree of saprophytic dependence.

In experiments on the early growth of some orchids, the length of saprophytic life has been shown to be dependent on external conditions. In the genus *Cymbidium* a prolonged period of saprophytism is usual. However, some species become green and partially photosynthetic in the light but may persist as subterranean saprophytic rhizomes in the dark. These develop aerial shoots and roots when illuminated. *C. macrorhizum* is a rootless saprophyte which always passes directly from the saprophytic phase to the reproductive phase. Burgeff (1936) has shown that the species of fungus infecting a particular *Cymbidium* species may alter the length of the saprophytic phase. Thus with a strain of *Rhizoctonia repens* the saprophytic phase was rapidly passed, but with *Corticium catonii* it was greatly prolonged. Such variation might explain the considerable discrepancy between the estimated periods from germination to flower production given by different observers for certain orchids.

This brief summary of the general points about the life-cycle of members of the Orchidaceae is enough to show that there are certain general problems common to them all. The occurrence of a passing or permanent saprophytic phase is striking. As in the Ericaceae, increased saprophytism is correlated with increased intensity and complexity of mycorrhizal infection. The question immediately arises as to whether the endophytic fungi are essential to success in saprophytic life, or whether for some reason saprophytes are more susceptible to attack. If we assume for present purposes that the Orchidaceae are functionally dual organisms and that host and fungus are physiologically interdependent, we must inquire how it happens that the fungi allow the system to absorb carbonaceous material more actively or successfully than the hosts alone; how it is that the fungi grow in association rather than alone; what physiological properties with regard to absorption the fungi have; and what structural features are possessed by the mycorrhizas that enable an exchange of material to take place between the component organisms.

XI

MYCORRHIZA IN ORCHIDACEAE— THE COMPONENTS OF THE SYMBIOSIS

The Fungi

The isolation of the fungal endophytes of Orchidaceae into pure culture has presented few special problems. Unlike those of the ectotrophic mycorrhiza of forest trees, they do not exhibit extreme sensitivity to cultural conditions, and many of them grow at a reasonably fast rate. Like other root-inhabiting fungi, some of them may be sensitive to competition from contaminants of the root surface, but, with ordinary precautions of surface sterilization, many have been isolated from orchid roots. It has sometimes been easier to use more elaborate methods than the placing of sterilized root fragments in agar. Sometimes groups of infected cells have been cut away and used as inocula. Coils of hyphae have sometimes been hooked out of infected host cells with sterile needles and inoculated into agar. N. Bernard isolated several species of *Rhizoctonia* from germinating seeds and from adult orchid roots. H. Burgeff also reported the isolation of several *Rhizoctonia* species in 1909, and since that time the number of known orchid fungi has increased.

The method used for confirming them to be endophytes of orchids has been that of back-inoculation into cultures of sterilized orchid seed. The stimulation of seed germination and development, together with the occurrence of seedling infection of a normal type, has been taken as evidence of the provenance of the isolate, although, as will be seen later, the conclusions so reached may not be fully reliable.

Burgeff, who has made so many of the successful isolations, summarized in 1936 what was known of the taxonomy, physiology, and specificity, of the orchid fungi. He described some twenty-three species or strains, comprised of fourteen imperfect fungi of the genus *Rhizoctonia* and nine clamp-forming Basidiomycetes of some of which the fruit-bodies are now known. The species of *Rhizoctonia* originated mainly from green orchids, whereas the clamp-forming Basidiomycetes were isolated from colourless saprophytes. In the course of work in culture, certain of the imperfect strains have been shown to be imperfect Basidiomycetes. A species of fungus isolated in 1929 by G. Catoni from *Cypripedium* sp. was referred to the genus *Hypochnus*. During cultural work this fungus developed, when grown in conjunction with a contaminating organism, *Cladosporium*, a fruiting-body with basidia. Cappelletti (1935) repeated the same experiment and clearly demonstrated the organism to be a Basidiomycete, which, though normally imperfect, was nevertheless capable of 'fruiting'. It is now called *Corticium catonii.* Spräu (1937) had a similar experience with an isolate from *Orchis mascula* which fruited to form a *Corticium* basidiocarp. A *Rhizoctonia* isolated in 1963 by S. E. Harley (now Mrs. S. E. Smith), also from *Orchis mascula*, was induced to fruit in culture by Warcup & Talbot (1966) who ascribed it to *Thanatephorus orchidicola*, this genus being a part of *Corticium* Pers. Indeed Warcup & Talbot (1962, 1963, 1966, 1967) are currently making a detailed study of the perfect stages of orchid mycorrhizal fungi.

A strain of *Rhizoctonia goodyerae-repentis*, isolated by Pérembolon & Hadley (1965) from *Goodyera repens*, was matched with numerous isolates obtained from various sources in Australia and one from Canada. These all gave basidial stages ascribed to *Ceratobasidium cornigerum* (Boud.) Rogers. In later work, Warcup and Talbot (1967) made many isolations of mycorrhizal fungi, identified on their production of hyphal coils in the cells, from some thirty species of Australian terrestrial orchids. They have successfully fruited about 50 of the 95 mycorrhizal strains obtained, and ascribed them to the Basidiomycete genera *Thanatephorus*, *Ceratobasidium*, *Tulasnella*, and *Sebacina*. *Rhizoctonia* (*Corticium*) *solani*, ascribed to *Thanatephorus cucumeris* by Warcup, is yet another mycorrhizal *Rhizoctonia* of which the basidial stage is known. This species was first proved to be mycorrhizal with *Orchis purpurella* by Downie (1957), and later confirmed as such by G. Hadley and his colleagues and by S. E. Smith.

Some of the clamp-bearing Basidiomycetes isolated from saprophytic orchids have also been placed in perfect genera or species on account of their production of fruit-bodies. An example is *Marasmius coniatus* var. *didymoplexis minor* (Burgeff, 1932). The clamp-free endophyte of *Gastrodia javania* is *Xerotus javanicus*, that of *Gastrodia elata* and *Galeola septentrionalis* is *Armillaria mellea* (Kusano, 1911; Hamada, 1940), that of *Galeola altissima* is *Hymenochaete crocicreas* (Hamada & Nakamura, 1963), and that of *Galeola hydra* is a species of *Fomes* (Burgeff, 1936).

The perfect stages of the endophytes of the European saprophytic orchids are not yet known. *Neottia nidus-avis* forms an association with a *Rhizoctonia* (*Mycelium radicis neottiae* of Wolff, 1926) which does not form clamp-connexions. Downie (1943) isolated both a clamp-bearing mycelium and several *Rhizoctonia* strains from *Corallorhiza*, whence previously Wolff (1932) had reported a strain of *Rhizoctonia* and Burgeff (1936) a clamped mycelium only.

In recent years, knowledge of the ecology and physiology of the fungi capable of forming mycorrhizal associations with the orchids has been greatly increased. Two species, *Armillaria mellea* and *Rhizoctonia*

(*Corticium*) *solani* are especially well-known because of their importance as destructive parasites.

Armillaria mellea is almost world-wide, being found in tropical and temperate climates where it is usually regarded as a somewhat specialized destructive parasite and saprophyte which spreads by rhizomorphs. It is a white-rot organism having powers of lignin and cellulose breakdown. It spreads readily from one food-base to another, and attacks living and dead substrates most readily if they contain sufficient quantities of starch or simple carbohydrates to give it a start. If it is well established on a decaying stump or tree base it may spread far, destroying litter and fallen or standing timber as well as living hosts. There have been many accounts of its ecology and physiology, especially in respect of its destructive activities (*see* Reitsma, 1932; Leach, 1937, 1939; Hamada, 1940*a*; Cartwright & Findlay, 1946).

Rhizoctonia (*Corticium*) *solani* exists in a number of strains that differ from one another in their physiology and in their parasitic potential. They parasitize a large range of plants including crops such as cereals, lettuce, tomatoes, cauliflower, etc. (e.g. *see* Daniels, 1963). Blair (1943) demonstrated that it was not a root-inhabitating fungus in the strict sense but could spread through the unsterilized soil even when fresh organic matter or living roots were absent. It is capable of utilizing cellulose (Garrett, 1962; S. E. Smith, 1966) and of translocating the carbonaceous material absorbed through its mature hyphae (Schütte, 1956; S. E. Smith, 1966).

These two examples alone stress the difference between the fungal associates of the partially saprophytic orchids and the autotrophic forest trees—those of the former group have a greater saprophytic activity when free-living than those of the latter. This difference will be still more apparent when the physiology of the orchid fungi has been considered.

The distribution of *Mycelium radicis goodyerae repentis* in pinewoods has been studied by Downie (1943*a*). She determined the presence of this fungus by sowing the seeds of *Goodyera* upon humus from different sources; when the endophyte was present, germination occurred. The superficial needles of the pine litter contained either mycelium, sclerotia, or spores, on their surfaces. Surface sterilization of fallen needles eliminated the fungus. The fungus was absent from buried cones and twigs, but brown needles that were still attached to the branches on the trees were infected, and the fungus was also present to a slight extent on green needles. It was not limited in distribution to localities in the woods where *Goodyera* was growing, but was more widely distributed. This interesting survey suggests a wide distribution of this fungus as mycelium as well as spores, for it is difficult to understand the data if we assume it to be present only as spores.

The work of Warcup & Talbot (1966) on *Ceratobasidium conigerum* (*Mycelium radicis goodyerae repentis*), which was isolated in various places from lupin, pine, oats, and wheat, as well as from soil, would seem to lead to similar conclusions, although Warcup himself inclines to the view that dormant spores are widely distributed and their presence sufficiently explains the occurrence of the fungus.

The relative ease with which green orchids are cultivated in glasshouses and gardens in all parts of the world suggests that suitable endophytes are widespread and that the endophytic fungi are perhaps not highly specific. In his work on orchids, Bernard obtained an impression that the fungi were highly specific—a view which is sometimes held of them even today. Bernard (1905, 1909) described experiments with *Phalaenopsis* seeds in which he made inoculations with different fungal strains from *Cattleya*, *Phalaenopsis*, and *Odontoglossum*. The first fungus infected *Phalaenopsis* seeds, was not digested, and killed them. The second fungus attacked the seeds, but arrived at equilibrium with them, so that attack and digestion became balanced; the seeds germinated. The third fungus entered the seeds but was completely eliminated by active digestion, and the seeds grew no further. These observations might be interpreted as evidence of specificity, or of variation in virulence of the fungi, or of varying susceptibility of the seeds. A similar and much more complicated series of reactions has been described by Burgeff (1936) who inoculated seeds of a *Laelia-Cattleya* hybrid with many fungal strains. He classified the reactions of the seeds into seven classes which ranged from absence of infection to complete parasitism. Amongst the species of *Rhizoctonia* isolated from orchid roots, a few have been found always to inhibit the growth of seedlings, while the remainder exhibit varying degrees of specificity in respect of the group whose germination they stimulate.

Burgeff (1936) isolated the fungi from many tropical orchids growing in natural habitats. He obtained some of the *Rhizoctonia* strains which are familiar as isolates from wild and glasshouse orchids in Europe, and, in addition, some new *Rhizoctonia* strains. Both he and Curtis (1939) have come to very similar conclusions about the specificity of *Rhizoctonia* to the green orchids. Amongst the green and partially saprophytic orchids, there is a tendency for certain of the fungal species to be most effective with certain taxonomic or ecological groupings of orchids; but any generalization which could be made would be highly imperfect. It often appears that the fungal strain isolated from one species of orchid is more effective in mycorrhiza formation with seedlings of another orchid species than with those of its original host. Tables of comparison are given by Burgeff (1936) and Curtis (1939) which illustrate these conclusions. Indeed their results are greatly reminiscent of the results of studying the specificity of the mycorrhizal fungi of ectotrophic forest trees and also of the specificity of *Rhizobium*.

Harvais & Hadley (1967) have come to essentially similar conclusions from an investigation of wild terrestrial orchids. They isolated root endophytes from surface-sterilized tissues of *Orchis purpurella*,

O. ericetorum, *Coeloglossum viride*, *Gymnodenia conopsea*, *Leucorchis albida*, *Listera ovata*, *Platanthera bifolia*, and *P. chlorantha*, growing in various habitats. They grouped the mycelia into twenty groups based on vegetative and chlamydosporic characters. Amongst the groups, *Rhizoctonia repens*, *R. solani*, *R. goodyerae-repentis*, *Hypha* sp., and *Rhacodium* sp., were recognized, but the remainder could not be ascribed other than to the genus *Rhizoctonia*. Of a total of 420 isolates, 244 belonged to the genus *Rhizoctonia*, *R. repens* accounting for 95 and *R. solani* for only 8. Harvais & Hadley concluded that, although certain strains are more consistently present in any particular species than are others, there is an annual fluctuation as well as a variation with habitat. When the isolates were tested by back-inoculation with the germinating seedlings of some of the hosts, a range of behaviour was observed. Some were non-infecting, others parasitic, others mycorrhizal, and yet others mycorrhizal but not stimulating to the seedlings of the host from which they had been isolated. Harvais & Hadley also observed that some strains, although not symbiotic with the seedlings of their host of origin, were symbiotic with others.

Curtis (1939) and Downie (1943) have pointed out that there is some evidence of a fairly direct effect of soil conditions on the distributions of some of the mycorrhizal fungi of orchids, but not on those of others. Curtis observed that *Rhizoctonia repens*, *R. lanuginosa*, and *R. mucoroides*, were not greatly affected by soil conditions, and indeed were repeatedly isolated from orchid roots in all kinds of soil. These three species were the most widely distributed orchid fungi under natural and glasshouse conditions, others being more limited in their tolerance. *R. sclerotica* and *R. stahlii* were found in open situations in neutral or slightly alkaline wet soil, while *R. borealis* occurred in acid soil such as that of coniferous woodlands.

Warcup has raised a further important point. He is at present investigating further the fact that, in the roots of some orchids, he has observed more than one *Rhizoctonia* at one and the same time. He has observed, in addition to a species forming coils, a further species forming no coils. On isolation, that forming no coils has proved to be the faster growing. It is therefore possible that many isolates from surface-sterilized roots are suspect and may not be true mycorrhiza-formers. They would be the analogues of the so-called pseudomycorrhizal fungi of ectotrophs.

The relative absence of specificity in these species of *Rhizoctonia* is very much in line with what is known of other members of the same genus. Several of them, including some which have been proved to be imperfect Basidiomycetes, are root-inhabiting forms with variable powers of pathogenicity. It may well be that the orchid roots and rhizomes with which others form such close associations are not their only ecological niches, and they may find suitable sites for growth on the root surfaces of other plants.

The Basidiomycetes of the holosaprophytic orchids appear to be more specific in respect of the range of hosts with which they can act as symbionts. This may, however, be a false conclusion arrived at in the absence of extensive work upon them. Certainly, *Armillaria mellea* can form associations with *Gastrodia* and *Galeola* species, so that it may have a wide host-range both as a symbiont and as a parasite.

The Physiology of Orchid Fungi

(1) *Nutrition*

The nutritive physiology of the orchid fungi has been much studied since the first decade of this century—especially by H. Burgeff, and later by his pupil S. Holländer and by others. It is worth emphasizing that Burgeff, quite early in his work, was unable to obtain evidence that any of these organisms was capable of fixing atmospheric nitrogen. Later, Wolff (1932) and others renewed the claim that the fungi of orchids were able to fix nitrogen, but more recently Holländer (in Burgeff, 1936) has, one supposes, finally disposed of this contention. In a series of extensive and careful experiments he was unable to show any significant fixation by any of the fungi that he studied. Since there are now no good ecological, cultural, nor experimental, grounds for believing in nitrogen fixation by orchids, we can assume this aspect of their physiology, like its counterpart in fungi of Ericaceae, to be resolved. As has been stated earlier, the assumption of nitrogen fixation seems to have clouded much of the thought on mycorrhizal phenomena; particularly in this case, it helped not at all in the explanation of the potential of orchids to live saprophytically on organic substrata for some period of their lives. The assumption that the fungi fixed nitrogen, simply had the effect of making all hypotheses seem highly complicated and improbable.

The orchid fungi usually grow readily in ordinary culture media. They are highly aerobic and are damaged, as are so many Basidiomycetes, by deficient oxygen supply. They may actively utilize many organic substances as a source of carbon, and in this there is especial interest, because typically the habitats of orchids are rich in decaying vegetable matter. The range of carbohydrates upon which orchid fungi will grow is very much wider than that which will support the growth of the fungi of ectotrophic mycorrhizas, and their growth-rates in terms of lateral spread on solid media are more rapid. All of them are able to use the simple sugars such as glucose, maltose, and sucrose. The last-named may be rapidly inverted and invert sugar released in excess of absorption, or it may be absorbed apparently as a single molecule. Starch is rapidly hydrolyzed and its products are an excellent source of carbon for growth. The strains of *Rhizoctonia* studied by Burgeff (1909) made between 1·3 and 4·5 cm. of growth in eight days on starch agar at 23° C., giving rise to a starch-free zone, or to zones of partial starch hydrolysis, in advance of the growing margin, as shown by iodine coloration. Some of the strains of tropical orchid fungi isolated in 1927–28 grew even faster on similar media, but the species of *Rhizoctonia* grew faster than the clamp-bearing types.

In contrast again with the fungi of ectotrophic mycorrhizas, these orchid fungi grew well on humus-decoction agar, on natural leaf-litter, and on poplar wood, all of which contained suitable sources of carbon and nitrogen. A multiplicity of carbohydrates and other organic compounds, glycosides, humic acid, glycogen, mannose, arabinose, xylose, and proteins and their decomposition products, all proved to be suitable nutrients for many of these fungi.

Pérembolon & Hadley (1965) further showed that five *Rhizoctonia* strains, whether symbiotic or pathogenic, could utilize pectins as carbon substrates. The production of pectinases was not correlated with the symbiotic or parasitic habit.

Holländer's experiments in which tannins were added to media containing glucose, showed that these substances were not suitable carbon sources for orchid fungi—in spite of certain claims to the contrary. Nevertheless his observations demonstrate that most of these fungi are capable of producing extracellular polyphenol oxidases, so that the tannins in the media become oxidized. The fungi of the tropical saprophytes were especially active in this respect. The production of extracellular oxidases which act on phenolic compounds, is a property usually correlated with ability of fungi to break down and utilize cellulose and other complex polysaccharides, and lignin. Indeed in subsequent experiments many of the fungi were shown to have this power. Table XLII gives a summary of some of Holländer's results on their utilization of cellulose and lignin.

The ability of strains of *Rhizoctonia*, derived from terrestrial European orchids, to use cellulose, was tested by S. E. Smith (1966) both in pure culture and in unsterilized soil. A loss of weight from pads of cellulose paper was observed in pure culture, the magnitude of which varied with the strain. *Rhizoctonia solani* from *Orchis* (*Dactylorchis*) *purpurella* gave the greatest loss, while some strains of *R. gracilis* and *R. repens* gave a negligible loss. In these cultures considerable breakdown had taken place, for the eluates of the pads contained soluble carbohydrates: glucose and trehalose in *R. repens* cultures, and glucose, mannitol, and cellobiose, in *R. solani* cultures. All the *Rhizoctonia* strains tested—*R. solani*, *R. gracilis*, *R. repens* (3 strains), *R. goodyerae-repentis*, and *Rhizoctonia x*—grew in unsterilized soil both in the presence and absence of added cellulose, showing them to be competitive saprophytes. Measurements of the rate of growth of mycelia of *R. repens* and *R. solani*, showed them to grow much faster in soil in the presence of cellulose than in its absence. This very striking ability of a strain of *Rhizoctonia solani* from an orchid to utilize cellulose, is in agreement with the work of Garrett (1962) on a strain of this species from swedes, and that of Daniels (1963) on many strains of varied pathogenic potential which were derived from various sources.

Armillaria mellea, the endophyte of *Gastrodia elata* (Kusano, 1911) and of *Galeola septentrionalis* (Hamada, 1940, 1940*a*), is also capable of utilizing lignin and cellulose. It is a white-rot fungus and exists in many different strains. It causes breakdown of woody tissue, of sapwood and sometimes of heartwood—also of both softwood and hardwood trees not only when they are standing but also when lying as wood and twigs on the forest floor. The fungus is highly aerobic and is capable of conducting oxygen through its rhizomorphs (Reitsma, 1932). There seems no doubt that it also conducts other materials by the same means, for its spread to a new site often depends upon one end of its rhizomorph being attached to a source of food.

TABLE XLII

CELLULOSE AND WOOD DESTRUCTION BY ORCHID FUNGI. Results of Holländer (from Burgeff, 1936).

Name and description	Pure cellulose	Cellulose in leaf-litter	Poplar wood	Lignin in poplar wood
M.r. petolae (*Rhizoctonia*)	++	+	++	—
M.r. schillerianae (*Rhizoctonia*)	+	?	++	—
M.r. neottiae (*Rhizoctonia*)	+	—	—	—
Corticium catonii	—	—	—	—
M.r. hydrae (*Fomes*)	+	++	+++	++
M.r. javanicae (*Xerotus*)	+	+++	++	+
M.r. callosae (Basidiomycete)	+	+++	++	—
M.r. pallentis (*Marasmius*)	+	+++	+	
M.r. paete (*Marasmius*)	—	+	—	—
M.r. nutantis (*Marasmius*)		+	++	

The orchid fungi use many kinds of inorganic and organic compounds as sources of nitrogen. Ammonium salts and simple organic compounds are good sources, nitrates not so good. Urea, amino-acids, peptones, and proteins, are satisfactory. The inorganic sources of nitrogen—ammonium salts and nitrates—when absorbed cause a pH change in the medium, such as occurs with many other fungi. These changes may greatly affect growth-rates, for the optimum acidity for growth of most of the fungi is between pH 5·0 and pH 7·0. Table XLIII shows the comparative growth-rates of a group of orchid fungi upon a number of nitrogen sources. The greater growth of many, especially those known to be Basidiomycetes (the last five in the table), upon the organic nitrogen sources albumin and nucleic acid, is interesting. Complex organic substances are also the best source of nitrogen for *Armillaria mellea*. This probably reflects the effect of some impurity in the source rather than the effect of the nitrogen source itself. It may well indicate that growth is greatly increased by some vitamin or growth factor in these substances.

TABLE XLIII

NITROGEN SOURCES AND GROWTH OF ORCHID FUNGI IN 2 PER CENT GLUCOSE MEDIUM. Results of Holländer (from Burgeff, 1936), expressed in mg. of dry mycelium per forty days' growth-period. Fungus: *Mycelium radicis*.

N. source	*petolae*	*schiller-ianae*	*neottiae*	*pallentis*	*javanicae*	*callosae*	*nutantis*	*hydrae*
$(NH_4)_2SO_4$	72·0	96·0	4·6	2·0	7·0	2·8	14·2	2·2
KNO_3	54·0	77·0	5·5	3·5	2·5	0·0	4·5	3·3
NH_4NO_3	129·0	71·0	6·4	4·5	6·5	3·2	7·5	3·8
$Ca(NO_3)_2$	88·5	96·0	7·0	11·2	7·6	8·3	9·5	6·1
Glycocoll	74·0	79·5	1·5	2·1	—	—	1·6	2·1
Alanine	110·0	—	1·5	2·5	—	—	1·2	—
Leucine	61·5	—	1·5	3·0	—	0·5	1·5	0·3
Asparagine	150·0	82	6·0	5·5	4·1	2·5	13·0	11·4
Urea	73·0	61	3·6	19·2	8·6	6·0	21·0	14·5
Peptone	93·0	101·5	11·1	10·5	8·4	6·8	12·4	8·0
Albumin	122·0	78·0	20·2	24·8	23·6	24·0	22·0	31·8
Lecithin	63·0	—	0	12·1	4·2	0·5	6·0	0
Nucleic acid	94·0	88·5	12·0	38·0	39·2	31·0	25·0	29·5

More recently Vermeulen (1946) has shown that the strain of *Rhizoctonia repens* which he isolated from *Orchis morio* was greatly stimulated in culture by the vitamins of the B complex, thiamin and biotin, and by other substances present as impurities in the maltose which he used as a carbon source in his medium.

Similarly, Pérembolon & Hadley (1965) record that all the strains of *Rhizoctonia* used by them (*R. solani* 3 strains, *R. repens* 1 strain, and *R. goodyerae-repentis* 1 strain) required yeast extract for good growth in culture, and de Silva & Wood (1964) have confirmed the stimulation of strains of *R. solani* by exudations from roots.

(2) *Translocation*

One of the most important recent advances in the study of the physiology of orchid fungi is the demonstration of their ability to translocate nutrients within their hyphae. A number of workers have shown, incidentally to their work on translocation in the fungi, that certain species, strains of which may be mycorrhizal with orchids, have the property of translocation. Schütte (1956) made an organized general survey of the ability of mycelia to spread from a medium containing a given nutrient to one which was deficient. Those that could spread in this way were described as translocators, and amongst them was *Armillaria mellea*. This method of testing was shown by Lucas (1960) and Thrower & Thrower (1961) to be essentially unreliable. Lucas used a split plate method in which a mycelium was made to grow across a diffusion barrier. Isotopically labelled nutrient was then applied to the mycelium on one side of the barrier; and movement through preformed mycelium to the deficient side was then observed. Lucas distinguished 'translocation' from 'transport by growth'. In the latter process, nutrient (especially ^{32}P), accumulated in the growing region, was transported in the tips of the hyphae as they grew across the barrier, whereas in translocation proper, the isotope was absorbed by and moved through the mature hyphae. *Armillaria* was indeed a translocator. Robinson (1963), working in Lucas's laboratory, showed that another potentially mycorrhizal species, *Rhizoctonia solani*, could translocate ^{14}C derived from [^{14}C] glucose and ^{32}P derived from $KH_2{}^{32}PO_4$. Other observers also have shown *Rhizoctonia* (*Corticium*) *solani* to be a translocator. Thrower & Thrower (1961) with ^{14}C compounds, and Monson & Sudia (1963) with a variety of isotopically labelled metabolites, all recorded translocation.

S. E. Smith (1966, 1967) has made a special study of translocation in proved mycorrhiza-formers. Using Lucas's method, *Rhizoctonia repens* was shown to translocate ^{32}P absorbed as $KH_2{}^{32}PO_4$ and ^{14}C absorbed as [^{14}C]D-glucose. Using a modification of this method, in which seeds of *Orchis* (*Dactylorchis*) *purpurella* were placed on the receiving side of the split plate, she showed that both isotopes ^{14}C and ^{32}P were translocated into the tissues of the seedling through the hyphae of the infecting fungus. Using Schütte's method (*see* above), Smith allowed mycelium of *Rhizoctonia repens* to spread from a medium containing full nutrients (including cellulose as the carbon source) onto a carbon-free medium. Seedlings of *Orchis* (*Dactylorchis*) *purpurella* or *O.* (*Dactylorchis*) *praetermissa*, placed on the carbon-deficient side, were stimulated to germinate and continued to grow after control seedlings (without cellulose in the donor side) had ceased to grow.

In her second paper, Smith extended these observations by analysing the compounds in the seedling which became labelled with ^{14}C as translocation into them proceeded. Previous analysis had shown that the main soluble carbohydrate compounds of *Orchis purpurella* tissue were glucose, fructose, and sucrose. Those of *Rhizoctonia repens* mycelium were trehalose and glucose with a trace of mannitol, those of *R. solani* (strain Da) were glucose and trehalose, and those of *R. solani* (strain RS10) were glucose, trehalose, and mannitol. Infected mycorrhizal seedlings of *O. purpurella* contained the expected soluble carbohydrates derived from both partners—according to which fungus was involved.

Infected seedlings were placed on sterile media so that fungal hyphae grew out of their tissues onto a new substrate. This mycelium was supplied with [^{14}C] D-glucose. The radioactivity per unit fresh-weight of the seedlings rose steadily. Fig. 19 shows two time-course experiments extending over 150 hours in which it is seen that, as translocation into the seedling proceeded, there was a conversion from the fungal carbohydrate, trehalose, to the host carbohydrate, sucrose, in each case. Translocation

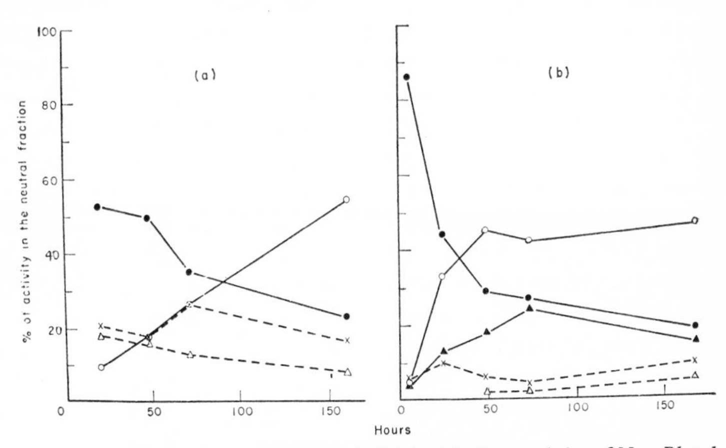

After S. E. Smith (1967). *By permission of New Phytol.*

FIG. 19.—Analysis of seedlings of *Dactylorchis purpurella* supplied with [^{14}C] glucose through their infective mycelium only. The percentage of the total radioactivity in the seedling as trehalose ●——●, glucose ×——×, sucrose ○——○, and mannitol ▲——▲, is shown. As time proceeds there is a conversion from the fungal carbohydrate, trehalose, to sucrose which is not found in the mycelium.

through the fungal hyphae into the endophytic mycelium, and from there into the seedling tissues, was clearly demonstrated. These experiments will be considered later, when symbiotic germination of orchid seeds is discussed.

SUMMARY OF POINTS ON FUNGAL PHYSIOLOGY

The outstanding properties of the fungi of orchid mycorrhiza which contrast them with those of ectotrophic mycorrhiza are their ability to utilize complex sources of carbon, their rapidity of growth, and their ability to compete as saprophytes with other organisms. Some may, in addition, act as aggressive parasites. Their ability to obtain nutrients and to translocate them through their hyphae, makes it credible that they may play an active and essential part in the nutrition of orchids—in both the saprophytic and autotrophic phases of the life-cycles of their hosts.

ASYMBIOTIC SEED GERMINATION

As was described in the last chapter, the seedlings of orchids, when germinated on natural unsterile substrates, are attacked by fungi very soon after the embryo has swollen by absorption of water. It is usually observed that those which have become infected develop further, whereas uninfected seeds fail to grow. The exceptions to this rule are *Bletilla hyacintha* which was first described in detail by Bernard, and *Sobralia macrantha*, which was later described by Burgeff as being similar. *Cattleya* and *Laelia* grow to some extent, and may become green but then grow no further in the absence of fungi. Even the two species which germinate alone were greatly improved in growth if inoculated with suitable fungi, the magnitude of the plantlets, the development of roots, and the size of the leaves, being all increased following fungal invasion. When these facts were first described, it was readily believed that the fungi in some way provided the stimulus to further growth of orchid seedlings, and for a time the method by which this stimulation was brought about was a matter of high controversy.

It should be realized that it has only relatively recently been discovered that many plant tissues may be grown in artificial culture. Cambial cultures, and cultures of root-tips and of stem-tips, are rather recent innovations. Their successful preparation has depended not only upon the perfection of techniques of sterilization and purification of media, but also upon a knowledge of the importance and action of accessory organic substances, vitamins, and growth factors, and of the needs of plants for trace elements. It has been shown that whereas adult plants may usually be grown successfully as saprophytes by providing them with suitable carbohydrates and inorganic materials, seedlings may depend initially upon stores of growth factors in their cotyledons, and only become self-supporting as regards these substances later. Root-tips and cell cultures require a permanent supply of vitamins. For example, Quendow (1930) grew *Brassica napus* in the dark as a saprophyte on inorganic salts and sugar, while Spoehr (1942) grew albino maize simply by feeding sugars to it through its cut leaves. By contrast, von Hausen (1948) showed that pea seedlings required the provision of vitamins when their cotyledons were removed. The media necessary for the growth of excised organs are still more complex, and the addition of vitamins and of small quantities of amino-acids as well as of trace elements may here be essential for success.

With the advantage of our present knowledge, it now becomes clear that many of the initial difficulties in growing orchid seeds aseptically have relatively simple explanations, and so we should pass over many of the controversies as being only an essential stage in the research. It will be best, therefore, to examine the experimental results closely rather than to bother about the hypotheses to which they at first gave rise.

It is not an aim of the work on aseptic seed-germination and seedling growth of the orchids to prove that the fungal infection is inessential to development, but rather is it to determine the conditions and the substances necessary for germination and growth. Then, knowing that fungal infection is the rule in natural habitats, we may be better able to estimate whether the fungal endophytes play an essential part in the process of germination. It must be remembered that the primary observation that most orchid seeds begin to grow only when the fungi infect them, at once strongly suggests a direct causation. We need to know what other conditions will bring about this commencement of growth, and whether, when it is biologically initiated, it is an effect specific to the endophytic fungi.

A second aim of work on asymbiotic germination and growth has been the more efficient production of seedlings for commercial and horticultural purposes. The two objectives have had much common ground, but the second does not necessitate such a precise control and knowledge of the constitution of media; hence its results, although often scientifically of great interest, are not always capable of full interpretation. Asymbiotic seed germination has recently been reviewed in detail by Arditti (1967), who gives full references and many tables of great value.

Bernard (1911) observed that the application of sugars to germinating *Bletilla* increased its growth. The plants so produced were more like those which were infected by fungi. But it was Knudson, in a series of papers (1922-30), who first described successful asymbiotic germination and growth of large numbers of Orchidaceae. He grew the seeds, after surface-sterilization with hypochlorite, on agar media of simple constitution containing glucose and inorganic salts. An inspection of his results shows certain very important points. First among these is that the primary swelling and increase in size of embryos which results in the rupture of the testa may take place in media lacking sugars, but that little further development occurs in such circumstances. These processes are, therefore, not certain indications of growth. Some of the embryos may go green in the light at this stage, but in the absence of an external source of carbohydrates they may show little or no further growth for as much as two years. The addition of sugars to the media may allow the development and growth of the protocorm and leaf rudiments. The full development of some plants may be possible on these simple sugar media, but the rate of growth may be greatly increased by the addition of decoctions of natural plant material such as those from beet and potato tissues. Germination of seeds and growth of seedlings on starch media, and on media containing insoluble organic matter, were negligible or slow under aseptic conditions. When the systems were inoculated by fungi, but not specifically by mycorrhizal fungi, growth was increased—perhaps as much as on sugar media. A *Phytophthora* species caused in one case about as good a growth of seedlings on starch media as inoculation by a specific mycorrhiza-former.

In the course of his work Knudson germinated seeds and produced healthy, well-developed seedlings of *Cattleya*, *Laelia*, *Cymbidium*, *Odontoglossum*, *Phalaenopsis*, *Ophrys* and *Dendrobium*, but with some genera and species his methods failed completely. Finally, in 1930, he reported the successful growth to the flowering stage of a *Laelia-Cattleya* hybrid in uninfected condition. The seed was germinated on 2 per cent glucose medium and the seedlings were grown for one year. A single seedling was transferred to a 250 cc. flask on a medium without sugar and grown for over a year, until it was 2 in. (5·08 cm.) high. It was then transferred to a 12-litre flask in which, after several years, it flowered. The plant was shown to be uninfected, although the medium was contaminated by a variety of extraneous organisms.

This series of experiments seemed to show that many orchids required a supply of sugar for the early growth of the seedling, and that one at least, *Laelia-Cattleya*, became self-supporting for carbon compounds when it became leafy.

After the publication of Knudson's first successful experiments, a great deal of work was done on asymbiotic germination. Clement (1924, 1924*a*, 1926), Baillon & Baillon (1924), Bultel (1926), La Garde (1929) and others, germinated orchids and grew the seedlings on sugar media with or without the addition of plant decoctions. As might be expected, various species were found to require slightly different cultural conditions, different constituents in the medium, and different pH values or temperatures. Bultel reported that he had grown a *Miltonia* under asymbiotic conditions right from seed to flowering.

Later experiments have verified and greatly amplified the results of the early work, and Arditti (1967), in his tables, gives a list of the effects of various carbohydrates on the germination and growth of numerous orchids under specified conditions. He also quotes the work of R. Ernst* which compares the value of various carbohydrates for one species, a cultivar of *Phalaenopsis*. It is of interest to note that the fungal carbohydrates, trehalose and mannitol, are both reported as supporting early growth of a species of *Dendrobium* during one year. They are not of much less value as carbon sources than sucrose, glucose, fructose, and mannose, for the *Phalaenopsis*, but are decidedly inferior for prolonged growth of *Dendrobium*. In experiments with *Orchis purpurella*, S. E. Smith (unpublished) also compared trehalose and mannitol with sucrose and glucose as carbon sources. Trehalose was of similar value to sucrose and glucose in the dark at 22°C. over 4 months, but mannitol supported growth no better than a carbohydrate-free medium. These investigations of the utilization of soluble carbohydrates that are common in fungal hyphae, are of importance in the consideration of symbiotic growth.

The common media for asymbiotic germination usually contain both ammonium and nitrate ions, and these inorganic sources of nitrogen are widely satisfactory although certain amino-acids and organic nitrogen compounds may in addition be stimulatory. In some cases amino-acids seem to be a requisite. For instance, the great effect of yeast extract upon the growth of the seeds of the saprophyte *Galeola septentrionalis* was ascribed to amino-acids by Nakamura (1962).

Although by the use of simple methods the list of orchids which could

* *See* Appendix §8.

be grown asymbiotically was greatly enlarged, there still remained a great number with which all attempts had failed. Very many, including many complete saprophytes such as *Didymoplexis* (Burgeff, 1932), and some green epiphytes such as *Vanda* spp., grew for a short time and then ceased to develop. Others gave erratic results, so that some seeds on the culture medium might develop whilst the majority did not. Following L. Knudson, many workers had added plant decoctions to media, with stimulating effects to seedling growth, and in 1934 Burgeff discussed this, suggesting that the seedlings required some vitamin that was contained in natural plant products. He pointed out that many species which formed mycorrhizas with *Rhizoctonia mucoroides*, such as species of *Vanda*, failed to develop in sugar in the absence of a suitable fungus. The young plantlet swelled and began to develop, but soon its cells became glutted with starch grains—a symptom previously observed by Knudson to occur with cultures in media that were too high in sugar content. As starch deposition occurred, the cells of the meristem of the plantlets differentiated until all the cells had become mature and cell division and growth ceased. These plantlets remained alive and might become green, but failed to enlarge further.

Burgeff (1934) noticed that plants which were grown on similar sugar media but with their fungi, developed quite normally; no excessive starch deposition occurred in their cells. Seedlings which were treated with fungal mycelia that had been killed by heat, also in some cases behaved normally. Extracts of all strains of *Rhizoctonia mucoroides*—whether inactive, stimulatory, or pathogenic—had a beneficial effect upon the growth of seedlings; yeast extract, too, had the effect of promoting normal growth. As a result of fractionation of plant decoctions containing the growth-promoting factor, Burgeff (1936) concluded that they contained vitamins of the B group. However, tests with pure thiamin, riboflavin, folic acid, and other substances, failed to give a clue to the exact identity of the growth-promoting factor.

The papers by Schafferstein (1938, 1941) have provided further information. Schafferstein showed that the symptoms observed were not peculiar to asymbiotically grown orchids which are usually associated with *Rhizoctonia mucoroides*. The seeds of members of the Vandeae, particularly, ceased to develop on sugar media, and became filled with starch, but those of many other orchids also behaved in a similar way. Schafferstein described these symptoms in four species of *Dendrobium*, in *Renanthera* sp., and in a *Laelia-Cattleya* hybrid, all of which may be associated with *Rhizoctonia repens*, and in *Oncidium grande* which may be associated with *Rhizoctonia lanuginosa*.

Schafferstein argued, as Burgeff had done, that the cessation of growth was not due to simple carbon starvation, as the cells were able to absorb sugar in excess. Moreover, since as little as 300 μg. of yeast extract per ml. of medium was enough to overcome the stagnation phase completely in most cases, the effect was probably due to vitamin deficiency. The initial stages of germination of *Phalaenopsis* seeds were in all cases normal, even without added yeast extract, and the abnormality only developed when food material was being absorbed saprophytically from the culture medium. Schafferstein extracted ungerminated seeds of *Phalaenopsis* and showed that the growth-substance was present in them in small quantities. It was present in much larger quantities in green seedlings raised on yeast extract media, and also in the green tissues of *Phalaenopsis amabilis* plants. After the holosaprophytic phase of seedling development was over, the green seedlings were again independent of external supplies of the growth-factor. There were therefore three developmental phases: (1) an initial phase when the food stores of the seed, including a store of the growth-factor, were utilized; (2) a saprophytic phase when the plants were dependent on external supplies of the growth-factor as well as of nutrients; and (3) a photosynthetic phase when the growth-factor was synthesized in adequate quantities. These observations explained the behaviour of seeds in culture—especially the point that, with many species, occasional seeds were able to develop much further than the average.

The active principle was found to be present in the green tissues of many plants. Besides stimulating the development of orchid seeds, it stimulated the growth of excised roots in culture—over and above the stimulation caused by thiamin. Its effects could be replaced by the addition of nicotinic acid and nicotinamide to the culture media. Noggle & Wynd (1943) followed up this work by testing the effects of pure vitamins on the seed germination of a *Cattleya* hybrid. They observed that germination occurred on a medium containing relatively impure maltose but not on one containing purified sugar. Although thiamin, ascorbic acid, and calcium pantothenate, had little effect on development in pure sugar media, nicotinic acid, pyridoxin, and riboflavin, had significant effects. The effects of these last three substances varied somewhat in magnitude and in kind. Only nicotinic acid had a very great effect on growth and development.

In a later paper, Mariat (1952) made the interesting observation that there is a range of vitamin dependence in the orchids. He pointed out that *Bletilla hyacintha* was vitamin-independent, and that *Cattleya* and *Laelia-Cattleya* were less dependent than *Vanda* on external supplies. In his actual experimental work he showed that these orchids were stimulated in the same manner, and by much the same factors, but to a lesser extent than those used by Schafferstein.

Arditti (1967) has tabulated what is known of the effects of vitamins and vitamin precursors on germination. He pointed out that, although there is variation between species with respect to known vitamins, nicotinic acid is commonly necessary. In some instances precursors of nicotinic acid may also have a stimulating effect (Arditti, 1966, 1967), but the co-enzymes NAD and NADP*, derived from it, are inhibitory if exogenously supplied. In a similar fashion, auxins and kinins may have stimulating effects on early growth. Doubtless the influence of a multitude of natural decoctions on the process of germination and early growth, is explicable by their content of growth-factors and trace elements.

It is clear from all this work that the minute orchid seeds may be, as other minute propagules often are, dependent on external supplies of food materials and growth-factors for any but the most limited amount of growth and development. They are similar in behaviour in many respects to many fungal spores. Some, such as the seeds of many epiphytic orchids, only require supplies of food materials and perhaps traces of vitamins; others, such as *Vanda*, cannot at first synthesize certain growth-factors at all and, in order to form essential enzyme-systems in their cells, require a supply of these also. The nicotinic derivatives are essential precursors of co-enzymes, and the symptoms of deranged carbohydrate metabolism are an expected effect of their deficiency. Synthesis of these active substances takes place normally in green tissue, and green plants are independent of direct supplies—except in the seedling stage, when the cotyledonary stores are normally sufficient to allow development up to the point where these plants become self-supporting. In the absence of stores in the seed, many orchids are dependent on external supplies for a time, and the complete saprophytes may be so dependent throughout their life.

Besides the large number of orchids whose seeds respond to the treatments described, there are yet many others which do not. Prominent examples of these are many of the terrestrial European and tropical orchids, and some saprophytes. Of these, many have not yet been germinated at all—even in the presence of apparently suitable fungi which have been isolated from their tissues. Others only give rise to good seedlings occasionally, while others again germinate after prolonged storage with soaking and shaking in water. There may be many possible explanations for this last effect. No doubt some of the results can be explained by the presence in the seed-coat of an inhibitory substance which is removed in the water. This is the explanation which has been suggested also in the case of the seeds of *Monotropa* and some Pyrolaceae (cf. p. 188). On the other hand the maintenance of seeds in a soaked, aerated condition may allow a slow synthesis of enzymic substances or enzyme precursors to take place, so as to make germination possible when suitable food material is provided.

The problems which have arisen over the non-symbiotic germination of orchid seeds are not essentially problems relating to them alone. They apply equally to many kinds of seeds and spores of angiosperms

* NAD=Nicotine adenine dinucleotide. NADP=Nicotine adenine dinucleotide phosphate.

and other green plants, and of fungi, and they will no doubt be gradually solved. It need occasion no surprise that the germination and growth of orchid seeds is stimulated by the presence of micro-organisms, for biological stimulation of germination is widespread in the plant kingdom (*see* R. Brown, 1946). Nor must it be assumed that the mechanism of stimulation is exactly similar throughout the Orchidaceae; there may be an interplay of many variables which affect certain cases differently. These may include:

(1) The effect of the age of the seed in decreasing viability.
(2) The necessity for an after-ripening period in seed maturation.
(3) The variability of seed structure, embryo development, and food stores.
(4) The quantity of enzyme and growth-factor precursors in the seed, and the rate of their synthesis in the seedling.
(5) The possible presence of germination inhibitors in the seed-coat.

To take one example: there appeared to be no specificity whatever in the fungal stimulation of germination of some seeds on starch media in some of Knudson's experiments; *Phytophthora* was as stimulating as *Rhizoctonia*. This would be expected if the seeds were simply dependent on the sugar supply, because any fungus whose activities resulted in an excessive release of simple sugar would stimulate growth. Such an experiment does not and cannot prove that all orchid fungi act solely in this way, nor that their penetration into the tissues is necessarily of no importance to the physiology of the orchids.

Symbiotic Germination and Growth

Although asymbiotic germination, growth, and even flower production, have been successfully brought about in culture in the case of some orchids, it is still as true as it ever was that infection usually occurs in the earliest stages of germination. Moreover, germination may occur in substrates such as peat, soil, potting fibre etc., if suitable fungi are present, but it does not occur in them if the fungi are absent. We know that the seedlings require simple carbohydrates and often in addition a supply of accessories, and we may ask in what way could fungal infection act to supply these needs.

In an experiment of Downie (1940) upon the early growth of *Goodyera repens*, the seeds in water alone showed the early signs of germination; but good development occurred only when salts, sugar, and potato extract, were also present. If the fungus also was provided, it had little additional effect during the first four months of development, except in those cultures containing salts and sugars but no potato extract. In these, infection gave the greatest stimulation of growth, and this might be ascribed to the formation of growth-factors by the fungus, or to a readier absorption of sugar through the fungus. Experiments on orchids germinated upon insoluble carbohydrate give different results. Aseptic germination on starch media is usually negligibly slow; the amount of growth differs little from growth on distilled water. In the presence of fungi, however, growth is as rapid as upon sugar media. Knudson (1925) showed that seedlings of a species of *Cattleya* grew equally fast upon starch media when infected and upon sugar media when uninfected. He ascribed this effect solely to the release of sugars into the medium, by the strain of *Rhizoctonia repens* that he used. Knudson observed a rise of sugar content in the starch medium, and also noted that uninfected seeds at some distance from the fungal hyphae germinated about as well as did infected seeds. Other seeds, lying on the glass of the tube and attached to the medium by hyphae, did not germinate. These observations, taken together with the fact that a great variety of fungi can stimulate *Cattleya* germination on starch media, convinced Knudson that fungal stimulation was due solely to the external enzymatic action of the fungi. He therefore held that orchid fungi were weakly pathogenic, and that penetration of the tissues of the orchids by them was not related to their growth-stimulating effects.

This is a view widely held by many people, and Burges (1936), as we have seen above, extended it to apply generally to mycorrhizal associations. However, it is clearly not applicable to ectotrophic mycorrhizas, for which the facts have been analysed in an earlier chapter. This does not mean that we must reject it in connection with the Orchidaceae, for they are greatly different in structure. There is in them no external sheath of fungal tissue through which substances must be absorbed; any soluble material produced in the substrate by the fungus could be absorbed directly by the host root. Knudson's and Burges's views are entirely credible as far as they go, but certain fairly obvious points arise and prevent a too ready acceptance of them as a complete general explanation; indeed the experimental work on translocation in the hyphae of orchid mycorrhizal fungi by S. E. Smith (1966), mentioned above, places them in a new light. In her experiments not only was carbohydrate translocated into the seedling from a glucose solution, but so were the products of cellulose hydrolysis as judged by seedling growth at a distance from the cellulose food-base. There was, moreover, a conversion within the seedlings of the fungal sugar, trehalose, to the angiosperm sugar, sucrose.*

There are therefore two means by which the fungi can intervene in the absorption of carbohydrate by orchids in the saprophytic state. They can release, by hydrolysis, sugars from insoluble carbon compounds in excess of their own utilization, so that free sugars are available for direct absorption into the orchid. By contrast they may take up all available carbohydrate and supply some to the orchid by translocation in the hyphae,

* *See* Appendix §8.

followed by release, or loss from the hyphae into the orchid tissues. The relative importance of these two processes requires to be assessed. Two experiments from Burgeff's laboratory (1936) seem to show that external digestion and direct absorption may not be generally important.

Two media containing, respectively, 2 per cent starch and 2 per cent sugar, were inoculated with orchid fungi. After a growth-period of two and a half months, these were killed by being maintained at a temperature of 54° C. to 62° C. for seven hours. Orchid seeds were then placed on the media and allowed to germinate. Three orchid fungi and three orchid species, of the genera *Dendrobium*, *Laelia-Cattleya*, and *Vanda*, were employed. After eight and a half months, each of the orchids on the sugar cultures had made several leaves and in one case roots also; those on the starch had hardly grown. This experiment cannot be considered unequivocal; nevertheless the external starch hydrolysis by these fungal strains

TABLE XLIV

RELEASE OF REDUCING SUBSTANCES BY *Corticium catonii* GROWING ON *Polypodium* FIBRE. Results of Holländer (from Burgeff, 1936).

Control before inoc.	279*
Inoculated 2 weeks	271
4 weeks	259
8 weeks	244
12 weeks	199
Control 12 weeks	282

* Relative reducing power

had not exceeded their consumption of sugars sufficiently to provide, in two and a half months, enough for germination and protracted growth of seedlings. Burgeff claimed that since, if the fungi and seeds had been inoculated together, growth would have occurred, this experiment proved that the hyphae conduct material into the infected seedlings.

In another experiment Holländer (in Burgeff, 1936) grew *Corticium catonii* on *Polypodium* fibre—a medium in which this fungus has great stimulating effects on many orchid seedlings. He analysed the reducing power of the extract of the fibre at intervals during growth. The results are shown in Table XLIV. No increase in reducing substances resulted. For comparison, reference may be made to Table XLVI, where an example of the effect of *Corticium catonii* on seedling development on *Polypodium* fibre is given.

In S. E. Smith's experiments (1966) each strain of *Rhizoctonia* released sugars into the medium from cellulose. Glucose, cellobiose, trehalose, and mannitol, were the soluble compounds identified in various fungal cultures. Here fungal utilization was smaller than production of free carbohydrates, but doubtless the quantities produced would be expected

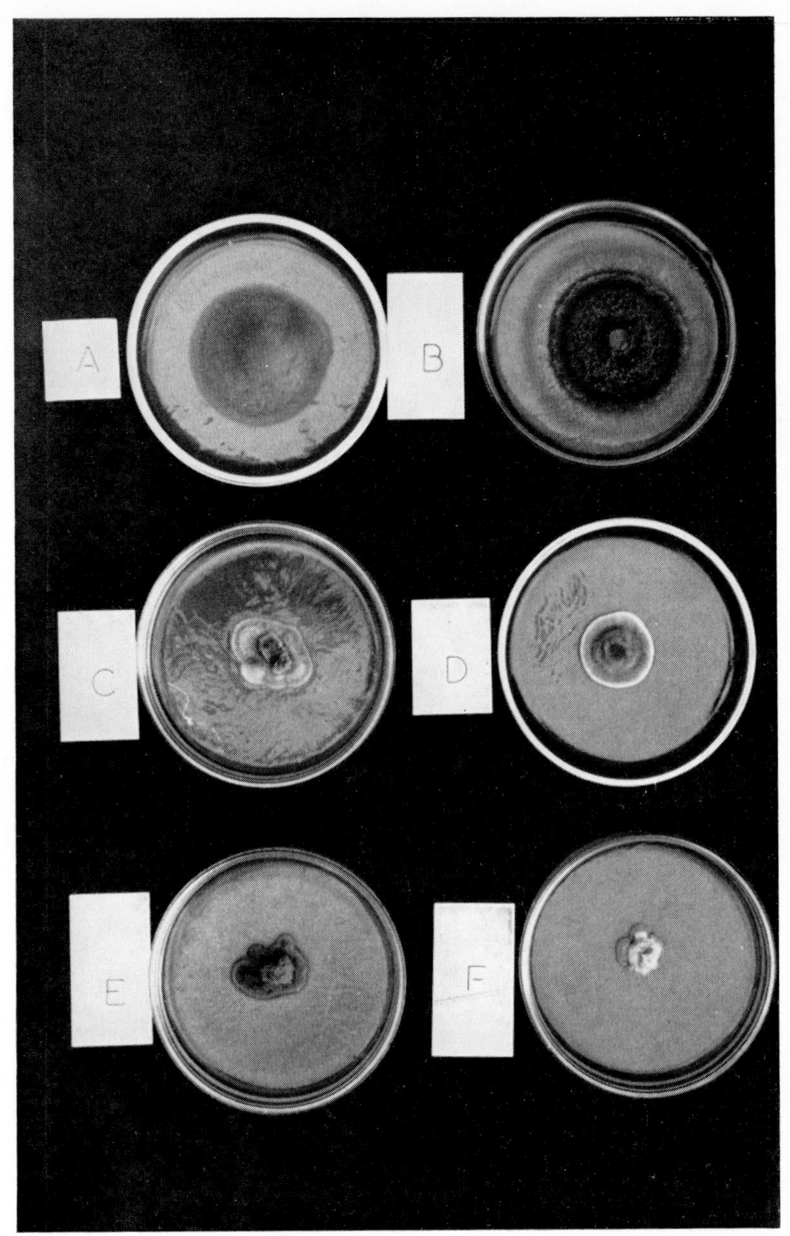

Photo by G. Woods

(*a*)

(*a*) Fungi from Ericaceae. A, *Mycelium radicis atrovirens*; B, *Phoma radicis callunae*; C, *Calluna* endophyte isolated by D. J. Read; D, *Erica tetralix* endophyte isolated by D. J. Read; E, *Mycelium radicis myrtilli* α isolated by Freisleben; F, *Mycelum radicis myrtilli* β, isolated by Freisleben.

PLATE 6

(*b*)

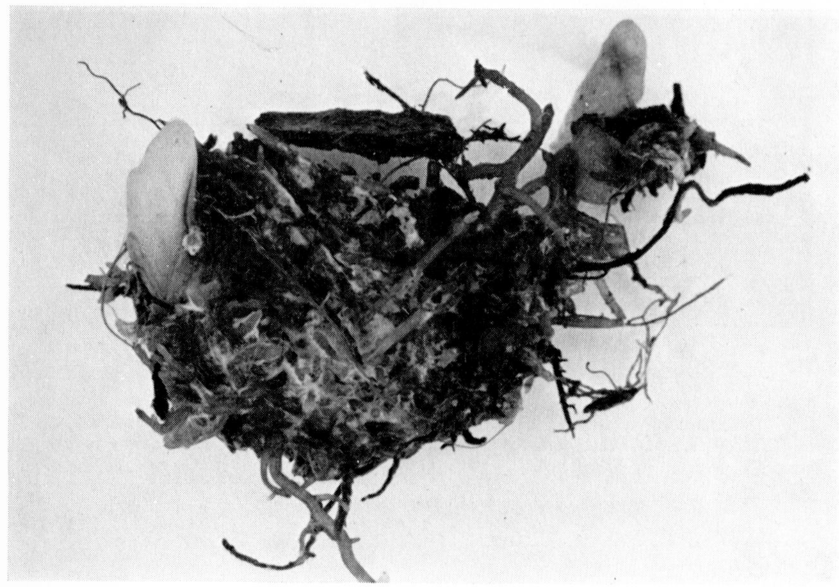

Photos by Mr. G. Woods

(*c*)

(*b*) *Monotropa* plants in woodland.

(*c*) Young *Monotropa* shoots growing from a mass containing roots of spruce and beech, fungal mycelium, and *Monotropa* roots.

PLATE 6

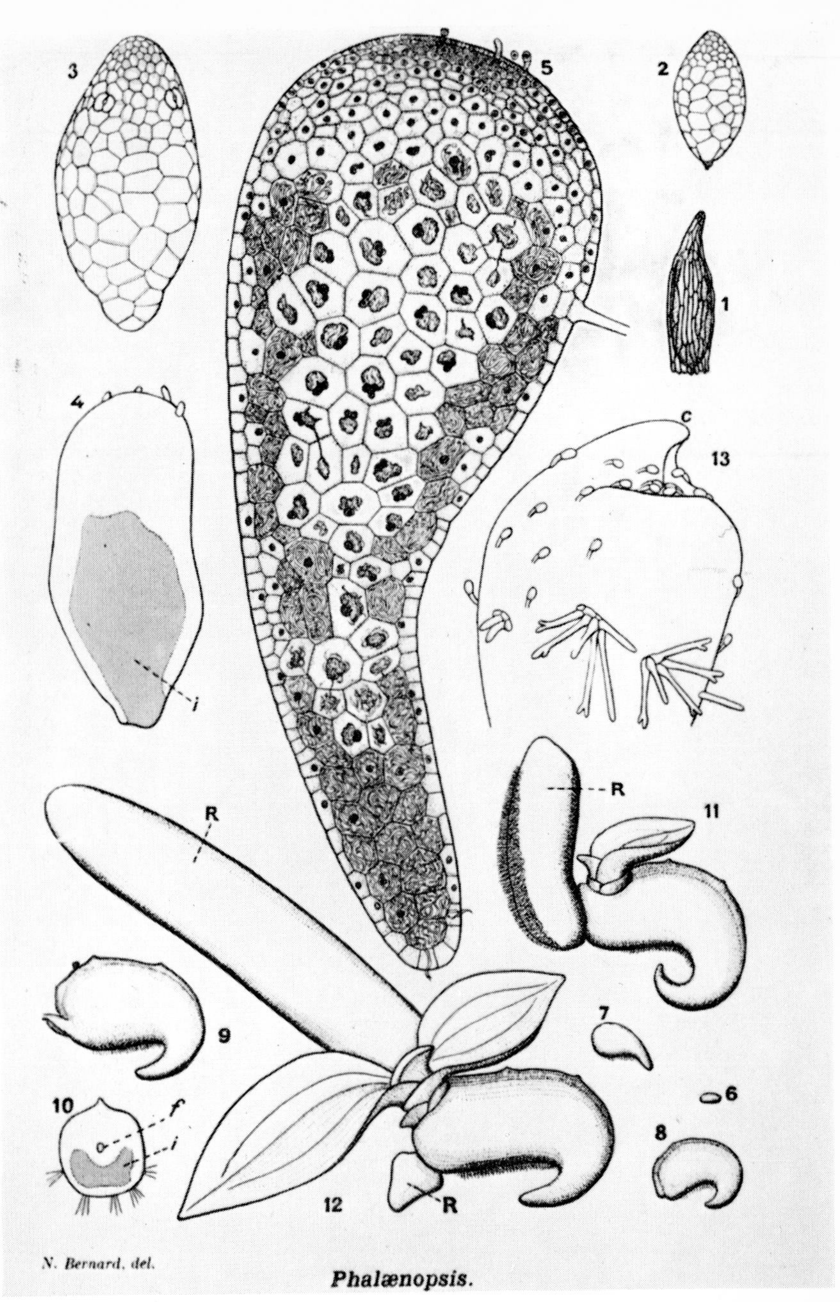

From Bernard (1909). By permission Ann. Sci. Nat. Bot.

Seed germination of *Phalaenopsis amabilis*. (1) Seed, × 100 (2) and (3) a few days and 3 months, respectively, after sowing without the fungus, × 100. (4) diagram of median section 6 days after infection with *Rhizoctonia mucoroides*. (5) drawing of the same 50 days after infection, × 100. (6) embryo after several months without fungus. (7), (8), (9), (11), and (12) plantlets respectively 1, 2, 3, 5 and 12 months after infection. (10) transverse section of infected embryo showing (shaded) the fungal region.

PLATE 7

to vary with the conditions for absorption and growth of the fungi. In any event it is doubtful whether the orchid would absorb these free carbohydrates in natural conditions in the face of competition from other soil denizens.

Certain observers (e.g. Beau, 1920) have reported, in contradistinction to Knudson (1925), that orchid seeds will germinate on damp glass if attached by infective hyphae to a nutritive medium. This is added proof that the conducting power of hyphal connexions between host and substrate is sufficient for the kind of transport of substances which is visualized by those who believe that infection is essential to the nutrition of orchids, and that it is essential to their proper physiological functioning under natural conditions. The system is envisaged as an external mycelium acting upon the insoluble organic compounds of the substrate and absorbing material which nourishes the whole mycelium—including that in the cells of the host. The hyphae within the host tissue are believed to lose substances to their host—either by being digested, or directly across the membranes of the living filaments. There is nothing improbable about this latter mechanism. We know that parasitic hyphae both absorb some materials from their hosts and lose others to them. The wilting substances produced by wilt fungi, e.g. alternaric acid by *Alternaria*, and gibberellic acid by *Fusarium* sp., are all examples of biologically active substances released in host tissues from which materials are also absorbed. There is also a loss of substances across the membranes of hyphae of ectotrophic mycorrhizas without any digestion occurring.

In spite of what has been said above, most people favour the view that the digestion of the hyphae by the host is the most important means by which substances pass from fungus to host. The visible is almost always more credible than the invisible, and therefore it is easier to believe that the endophyte loses food material to its host as it disintegrates in the host tissues. However, digestion will result in the gain to the host of only that material which is present in the hyphae when digested.

This is a matter upon which experimentation is required. It is necessary to determine the relative importance of digestion and direct passage though intact hyphae in the nutrition of orchids. Burgeff (*see* his article in Withner, 1959) recognizes two forms of digestion of fungus in orchids. In the common 'tolypophagy', coiled fungal hyphae in the cells are killed as a whole, their contents are lost to the host, and their remains persist as clumped bodies. In 'ptyophagy', the apices of the hyphae release or reject protoplasmic masses which undergo dissolution. The latter, found in some complete saprophytes such as *Galeola* and *Gastrodia*, results in a more continuous release of material to the host.

The rapid and continuous manner in which carbohydrates moved from fungus to host in S. E. Smith's labelling experiments (1967) on the tolypophagous system of *Orchis purpurella*, is slight evidence in favour of steady release rather than release by digestion of carbon compounds. On the other hand the fact that the fungus can utilize sucrose (the soluble sugar of the host), as well as trehalose, is evidence that the one-way mechanism of transfer into a 'sink', suggested by D. H. Lewis & Harley (1965*b*) for ectotrophic mycorrhizas, is not applicable in this case. Both the fungus and the orchid can it seems utilize the common soluble disaccharide of the other.

In an unpublished experiment, S. E. Smith tested the possibility that photosynthetically produced carbohydrates could find their way into the fungus. *Orchis purpurella* seedlings were allowed to assimilate $^{14}CO_2$ in the light or dark for five hours. In the light, 50 per cent of the radioactivity was found in the neutral (carbohydrate) fraction, but only 6 per cent of the dark-fixed CO_2 was carbohydrate. The remainder was in organic acid and basic (amino-acid) fractions. The analysis of the neutral fraction labelled in photosynthetic carbon-dioxide fixation showed 94 per cent in sucrose, 3·3 per cent in glucose, 1·0 per cent in fructose, and only 1·7 per cent in the fungal carbohydrates trehalose and mannitol. It would seem from this that the reciprocal flow of carbohydrates from host to fungus is very slow, and this could be interpreted as evidence that digestion might be the most important means of transfer of carbohydrate. The problem is, however, far from being solved, and it is to be hoped that further experiments will be available soon.

Whatever the actual means by which the fungus is supposed to lose material inside the host tissue, the assumption that it is lost there involves an additional assumption, namely, that there are sufficient hyphal connexions for translocation to occur. Controversy on this point is confused. Burgeff (1936) argued from his estimates on *Platanthera* that there are enough such hyphal connexions, but others claim that there are not. Burgeff estimated that some 32,000 hyphae connected an adult *Platanthera* plant to the soil. Taking the diameter of the hyphae to be on an average 4 μ, this would give, or so Burgeff thought, a total cross-section for conduction large enough to explain absorption of food material. One must remember, too, that *Platanthera*, like many other European terrestrial orchids, is capable of considerable photosynthesis and may, like the plant which flowered asymbiotically for Knudson (1930), be wholly or partly autotrophic for carbon compounds in the adult state.

The critical phase of absorption of external supplies of carbonaceous material for most orchids is during early growth of the seedling in the saprophytic phase—a time when the proportionate quantity of hyphal connexions with the soil is often very great. Moreover, many permanently saprophytic forms, and especially the large tropical types, are connected with the soil or to living or dead hosts by large aggregates of hyphae, or by rhizomorphs. These may run out into the substrate of the bark or wood of their supports and provide macroscopic evidence of connexions with the food-base which are known to be capable of effecting translocation. Many years ago, de Cordenoy (1904) described the outgoing hyphal strands from *Vanilla* roots. He pointed out that not only did these connect the roots with the supporting plant, but that they penetrated both the support and the orchid. The nature of the support, too, was critical for good growth. *Casuarina equisetifolia* was a bad host, but *Pandanus utilis* and *Jatropha curcas* were very good hosts.

The behaviour of the orchids associated with *Armillaria* also throws some light on this matter, for here a good deal is known about the properties of the rhizomorphs which connect the plants with their substrate. In these orchids the underground organ may be simple; it is especially so in *Gastrodia elata* as described by Kusano (1911). This orchid exists as an uninfected underground tuber which remains quiescent until it is attacked by *Armillaria*. Then part of the tuber is at first digested, but soon the tissues of the host control the fungus. The host then flowers, giving rise to an aerial chlorophyll-free scape, and forms secondary tubers. Its only apparent means of absorption are the rhizomorphs of *Armillaria*.

Highly interesting and very similar points arise from a valuable paper by Ruinen (1953), who worked on epiphytes in the East Indies. Amongst those which she observed in detail were orchids of the genera *Taeniophyllum*, *Thrixospermum*, *Dendrobium*, *Oberonia*, and *Eria*. By direct observation as well as by experimental cleaning or artificial infestation of living supports, she was able to demonstrate, quite unequivocally, that these epiphytes exert a deleterious effect on their hosts. The effect is not simply one of suffocation or mechanical suppression, but, as she put it, 'the mycorrhizal fungus of the epiphyte is potentially parasitic on the supporting trees'. The contact between the orchid root or rhizome and its support may be solely the line of contact between the two organisms; but the fungal symbionts cross over this line of contact, forming an effective connexion between the two. This Ruinen has illustrated by means of detailed figures of hyphal distribution. The cells of the support show disease symptoms due to fungal activity, while the epiphyte remains vigorous and healthy. In these cases there can be little doubt that the epiphytes are nourished, through the fungal hyphae, at the expense of the supporting trees.

Indeed the position of these epiphytes, and of many saprophytic orchids such as *Gastrodia* and *Galeola*, is reminiscent of the condition of *Monotropa* as described by Björkman (1960). They are epiparasites—parasites on fungi which are themselves parasitic or mycorrhizic upon other hosts. In a recent paper, Campbell (1963) has described the New Zealand saprophyte, *Gastrodia minor*, as an epiparasite on *Leptospermum scoparium*. *G. minor* arises from a subterranean tuberous rhizome upon which coarse septate hyphae bearing clamps ramify as single hyphae or strands. Little infection occurs except in juxtaposition to a root of *Leptospermum*, when extensive penetration of the cells together with the onset of ptyophagy is observed in the orchid. The root of *Leptospermum* also develops an external sheath of hyphae, a Hartig net, and some intracellular penetration. The root of *Leptospermum* is clearly parasitized, so that its cortex is disintegrated whilst the *Gastrodia* develops and flowers. The parallel with *Monotropa* is not exact but is sufficiently close to be of great significance. Indeed it will be very interesting to obtain further information about the permanently saprophytic angiosperms, for there are indications that epiparasitism of these kinds may be much more common than has hitherto been suspected.*

Altogether, the balance of evidence is in favour of the fungus acting, in all the partial and complete saprophytes, as an absorbing and conducting system. It would seem highly unlikely that their effect is purely external, and that the absorbing organs, whether roots or rhizomes, of these orchids, with their relatively small surface area, are the sole absorbing structures. What may be the mechanism whereby material is lost by the fungus to its host is still undecided. The digestive action of the host is, at least in part, an efficient defensive mechanism. It may also be nutritive.

Quantities of Material Absorbed

The quantities of materials absorbed saprophytically by various orchid species are highly variable. Some pass quickly through the saprophytic stage; others remain saprophytic throughout life. The large saprophytes may be extremely massive plants, absorbing all their food substance, including all the material lost as CO_2, by saprophytic means.

TABLE XLV

GROWTH OF *Cymbidium* ON *Polypodium* FIBRE FOR TEN MONTHS

	Length cm.	Fresh weight gm.	Dry-weight gm.
Rhizome	101	16	0·745
Long shoots	688	64	2·247
Roots	220	25	1·052
Total	—	105	4·044

These may form and maintain plant bodies many kilos in weight. Little is known of the time-period which these may pass in attaining their full size. Burgeff and his colleague Holländer (cf. Burgeff, 1936) have given experimental estimates of the quantities of material which may be absorbed by some of the smaller orchids in reasonable experimental times. In one experiment a small rhizome of *Cymbidium*, infected with *Corticium catonii*, was grown for ten months in *Polypodium* fibre. The rhizome measured 10 mm. in length initially, and it was placed on the fibre and irrigated with salt solution in the dark. Table XLV gives the estimates of the size and weight of its various parts at the end of the growth-period.

* *See* Appendix §8.

This dry-weight of about 4 gm. was attained by the breakdown of the *Polypodium* fibre and the absorption of its products. The total weight absorbed, allowing for CO_2 emission, must have greatly exceeded this value. *Corticium catonii* is not known to cause a breakdown of cellulose; hence the starch, proteins, and simpler carbon compounds, must have supplied the carbon used. The weight of fibre presented initially was 56 gm., of which only about 1·7 gm. was estimated to be starch.

In another experiment the growth of *Cymbidium*, which is capable of both saprophytic and autotrophic existence, was compared in light and dark. *Cymbidium* seeds were grown for one year on nutritive agar in the dark. Some were then transferred to fungus-free *Polypodium* fibre and kept in the light, where they became green and photosynthetic; others were placed on *Polypodium* fibre inoculated with *C. catonii* and kept in the dark. The results are shown in Table XLVI.

This experiment indicated the great efficiency of infected seedlings in absorbing materials and in growing, even in the dark. In this experiment, something like ten times as much weight was made by them—as

TABLE XLVI

GROWTH OF *Cymbidium* ON *Polypodium* FIBRE IN LIGHT AND DARK

	Uninfected, in dark 1 month, in light 2 months	Infected, in dark 3 months
Dry-weight mg.	5·33	52·0
N content per cent	1·60	5·2
Ash per cent	12·48	10·15
N mg. equiv.	0·297	7·275
Bases mg. equiv.	0·785	3·236

compared with uninfected seedlings with full photosynthetic ability. In the writings of Burgeff as well as in those of Ziegenspeck (1922, 1936), much is made of the relative quantities of bases and nitrogen absorbed. The argument put forward, in its essential form, is that when nitrate is absorbed, bases and nitrogen enter equally into the plant. When ammonia is absorbed, nitrogen may enter without equivalent quantities of bases. Hence if the gram equivalent weight of nitrogen in the tissues exceeds that of the bases, it proves that nitrogen is absorbed as ammonia or as organic compounds. I do not think that this is acceptable. Even if, during salt-absorption, anions and cations did enter in amounts that were nearly equivalent, this could be true only of a relatively short period of primary absorption. Very soon after, nitrate would be reduced and rapidly assimilated into organic nitrogen compounds; nitrates do not remain long in the tissues unchanged. There would therefore soon be a cation excess capable of being satisfied by further anion absorption.* It is not now possible to think that there must be balance between the products of inorganic anions and cations in plant tissues. Hence these

* Or by anion synthesis as organic acids.

figures do not seem to prove anything about the nature of the compounds of nitrogen which are absorbed by either infected or uninfected plants.

The greatly increased nitrogen absorption in proportion to the ash absorption and dry-weight increment must be explained by other means. It is a general observation that mycorrhizal tissues have a higher nitrogen content than do uninfected tissues, but the differences are usually not so large as the ones observed here. It seems to me that a comparison is being made between a rapidly growing, infected plant on one hand and a stagnating, uninfected plant on the other, and that the nitrogen content perhaps reflects the greater proportion of actively growing tissue—as well as such greater availability of nitrogen to the infected plant as may be caused by fungal activity breaking down protein in the fibre.

This raises an important final point about the kinds of materials absorbed by orchids in the infected condition. All through the preceding account, emphasis has naturally been placed upon the absorption of organic matter—in view of the temporary or permanent saprophytic habit of orchids. It is important, however, not to ignore the possibility of a more efficient absorption of inorganic compounds. We have already obtained good evidence of this in other types of mycorrhizas. Even allowing for differences in structure, there is no reason why the properties of salt absorption possessed by ectotrophic mycorrhizal hyphae could not equally be possessed by orchid fungi. Admittedly there is, as yet, no evidence for this which equals in weight that obtained with forest trees, although the experiments of S. E. Smith (1966) have shown certain orchid fungi to be capable of absorbing phosphate and translocating it into the tissues of developing orchid seedlings. It must be realized too, that the total surface area of contact between orchid roots and their substrates may be relatively small, and that their hyphal connexions materially increase it. I can see no reason to doubt that similar mechanisms of inorganic salt absorption might apply to these symbiotic associations as apply to ectotrophic mycorrhizas, and they might be of ecological importance especially in green, carbon-autotrophic stages.

Summary

Typically, orchidaceous plants either pass through a saprophytic stage or are saprophytic throughout their life. All are infected by fungi for part or the whole of their lives—always during the saprophytic stage. Asymbiotic growth is possible when suitable simple carbohydrates and accessories are supplied during the saprophytic stage; but in the absence of these, on natural plant remains and humus, asymbiotic growth does not occur. The properties of orchid fungi, as exemplified in their carbohydrate and carbon nutrition, are such as to be in keeping with the hypotheses that they supply carbon compounds to their hosts. It seems most probable that this is brought about by their digestion of complex organic compounds, by absorption of the products of this digestion, and by conduction of them throughout their mycelium. Part is lost to the host within the host tissues. The digestive action of the host, which causes a breakdown of the fungal hyphae within the infected zones, is undoubtedly a defence mechanism against complete parasitic invasion; it also results in some gain of material by the host. It seems doubtful if this is the only form of exchange between the two partners. Exchange of material across intact fungal membranes is a possible and, on comparative grounds, actually a probable route of exchange. The orchid fungi appear to have properties of absorption and translocation of inorganic compounds similar to those of ectotrophic mycorrhizas, but their intervention in inorganic nutrition has not yet been thoroughly investigated.

XII

OTHER MYCOTROPHIC PLANTS WITH SEPTATE ENDOPHYTES

BESIDES those already described, there are very many other plants in the roots of which septate fungi are normally found. A number of these kinds of mycorrhizas have been described; details of their morphology and histology have been observed, but little valuable experimental work has been carried out on them. There is a great tendency to extrapolate results obtained by experimentation with the Orchidaceae and Ericales into an interpretation of the physiology of these other mycorrhizas—often in a most uncritical manner. Before beginning a detailed account of some of them, it is therefore important to state clearly what grounds there are for generalization.

Within each of the two groups just described, the Ericales and the Orchidaceae, there is some uniformity in physiological or ecological properties, or at least some common pattern of behaviour. In each case, their mycorrhizas may be viewed as variants of a common type; further, there are grounds for suggesting that the two groups have something in common in the functioning of their mycorrhizas. It is, perhaps, also dimly possible to see some similarities in function and behaviour right through all the kinds of mycorrhizas described—whether ectotrophic, ericalean, or orchidacean. In particular, a progressive range of behaviour of the fungi in each of the types of root system can be seen.

The fungi of ectotrophic mycorrhiza, so far as the proved mycorrhizas are concerned, are usually Basidiomycetes. (It may be that some Ascomycetes, other than those mentioned in Chapter IV, e.g. Tuberales, are also capable of forming mycorrhizas, but this has not been clearly proved.) Within these Basidiomycetes there occurs a range of physiology from those with restricted carbohydrate nutrition to those few which show active cellulose and lignin decomposition. Both monocaryotic and dicaryotic mycelia are capable of mycorrhiza formation. In addition, numerous kinds of strains of sterile root-inhabiting fungi such as *Rhizoctonia sylvestris*, *Mycelium radicis atrovirens*, and *Cenococcum graniforme*, are widely present on the surfaces and in the tissues of the roots of many host plants, although they may not always form morphologically and histologically normal mycorrhizas. The last-named fungus, *Cenococcum graniforme*, is also capable of forming mycorrhizas of normal structure with a large variety of plants. The other two fungi may be classed as pseudomycorrhiza-formers or ineffective mycorrhiza-formers. These names indicate that the composite organs produced by them with the roots, do not seem to function as do normal mycorrhizas. These sterile imperfect fungi have much in common structurally with some of the mycorrhizal fungi isolated by Doak (1928) and Bain (1937) from Ericales, in the roots of which they formed a morphological and histological pattern similar to that found in naturally occurring mycorrhizas of Ericales. Again, they are similar in some ways to the *Rhizoctonia* strains that have been isolated from orchids, some of which strains are known to be imperfect Basidiomycetes. Moreover some ineffective forms are found amongst orchid fungi. Downie observed that the *Rhizoctonia* isolated by her from *Corallorhiza* was ineffective; Burgeff found *Mycelium radicis* (*Ophrys*) *tenthridiniferae* and *M. r.* (*Cattleya*) *labiatae* to be parasitic or inactive in mycorrhiza formation.

In an earlier chapter it was emphasized that the fungi associated with the roots of plants are very diverse in kind and activity. Amongst the fungal population on most root systems there occur sterile mycelia which are not, in normal conditions, parasitic; they include many which are often called 'sterile dark mycelia' or *Rhizoctonia*. For instance, Hildebrand (1934), Truscott (1934), and Hildebrand & West (1941), have isolated them from strawberry roots, while Katznelson & Richardson (1948) have obtained them from tomato-roots. There can be no doubt that these sterile mycelia are essentially root-inhabiting fungi in an ecological sense (even if not in Garrett's restricted sense). On the root surface and in the subjacent tissue, some of them may cause minor lesions in some conditions of host growth. Waid's work (1959) on the fungi of the deep-seated root tissues of grasses, show them also to be of common occurrence as denizens of ageing organs.

Students of mycorrhiza have observed and isolated sterile imperfect fungi from a great variety of root systems. Burges (1936) suggested the term 'casual mycorrhiza' for such cases as that of *Schelhammera undulata*. This liliaceous plant has roots which are said to be normally uninfected, but occasionally the cortical cells of the lateral roots are occupied by coiled masses of hyphae or chains of oidia such as are found in the cells of Orchidaceae and Ericaceae. These hyphae were described by Burges as being of the '*Rhizoctonia* type'.

In the nineteen-twenties, Peyronel (1922, 1923, 1924) put forward the idea that many plants had a double infection in their roots. Thus two kinds of hyphae were to be found in the cells: septate hyphae and also hyphae without cross-walls. The latter were of a type that is common amongst Phycomycetes. This kind of phycomycetous infection is the subject of the next chapter. The septate mycelia were fairly easily isolated into pure culture, but not so the Phycomycetes. Peyronel obtained cultures, which he classified as *Rhizoctonia* and which were similar to the *Rhizoctonia* species of orchids, from a wide variety of wild and cultivated flowering plants. These included *Triticum*, *Zea*, *Hordeum*, *Solanum*, *Beta*, *Arum*, *Euphorbia*, *Circaea* and *Saxifraga*. In such cases the *Rhizoctonia* seemed to attack the roots later than the phycomycetous mycelium, and it appeared to colonize in particular the outer cortical cells of ageing roots. From the various accounts these *Rhizoctonia* spp. appear to behave as weak parasites.

Peyronel's concept of double infection, and his demonstration of the frequency with which Rhizoctonias occurred in root tissues, ought to have had a more steadying effect on speculation about mycorrhizal phenomena than in fact it did. Many people working on the phycomycetous infection have described, without due attempt to make sure of their ground, how a single mycelium in their root material was septate on the root surface and in the outer tissue but aseptate within.* Many have made attempts to isolate the phycomycetous endophyte, have obtained sterile mycelia or *Rhizoctonia*, and claimed to have cultured the phycomycete—often in the face of direct evidence that back-inoculation did not produce the normal histological pattern of phycomycetous infection.

It now seems that Peyronel's concept of double infection, useful though it was even allowing for the subsequent errors described above, is really a great simplification of what actually occurs in a very large number of plants. A particularly instructive case of multiple rather than double infection is provided by the work of Lihnell (1939) on *Juniperus communis*. In one of his sets of experiments, Lihnell isolated fungi from surface-sterilized roots on malt agar. Eighty-seven sets of roots were sampled. From all of these *Mycelium radicis atrovirens* was isolated, and out of a total of about 1,000 roots examined some 650 gave cultures of this fungus. A species of *Rhizoctonia*, *R. juniperi*, was obtained from roots in thirteen samplings, and eight other sterile mycelia, some of which were related to *M.r. atrovirens* and one of which was a Basidiomycete, were also obtained. In addition, the expected Mucorales and Fungi Imperfecti appeared in the cultures. All these isolations were obtained from juniper roots in which a phycomycetous fungus with aseptate hyphae, and bearing its normal mycelial organs (*see* next chapter), was present, although it was not isolated. Moreover, on occasion, juniper roots were observed to form ectotrophic mycorrhizas with *Cenococcum graniforme*.

In synthesis experiments, most of the isolated fungi, when grown in association with sterile *Juniperus* roots, behaved as weak parasites or were harmless to the roots. Some of them gave histological patterns in the cells, penetration with coil formation, etc., producing conditions not unlike those found in other plants and that are often called mycorrhizas—the 'casual mycorrhiza' of Burges.

We could hardly look for a better example to emphasize the fact that a single plant may associate with a great variety of fungi. Nor do

* For conditions under which this is to be expected *see* Chapter XIII.

we need further confirmation that the theory of double infection is too simple.

Infection of roots by root-surface and rhizosphere fungi grades through a wide range of 'casual mycorrhiza', which are weakly parasitic associations, to strongly parasitic and mutualistic relationships between the roots and their fungi. The first, the rhizosphere and root-surface associations, are a relationship comparable with that between adjacent plants of any community. The others bear a special relationship with one another which may be one of parasitism in a harmful sense, or may result in a specific mutual stimulation of physiological activity.

Each taxonomic grouping of these fungi comprises a variety of related species or strains and may therefore have a variety of effects on a single or a different host. In the genus *Rhizoctonia* there are for any host parasitic forms, such as those of *R.* (*Corticium*) *solani* and *R. sylvestris*, and harmless or mycorrhizal forms. Studies on specificity in mycorrhizal formation show that it is not possible to predict whether a particular strain will be symbiotic, parasitic, or inactive, with any given orchid. Most species or strains are not always symbiotic nor always parasitic, but vary in their activity with the environmental conditions and with the host plant. Strains of *Rhizoctonia solani* may, for instance, be parasitic on one host and mycorrhizal with another (Downie, 1957).

The example of *Armillaria mellea* is particularly instructive, for this fungus exists in a variety of physiological strains. Some of these form mycorrhizas with species of tropical saprophytic orchids; others are intensely parasitic, and the range of potential hosts for the species as a whole is exceedingly large. Again, *Fomes*, *Xerotus*, and *Marasmius*, are essentially humus- or wood-destroying or parasitic genera but some species form mycorrhizas with orchids. Similar variations are found within the genera of fungi which form mycorrhizas with forest trees. *Tricholoma* contains both mycorrhizal and wood-destroying species (e.g. *T. rutilans*). *Boletus* has some parasitic members, e.g. *B. parasiticus*.

Gosselin (1944) has described a type of ectotrophic mycorrhiza of spruce which is said to be formed by *Polystictus circinatus*. This fungus appears also to be the cause of butt-rot of the trees. In this case, if Gosselin were correct, one and the same fungus would be both mycorrhizal and dangerously pathogenic with the same host at different stages of growth in the same habitat. However, Whitney (1965) has reinvestigated this fungus under its name of *Polyporus tomentosos*, var., *circinatus* and found it always to be both non-mycorrhizal and virulently pathogenic in inoculation experiments.

The noteworthy conclusion which comes out of this discussion is that it is of primary importance to distinguish 'casual mycorrhiza' from kinds of infection which the plant habitually suffers. Only where an infected condition is a normal attribute of a host growing under natural ecological conditions, and where a consistent histological and anatomical structure in the infected condition is possessed by the infected organs, may we properly apply the name mycorrhiza. In the three groups of plants which were studied in detail in previous chapters, an infected condition is, so to speak, a built-in part of their structure and habit, when they are functioning normally in natural conditions. If we see similar infection in other plants, we may use the similarity to help us ask the right questions, and to construct working hypotheses which will simplify experimental work. When we find infections of different structure, the first step is to ensure that they are a normal attribute of the species of host on which they are observed, after which we may consider the functions of their composite host-fungus organs without inadvisedly extending to them the conclusions reached on the functioning of other types of mycorrhizas.

Below are given descriptions of a few additional mycorrhizas, to illustrate the range of septate infections.

Mycorrhizas of the Orchid Type

(1) Aneura

The accounts of M. C. Rayner (1927) and Nicholas (1932) have given most of the available information on the association of Bryophyta with fungi. Leaving aside obvious pathogenic infections, most of those consistently infected contain non-septate hyphae in tissues of their thalli. A few species are, however, inhabited by septate fungi. Amongst these last are a very few in which the structure of the host-cells, and the behaviour of the fungi within them, resemble those in the Orchidaceae.

It is interesting to note that amongst the species so infected are two examples of saprophytes. These are the chlorophyll-free saprophytic varieties of *Aneura pinguis*, described by Denis (1919), and *Cryptothallus mirabilis*, an obviously-related subterranean species described by Malmborg (1933, 1934). Strikingly enough, M. Stahl (1949) has shown that there is similar infection in a number of green species of *Aneura*. In this genus we have *Aneura palmata* which is usually uninfected, *A. pinguis*, *A. multiflora*, *A. sinuata*, and *A. maxima*, which are green and infected, and the chlorophyll-free varieties of *A. pinguis* which are saprophytic and infected.

The septate hyphae pass through the rhizoids and infect the cell layer next below the outermost. The fungus may later spread to several cell layers, and in the bulky forms, such as the tropical *A. maxima* and large specimens of *A. pinguis*, the fungi may occupy all cell layers of the central region of the thallus except the upper and lower epidermis. The wings of the thalli are usually free from fungus, as also are the growing-points. In normal individuals the reproductive organs and sporogonia also remain uninfected. Malmborg has, however, reported hyphae in the sporogonia of *Cryptothallus*.

The fungus forms hyphal coils in the cells, and digestion of these coils occurs in much the same manner as has been described for many orchids. This kind of digestion of hyphal clumps has been termed 'tolypophagy' by Burgeff.

M. Stahl sterilized the thalli of *Aneura* with hypochlorite and observed that the endophyte grew out of the cells on to culture media. In this way a series of fungi were isolated from *A. pinguis*. Some of these were determined as species of *Rhizoctonia*, and one proved to be a clamp-bearing Basidiomycete. Stahl compared the forms of *Rhizoctonia* with those which have been isolated from orchids and classified by Burgeff. *Rhizoctonia aneurae pinguis* '*A*' was attributed to *R. asclerotica*, *R. a. pinguis* '*E*' to *R. repens*, and *R. a. pinguis* '*H*' to *R. subtilis*—all three being species which are capable of forming mycorrhizas with orchids. The determination of Stahl's strains was described with some reserve, but at least they bore a structural and cultural similarity to orchid fungi.

M. Stahl succeeded in producing mycorrhizas in *Aneura pinguis* on *Polypodium* fibre by inoculating with *R. a. pinguis* '*A*' and *R. a. pinguis* '*H*'. These two fungi grew well on the substrate, the latter especially growing fast and permeating the fibre very rapidly. The infection caused by *R. a. pinguis* '*H*' was also very extensive, being more so than that found under natural conditions; but in both, infection resulted in the expected structures. *R. a. pinguis* '*H*' was reisolated from the artificially infected thalli and taken again into culture, thus demonstrating the reliability of the technique.

Stahl observed in these cases that infected plants appeared to grow faster than uninfected plants.

A further strain, *R. a. pinguis* '*F*', also infected the rhizoids of *Aneura* in pure culture but did not penetrate farther. The other isolates were ineffective.

The infection of these Bryophytes is sufficiently similar to that in orchids for working hypotheses about the bryophyte infection to be based on what is known of orchid mycorrhizas.

(2) *Gentianaceae*

In the Gentianaceae there are a few species, such as the aquatic *Menyanthes trifoliata*, which have been reported to lack mycorrhizas (a condition not uncommon in plants of very wet places). Most of the other genera and species form mycorrhizas, and some of them are complete saprophytes. Most mycorrhizal forms have small seeds with little food reserve in them, and which have proved to be very difficult to germinate in artificial conditions. The seeds, in germination tests, may show some similarities of behaviour to orchid seeds, such as 'patchy' germination—a behaviour ascribed to vitamin deficiency in small seeds. The adult plants of many species are usually uncommon, being restricted edaphically to peculiar sites.

Although not much is known about the endophytes of these plants and almost nothing about the physiology of their absorbing organs, there are some indications that their mycorrhizal infection resembles that of orchids. Neumann (1934) has given the most recent account of them and has confirmed the older descriptions of green and saprophytic forms.

Neumann described the roots of twelve species of European *Gentiana* and especially of *G. clusii*. The roots were infected by a septate fungus which occupied one or more layers in the cortex. The hyphae in the cells formed coils which later were digested by the host. The apical growing-points remained uninfected, and frequently the distribution of infected regions in the cortex of the root indicated that repeated local infection took place from the soil.

The septation of the hyphae, the formation of coils, and tolypophagous digestion, are all reminiscent of orchid mycorrhizas. Neumann, in comparing this kind of mycorrhiza with others, stated that it was above all like that of orchids.

Neumann succeeded in isolating two fungi from *G. clusii* and three from *G. walugewi*, of which one of the former was a *Rhizoctonia* similar to some orchid endophytes. She was unable to accomplish successful synthesis.

Both Neumann, using *G. pneumonanthe*, and Favarger (1953), using many species of *Gentiana*, have succeeded in bringing about asymbiotic seed germination. Favager used a cold treatment to activate the seeds. Hence, claims that the seeds of this genus are obligately dependent upon fungi for germination, are of similar value to the like claims made many years ago for orchid seeds.

The similarities of mycorrhizal habit and endophytic fungi in Gentianaceae and Orchidaceae, and the presence of saprophytes in both families, all suggest a likeness between the mycorrhizas of Gentianaceae and Orchidaceae which can only be verified by further study.

(3) *Saprophytes*

Saphrophytic plants are found in the Burmanniaceae, the Triuridaceae, and amongst the prothalli of pteridophytes. All of these are known to be mycorrhizal. In the Burmanniaceae and the pteridophytes, some species have been described as containing septate hyphae. From some of these, species of *Rhizoctonia* have been isolated, and were it not for the warnings which the distribution of *Rhizoctonia* gives us, it might be attractive to suggest that their mycorrhizas had similarities with those of the Orchidaceae.

Many earlier observers who worked upon tropical saprophytic Burmanniaceae described their endophytic fungi as septate, and their accounts seem to indicate some resemblance to those of the Orchidaceae. Van der Pijl (1934), who has reviewed some of this work, found aseptate

Phycomycetes in both the saprophytic and green Burmanniaceae with which he worked.

Similarly, *Psilotum* gametophytes have been reported to contain both septate and aseptate fungi. Bierhorst (1953) had the opportunity of seeing many specimens, and has shown that two endophytes may indeed be present. One without septa is consistently found in the cells, and a second, septate, fungus is present less frequently.

Although it may be tempting to assume that saprophytic behaviour is always dependent upon infection by fungi like those of orchids, these examples show that this is certainly not the case. If the results of anatomical work on saprophytic groups is reliable, we must accept it as true that the chlorophyll-less habit is not solely correlated with one particular type of fungal infection but with at least two, an orchid-like infection and a phycomycetous infection and perhaps with others. It is solely the observation that saprophytism only occurs where mycorrhizal infection also occurs, that really has given rise to the view that the two are *always* causally related.

Other Types of Mycorrhiza with Septate Endophytes

(1) *Bryophyta*

Although *Lophozia excisa* seems to have an endotrophic infection with septate hyphae not very unlike that of *Aneura*, other species of *Lophozia* appear to be very different in this respect. Long ago, Němec (1899, 1904) described fungal infection in *Calypogaea trichomanes*. Here, septate hyphae enter the host through the rhizoids and pass up to the cells of the thallus. In the swollen rhizoid-base a kind of pseudoparenchyma may be produced against the wall adjacent to the main body of the thallus. This pseudoparenchyma may be quite extensive and can even fill the basal swelling of the rhizoid. Haustorial processes are given off from its surface, penetrating upwards as simple pegs into the adjacent thallial cell. There may be slight variations in the extent of the pseudoparenchyma and in other details, but an infection of essentially this kind has now been described by M. Stahl (1949) in *Lophozia lycopodioides*, *L. ventricosa*, *Haplozia sphaerocarpa*, and *Diplophyllum albicans*. Stahl also described a somewhat similar infection in *Mylia taylori*, *Gymnomitrium concinnatum*, and *Anthelia juratzkiana*. In these last three, the pseudoparenchyma was less well developed than in the others, and long haustoria penetrated the thallial cells to form loose aggregates. Later on these haustorial aggregates degenerated to form an undifferentiated mass, which seemed to indicate digestive activity by the host.

From this kind of mycorrhiza, M. Stahl isolated a slow-growing dark-coloured mycelium. She gives no account of experimental synthesis of the association. Němec thought that this type of infection was due to one of the Pezizaceae, *Mollisia jungermanniae*.

There are no other known examples of infection resembling this remarkable type which seems, according to M. Stahl, to be quite common in leafy liverworts. Without further work it is impossible to suggest anything worthwhile about the physiology of the association.

(2) Lobelia gibbosa *and* L. dentata

An interesting type of infection in which there is no intracellular penetration has been described by Fraser (1931) in two Australian species of *Lobelia. L. gibbosa* is an erect, unbranched, somewhat fleshy herb which occurs in every Australian state. It has long cauline leaves which may be reddish owing to anthocyanins which they contain. It possesses great powers of drought resistance and may wither, except for the stalk of the inflorescence, and yet open functional flowers. *L. dentata* is a branched herb with a large spreading root system of fairly coarse roots, which lack root-hairs. In rare instances it lacks chlorophyll and appears to grow saprophytically. The seeds of both species are small. Those of *L. dentata* average about 10 μg. in weight, which is the size of a large orchid seed, and contain a small, little-differentiated embryo set in endosperm. Fraser was unable to germinate the seeds of either species in the laboratory.

Seedling stages of *L. dentata* were found in natural stands. Early in life the young plantlet was infected. It consisted of a swollen hypocotyl bearing very small cotyledons, a long chlorophyll-free shoot with scale-leaves, and a single large smooth root. Rhizomorphs of the infecting fungi attack the hypocotyl, forming locally a hyphal mat over the surface. The cells of the cortex are transversely much elongated and the hyphae penetrate between them. An extensive saprophytic growth occurs to produce a plant that is many times greater in weight than the seed.

The root system becomes extensive in well-grown plants, and the fungus always remains intercellular. Fraser observed, however, that the roots of plants grown in good conditions in the greenhouse were little infected—a point reminiscent of mycorrhizas in forest trees. Actively growing roots of young plants in natural conditions were heavily infected and showed different conditions along their lengths. Less than a centimetre from the apex, hyphae in the cell-walls and intercellular spaces were relatively sparse and seemed to be growing towards the root apex, keeping pace with the growth of the root. Behind this zone there was a second zone, of intense development of fungus in the cortex and with the hyphae containing oil. Farther back still, the quantity of oil stored in the hyphae increased and was particularly abundant in the middle cortex. Behind these active fungal zones, a progressive collapse of the hyphae took place. At first the hyphae in the inner cortex became depleted of oil, and eventually they were crushed as the host cells simultaneously increased in size. This process continued until all the hyphal reserves were depleted, and finally living fungus was to be found as slender hyphae only in the outer cortex. In the oldest parts of the roots, where fungal disorganization was most advanced, the latex of the host was plentiful and stained dark with osmic acid.

Fraser believed that these annual plants depended in their early growth upon infection: the rhizomorphs attacked the tissues, their activities were controlled and thereafter, as they collapsed, organic matter in the form of oils was made available for growth of the host.

This type of infection does not seem to occur elsewhere. Clearly, to be certain of the physiological conclusions reached, a direct analysis of the tissues would have to be made and something of the physiology of the fungus would have to be learned.

(3) Helianthemum chamaecistus *and other Cistaceae*

Boursnell (1950) has described the results of some of her investigations into fungal infection of the Cistaceae. The interest of this work lies particularly in the fact that the infection is described as an example of cyclic symbiosis. The root-infecting fungus appears to exist not only in the roots but throughout the tissues of the host plant, so that it becomes established on the seed-coat and seems to be matrilinearly transmitted by this means.

Seed-borne pathological infections are of course well known and have been mentioned in Chapter II. *Lolium* is a particular example of a genus in which a symptomless stem-infection is transmitted by seed. There was a time, indeed, when it was believed that the stem infection of *Lolium* might be a continuation of the mycorrhizal infection of the roots. But this suspected case of cyclic mycorrhizal infection was shown to be false, and the root infection was found to be caused by aseptate phycomycetous fungi while the stem and seed infection was by sterile septate fungi. It has now been shown that a number of species of fungus, all septate, may inhabit the stem tissue or seeds of species of *Lolium*.

Cyclic mycorrhizal symbiosis was also claimed for many Ericaceae; but recent work has failed to confirm this, as was explained in an earlier chapter. Hence there are now no other fully credible examples of this condition known; that described for the Cistaceae appears to be unique.

As described by Boursnell (1950), the root systems of *Helianthemum chamaecistus* are differentiated into long and short roots in a manner somewhat reminiscent of *Arbutus* roots. The short roots become somewhat swollen and tuberous as they develop, and have in their tissues a different distribution of fungal hyphae from that in the long roots. Both kinds of root are inhabited by septate hyphae, but the short roots may also bear, especially near their apices, a dense surface weft or sheath of hyphae.

The long roots are of two ages, young (first-year) and old (second-year). Those in their second year have branching hyphae penetrating the cortical cells in varied frequency along their length; they are especially densely inhabited in the middle regions. The cortex is separated from the stele by a tannin-impregnated endodermis, which appears to be a barrier to the spread of the hyphae. Nevertheless, hyphal filaments seem to occur in the stelar tissues, into which they may pass below the region where tannin-impregnation is complete. The very active mycelium in the cortex forms coils within the cell lumina and, later in the year, undergoes vigorous digestion. The hyphae increase in width, their membranes become thinner, and eventually they lose their integrity and become broken up into small fragments. By autumn the empty hyphae are so disorganized that it is difficult to trace them from cell to cell.

During the course of their second season, the long roots give rise to two kinds of branches. Those which are not infected as they pass through the cortex develop as long roots, which, during their first season, are infected by hyphae growing into them from the cortex of their mother-roots. The hyphae grow through their tissues from base to apex, and remain undigested and active in the cortex until the following year. If, on the other hand, the root initial becomes directly infected as it passes through the cortex of the mother-root, a short tuberous root develops. Such roots show two zones—an apical zone with dense intercellular cortical infection having a wide ectotrophic sheath on the surface, and a proximal zone where the infection is mainly intracellular and has a narrow sheath. In the proximal zone of these tuberous roots the fungus undergoes disintegration in the cells. The tuberous roots are of one season's duration and are abscissed by cork formation at the end of the season.

As was described in the case of *Calluna*, infection of the stem is difficult to see. Boursnell describes the fungus as occurring in the pith, cortex, and phloem, and states that it is digested in the tissues. Although it appears to be active and to grow in the younger regions, it is absent from the older parts, and 'there is no apparent connexion between the mycelium in the stem and root of adult plants'. The fungus also occurs, in greatly attenuated form, in the vascular parenchyma of the leaves, and in the mesophyll (though not in the chlorophyllous tissue).

The seed, which is equipped with a mucilaginous coat formed from the outer integument, is penetrated by the fungus. However, the hyphae do not pass through the inner integument nor the tannin layer which lies within. On germination, the embryo appears to become infected from the seed-coat.

Boursnell observed that surface-sterilized seeds germinated abnormally, without the fungus. In distilled water, if previously punctured, they swelled but growth soon ceased. Little improvement was obtained by the addition of nutritive salts or sugars to the culture medium. Good germination and growth were, however, obtained by provision of thiamin, pyridoxin, or nicotinic acid; when all three of these substances we represent, excellent growth occurred.

Boursnell also described experimental results which showed that calcium oxalate deposition occurred in the cells of infected seedlings, whereas this did not occur in uninfected seedlings. She therefore believes that the fungus has the effect of protecting its host against excessive calcium absorption from the calcareous soils on which it grows.

One notes that the two physiological effects reported both arise out of activity of the fungi within the tissues. Little importance is attached to the possible absorptive function of the infected roots. The ectotrophic condition in the tuberous roots is thought to be too short-lived to be important.

This example of cyclic symbiosis is very reminiscent of the description of the Ericaceae given by M. C. Rayner. The figures of Boursnell (1950) do not show in sufficient detail all of the points concerning the distribution of the fungus which she describes, so that it is difficult to make a full assessment of the case. Without wishing to be too critical it may be suggested that, as the other described cases of cyclic infection are so open to doubt, a very thorough extension of Boursnell's work should be made. The cyclic symbiosis aside, the condition of infection of the roots is of great interest because of obvious similarities with *Arbutus* in particular.

Conclusion

These few examples are enough to show the range of form and structure of mycorrhizal organs inhabited by septate hyphae. They serve to stress the inadvisability of generalizing too widely about the physiological and ecological properties of these kinds of fungal infection.

XIII

MYCORRHIZAS CAUSED BY ASEPTATE MYCELIA

THE mycorrhizal infections caused by fungi with aseptate mycelia have sometimes been called phycomycetous mycorrhizas and, sometimes, vesicular-arbuscular mycorrhizas. The first name is derived from the probability that the causal organisms are Phycomycetes. The second is given because two kinds of organ, vesicles and arbuscules, are often formed within the infected tissues.

In these infections,* which are by far the most frequent of all types of mycorrhiza, the infected organs do not always show such marked morphological peculiarities as are observed in other mycorrhizas. Some of them may look superficially like uninfected roots, be quite undistorted, and bear root-hairs. Others may exhibit specialized growth-forms and be deficient in root-hairs. So many different plants, of so many diverse phyla, families, and genera, are infected by phycomycetous endophytes, that the list of those reported to be free from infection is shorter than the list of those reported to be infected. Of course, as yet, only a small part of the plant kingdom has been investigated, but up to now it has proved more usual to find this kind of infection in plants not otherwise infected by mycorrhizas, than to find an absence of infection. B. Boullard (1956), in his review of works on endotrophic mycorrhizas, has given the families Juncaceae, Cyperaceae, and Cruciferae, as groups in which mycorrhizas are still reported to be absent. I feel, and I think Professor Boullard would agree, that even in these families some species or genera may be found to form mycorrhizas.

On the whole, aquatics and plants of very wet places are most likely to be non-mycorrhizal, although plants which grow in areas of periodic inundation, such as salt marshes, are often infected (Boullard, 1958). Some liverworts and many mosses are usually non-mycorrhizal. The pteridophytes are mostly infected, but not the aquatic ferns and lycopods. In the Equisatales, mycorrhiza is not well developed and its intensity varies with habitat (Boullard, 1957). The Gymnosperms comprise a mycorrhizal group and they all have either ectotrophic or vesicular-arbuscular phycomycete mycorrhiza.

In the present chapter, the kinds of mycorrhizas to be described

* *See* Appendix §9.

comprise a group of similar infections, present throughout the living plant kingdom and also described in fossils. It is highly regrettable that so little is known about them from a functional standpoint. As Dr. T. H. Nicolson wrote, in his excellent account of the mycotropic habit in grasses (1955), 'If the . . . fungi causing these infections do aid the nutrition of the invaded plant, even to the smallest degree, the importance of the phenomenon cannot be escaped in view of its world-wide distribution in the Gramineae and in other families of plants'.

The history of the discovery of phycomycetous mycorrhizas, accounts of controversies about the function of the mycelial organs, and lists of the known hosts, have been given by many previous writers (e.g. M. C. Rayner, 1927; E. J. Butler, 1939; A. P. Kelley, 1950). Here, after a general description to show the constancy and degree of variation of structure in this type of mycorrhiza, the very important recent developments in their study will be discussed.

General Description

Within the tissues of host plants infected with phycomycetous mycorrhizas, aseptate hyphae, often of relatively large diameter, are observed. These pass from cell to cell, or between the cells, of unspecialized parenchymatous tissue such as the cortex of the root. They do not enter meristems, stelar tissues, chlorophyllous tissues, nor other specialized systems of healthy organs. Within the cells thay may form coils of one or many turns which in extreme cases fill the lumina. In other cells, coils may be absent. Branches of the main hyphal systems, usually of smaller diameter, pass into or are formed in the cells. They grow either to produce simple haustoria, or, more frequently, to produce complex hyphal branch-systems, like small bushes, which are called 'arbuscules' (Plate 9, Fig. 20). The branching of the arbuscule is often approximately dichotomous, and the apices of the branches are often narrow and acute. These branches may become swollen at their ends into round fungal bodies, often called 'sporangioles' (by analogy with the sporangioles of Mucorales) or 'ptyosomes'. The sporangioles may become free within the cells, but degenerate there. Intercalary or apical swellings are also formed upon the main hyphae. These 'vesicles' (Plate 10, Figs. 21, 22, 23) are often very large and thick-walled, so that they may distort the cells or intercellular spaces in which they develop.

The hyphae are multinucleate, with the nuclei usually scattered but sometimes appearing as if they were arranged in pairs. The vesicles are also usually multinucleate, and also sometimes heavily charged with food reserves in the form of fat or oil.

In the forms of infection where complex arbuscules are developed, the whole arbuscule-sporangiole complex may disintegrate in the cells. This has been called by H. Burgeff 'thamniscophagy'—an uncomfortable word describing the uncomfortable concept of digestion of little bushes.

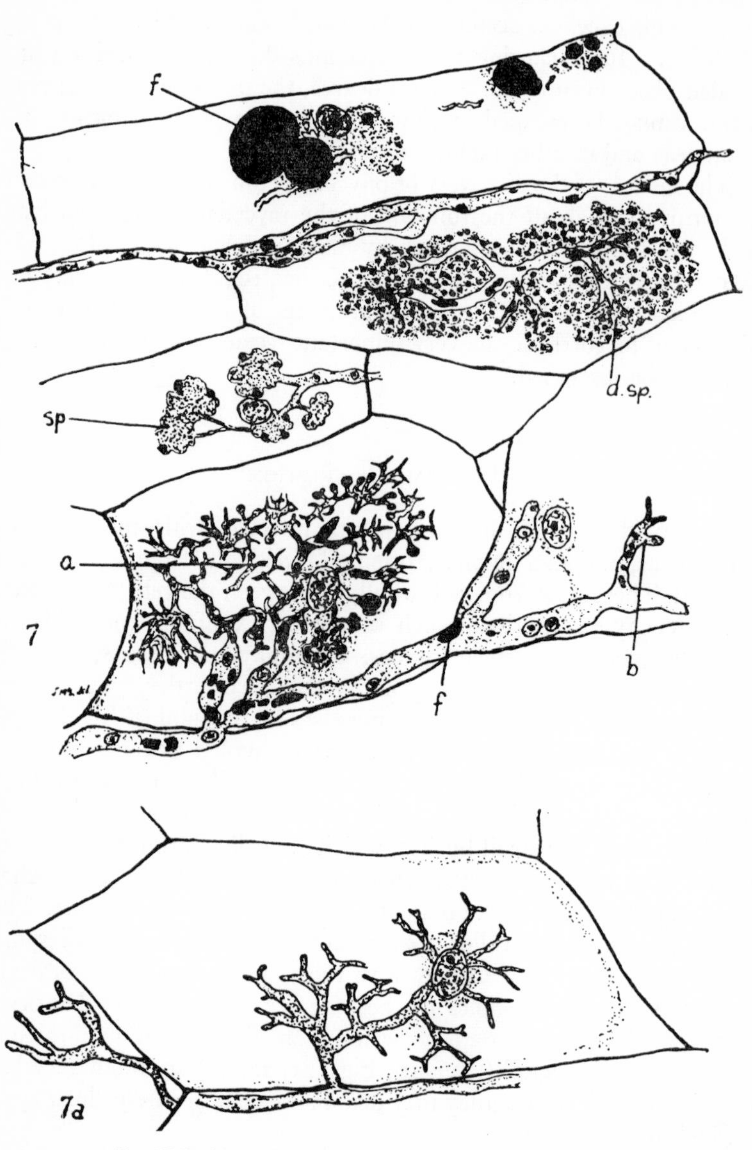

After McLuckie & Burges (1932). *By permission of Proc. Linn. Soc. N.S.W.*

FIG. 20.—Parts of longitudinal sections of the root of *Eriostemon crowei* showing arbuscules in various stages of development.

In those forms without arbuscules, or without arbuscules for much of the year, such as are found in Araliaceae and Cornaceae (Boullard, 1953, 1953*a*), in some pteridophytes (Burgeff, 1938), and in some liverworts

(M. Stahl, 1949), hyphal coils may be digested. This process is called by Burgeff 'tolypophagy'. Other, less common digestive processes will be described below.

The detailed arrangement of the fungus in the tissues may be different in different species of plants. From taxonomic considerations it does not

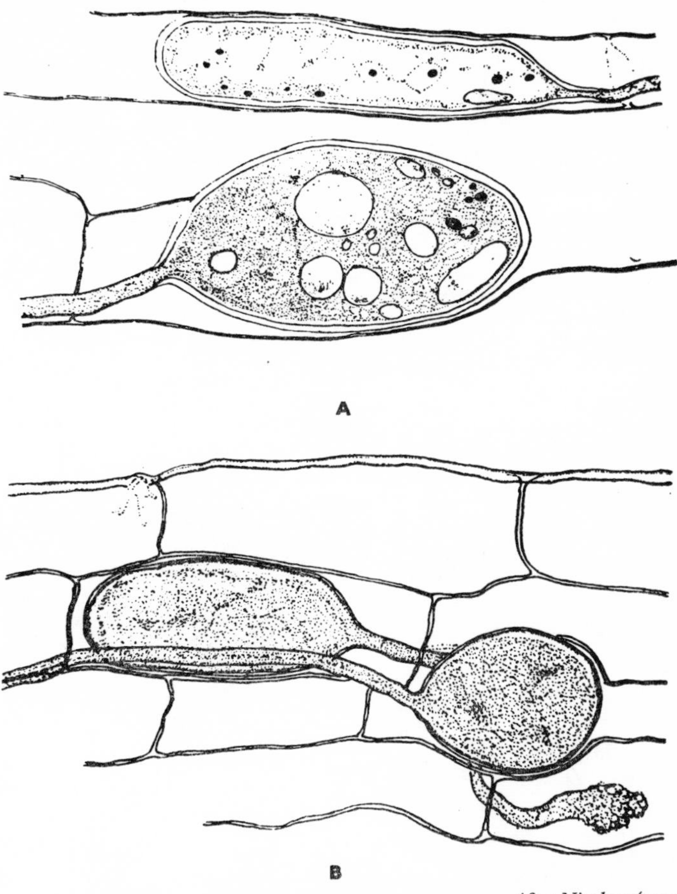

After Nicolson (1955).

FIG. 21.—(A) Intracellular, and (B) intercellular, hyphae and vesicles in *Avena sativa*.

seem to be possible to predict which kind of arrangement will be present. Gallaud (1905), who described the presence of vesicular-arbuscular mycorrhizas in many angiosperms, classified the kinds of fungal behaviour into two main classes.

The first was called the *Arum* type, because *Arum maculatum* typifies it. The main hyphae enter the cells of the root surface but soon become mainly intercellular. They give rise to branches which penetrate the cells to form relatively simple arbuscules that are scattered about the infected zone—not localized into hyphal and digestion layers.

In the second or *Paris* type, named after *Paris quadrifolia*, the mycelium is always intracellular. The arbuscules are very complex, and are localized in definite cell-layers within the tissues where digestion occurs.

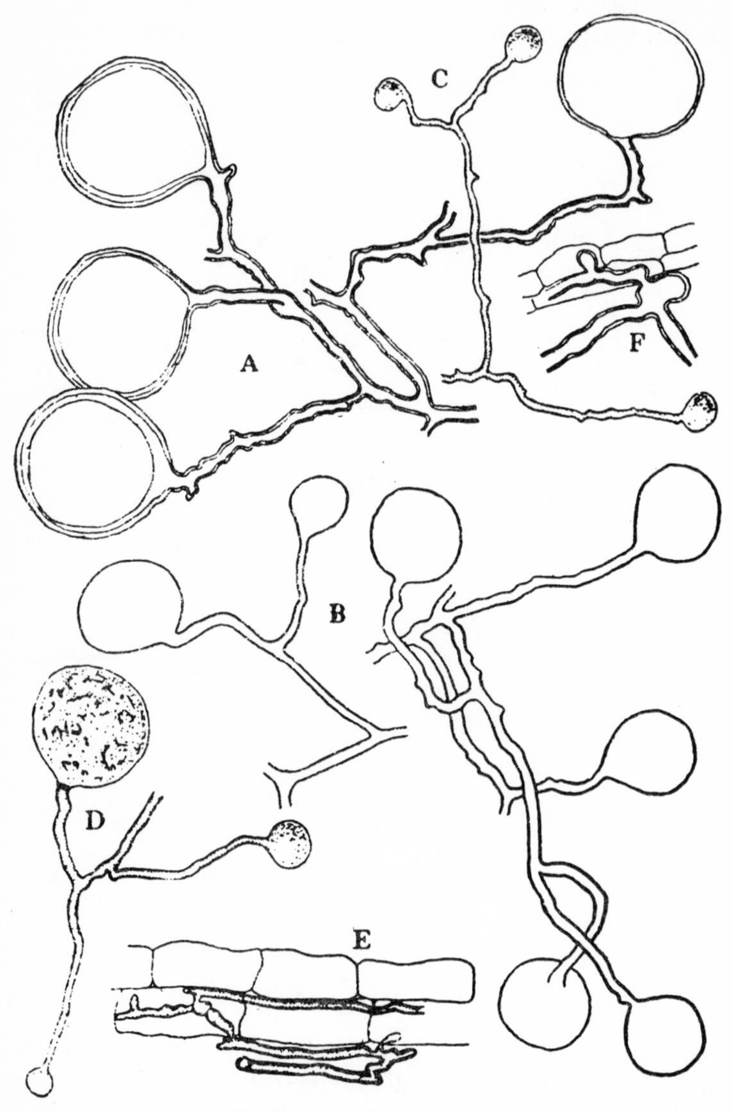

After Butler (1939). *By permission Trans. Brit. Mycol. Soc.*

FIG. 22.—Extramatrical hyphae in cotton (A, B, C and D). Note the angular projections on the hyphae. E and F: penetration of root tissue in cotton and *Abutilon* respectively.

Later descriptions of other digestive patterns have complicated the simple classification of Gallaud. Amongst those infections where arbuscules are found, M. Stahl (1949) has described in some liverworts a

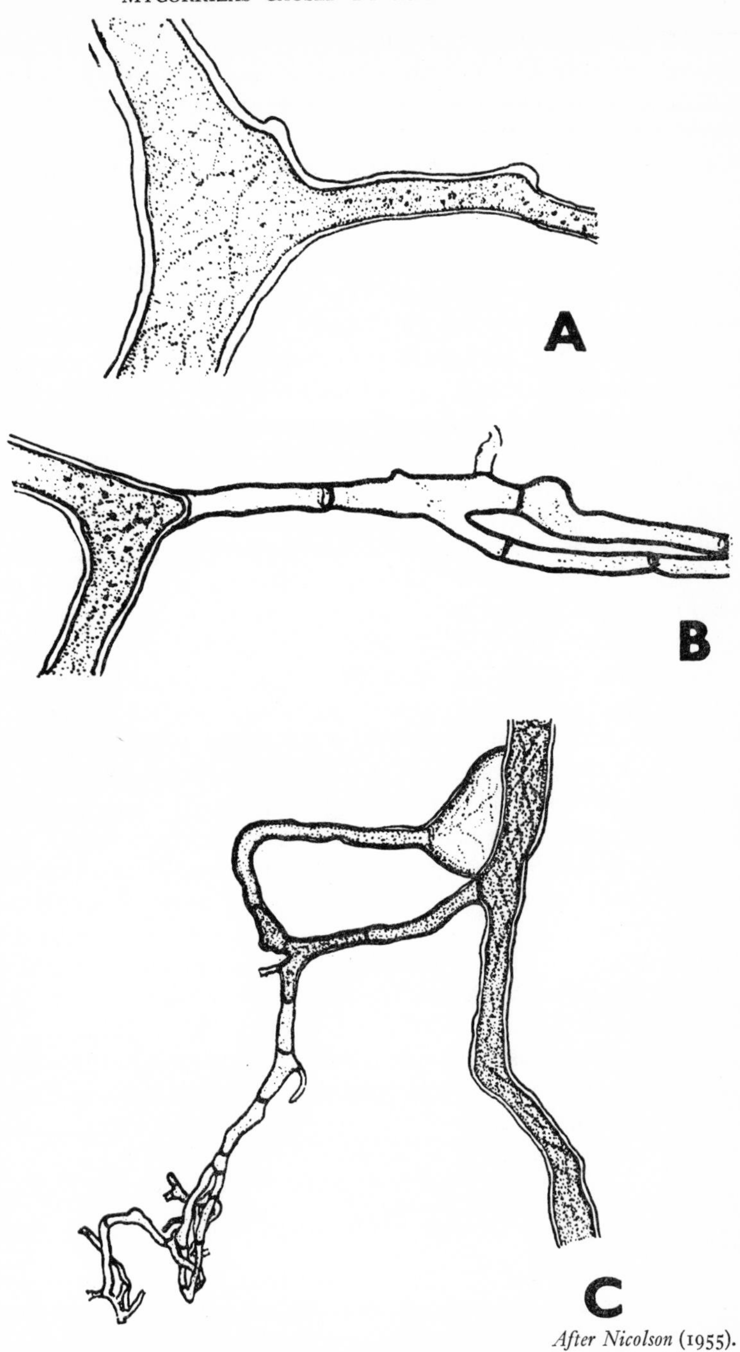

After Nicolson (1955).

FIG. 23.—Extramatrical hyphae in Graminae. A. Thin-walled hypha in direct protoplasmic connexions with main hypha in *Festuca rubra* var. *arenaria*. B. Septate thin-walled hypha in *Dactylis glomerata*. C. Septate thin-walled hypha bearing a vesicle. At the base the thin-walled hypha is filled with protoplasm and the basal septum is newly formed.

digestive pattern where arbuscules and vesicles become disintegrated together; this is called 'thamniscophysalidophagy', a word which I shall use as rarely as possible! In those infections where no arbuscules are found, tolypophagy is usually the form of digestion; but sometimes the haustorial hyphae do not suffer disintegration and, on the assumption that an exchange of substances in solution takes place, Burgeff has given the name of 'chylophagy' to the supposed nutritional mechanism.

In phycomycetous mycorrhizas there is rarely any sort of sheath of fungus formed on the root surface.* The nearest approach to a sheath is a loose external weft of hyphae which lies in the soil and is connected with the internal mycelium by connexions through the epidermal cells of the host. The degree of development of the extramatrical mycelium is very varied, and some observers have denied its existence.

The nature of the fungi involved in phycomycetous mycorrhizas has been the subject of much discussion, which has been confounded by the frequent presence of sterile septate mycelia on or in the cells of the roots—in addition to an aseptate organism. As has been noted in previous chapters, many sterile septate fungi are widespread inhabitants of roots and grow readily in culture. Setting these aside, two well-known genera have for a long time been considered as possible endophytes: *Endogone* and *Pythium*. B. Peyronel many years ago observed that fruiting-bodies of *Endogone* were attached to many mycorrhizal roots, and more recently L. E. Hawker has revived earlier views by repeatedly isolating species of *Pythium* from the roots of *Allium* and other mycorrhizal species. Barrett (1961) isolated a number of slow-growing sterile forms by special methods, and revived for them the name *Rhizophagus* which was originally given by P. A. Dangeard to the endophyte of *Populus* roots. To avoid the complexity of these names when there is no taxonomic certainty, and to avoid the term mycorrhizal fungus when the case is not proved, Boullard has (1957) used the term 'rhizomycete' for all fungi isolated from or observed in endotrophic mycorrhizas. We may have cause to use it!

In the course of this chapter a few examples of phycomycetous mycorrhiza will be considered first; these will be grouped mainly taxonomically. Then the fungi will be further discussed. Finally, in Chapter XIV, the physiological effect of this type of mycorrhizal infection on the growth of the host plant will be considered.

Fossil Mycorrhizas

Mention of some of the fossils which have been considered to be phycomycetous mycorrhizas will help to emphasize the widespread nature of such infections. If they are acceptable as genuine, then it must surely follow from their existence for millions of years that this type of mycorrhiza is not only of no selective disadvantage but very probably generally of selective advantage. E. J. Butler (1939) and A. P. Kelley

* *See* Appendix §9.

(1950) both give accounts of such fossils. Kelley emphasizes the difficulties of interpretation and advises caution especially where hyphae and vesicles, reminiscent of present-day mycorrhizas, are found in stelar tissue or in other parts of plant organs where mycorrhizal fungi would not be expected.*

The earliest record geologically is of *Palaeomyces asteroxyli*, found in the inner cortex of the rhizomes of *Asteroxylon mackiei* from the Devonian rocks of Scotland. It formed aseptate hyphae which gave rise to fine branches and terminal vesicles in large numbers in the rhizomes, and so is reminiscent of the mycorrhizas of the modern *Psilotum* and *Tmesipteris*. In the same chert beds, other fungi of similar types are found associated with what appear to be decaying stems of *Rhynia* (Kidston & Lang, 1921).

From the Carboniferous, Weiss (1904) described, under the name of *Mycorrhizinium*, roots or rhizomes showing obvious digestion phases, and Seward (1898) gives examples of vesicle-bearing aseptate hyphae in the tissues of *Lepidodendron*. The best example is, however, that of *Amyelon radicans*, the rootlet of *Cordaites*, which was described by T. G. B. Osborn (1909) from Lancashire coal measures. Here the roots bore short coralloid branches which were perhaps reminiscent of those of *Taxus* and other gymnosperms. In these roots, large but unbranched, almost aseptate hyphae had exploited the cortex and had given rise to clumps in the inner cortical region. In the older cortex of roots where periderm initiation was apparent, vesicles borne on the large hyphae were found. Later, Halket (1930) confirmed Osborn's findings (except that he found more frequent septa), showing that the infection was of common occurrence and that structures like arbuscules were to be seen.

Similar aseptate hyphae and vesicles have been recorded in fossil lycopods, ferns (including coenopterids), and gymnosperms, but again in many cases the preparations are not immediately convincing—sometimes because of the absence of any clear connexions between hyphae and vesicles, and sometimes because of the position of the fungus in the organs of the plant.

An interesting paper by Dowding (1959) considered the modern ecology of *Endogone*, a known mycorrhiza-former with many plants. She showed that *Endogone* is a common endophyte of living and subfossil roots in the recent and glacial soils of Alberta, that it may be found as clusters of spores and mycelium attached to plant remains in the blue-clay layer below peat-bogs, that it is found as mycelium and spores among newly-dead plant remains and living roots at the surface of swamps and bogs, and that it is associated with living mycorrhizal roots and may, to a restricted degree, be made to grow from them in water culture and to form external vesicles or chlamydospores. Her description of the ecology of *Endogone* shows it to be intrusive in newly-formed plant detritus, and allows one to conclude that many of the supposed fossil records of

* *See* Appendix §9.

phycomycetous mycorrhiza from the Devonian onwards may be accepted as reliable; indeed the presence of aseptate vesicular fungi in the associated plant detritus is to be expected.

Bryophyta

In the liverworts there is a large range of types of mycorrhiza. Those with septate hyphae have already been mentioned, while M. Stahl (1949) has described others with phycomycetous endophytes in which the digestive pattern may be any one of several different kinds.

Amongst the thallose liverworts, M. Stahl described many which show a normal thamniscophagy. In some of these the degree of infection varied greatly with cultural conditions, while in others infection was always intense except under special artificial conditions. She has classified the liverworts with this kind of hyphal digestion into several categories, as follows.

Pellia, *Monoclea*, and *Lunularia*, have infection of very varied intensity. Usually the penetrating hyphae pass through the rhizoids and exploit only the lowest layer of the thallus. Occasionally they may spread throughout all the tissues except the upper epidermis and the green cells, which are avoided even when infection is otherwise complete. In these plants there is no distinction between layers of cells with arbuscules and layers with coils. The digestion cells are mixed in the same zone as the cells with living hyphae. Vesicles, too, occur seemingly at random in the cells of the infected region.

The genera *Conocephalum* (*Fegatella*), *Preissia*, and *Marchantia*, with considerable tissue-differentiation in the thallus, have a much more localized zone of infection. The fungus occupies a zone above the 'conducting strands' and below the assimilating region, which zone in cross-section appears as a semicircular area in the middle of the thallus. The wings of the thallus, like the assimilating tissue, are not infected. The tissues containing fungi are divided into two zones. Below, one or two cell layers are occupied by wide, deep-staining hyphae which produce vesicles in abundance, and above are one or more layers in which arbuscules are formed. These arbuscules undergo typical digestion with sporangiole formation.

Anthoceros differs from both the above types in having intercellular hyphae which give rise to intracellular arbuscules in the cells. Nevertheless, in the experiments of M. Stahl (1949) which will be discussed later, the endophyte of one of its species was capable of forming a mycorrhizal association with a species of *Marchantia*.

The fungi described as mycorrhizal with the bryophytes include a species of *Pythium* (Carré & Harrison, 1961) which was brought into culture from, and back-inoculated into, *Conocephalum conicum*; and also a slow-growing fungus, resembling the *Endogone* of many workers, with which M. Stahl (1949) obtained successful cross-inoculations between species of a number of genera.

Digestion in most species is thamniscophagous or tolypophagous, but a strikingly different type of infection in the liverworts was that described by M. Stahl in the genera *Symphyogyna*, *Pallavicinia*, *Treubia*, and *Calobryum*. In the digestive cells of these, both arbuscules and vesicles were formed. Both of these types of organ were involved in the digestion process, which resulted in the release of their contents into the host. This type of digestion was called by the twenty-three letter name given on p. 248.

Pteridophyta

The subject of mycorrhiza in pteridophytes was reviewed by Burgeff (1938), who pointed out that phycomycetous mycorrhizal infections were of general occurrence in the group. We owe, however, to Boullard (1957) a valuable monographic study so comprehensive that we know more of the structure of the phycomycetous mycorrhizas in the pteridophytes in an organized way than we do of those in other phyla. Boullard considers previous work together with his own extensive investigations, and concludes that symbiosis is of general occurrence in all families of pteridophytes except Marsileaceae, Salviniaceae, and Isoëtaceae, although it may be variable or sporadic in some of the others. Table XLVII gives a summary of his conclusions. As will be seen, Pteridophyta with massive or saprophytic gametophytes are typically heavily infected at that stage, but the constancy of infection in laminate or filamentous green gametophytes is much less. In the sporophyte stage certain groups, amongst them some that are often considered primitive, are heavily infected with fungi. All the known forms of digestion are recorded from the group, but thamniscophagy and tolypophagy are commonest.

The difficulties of germinating the spores and growing the sporelings of the species with massive gametophytes, have been well known for a long time. Indeed the prothalli of many of them have been incompletely studies for that reason. Since some, such as species of *Ophioglossum*, are capable of regenerating the sporophyte from adventitious buds on roots in a manner reminiscent of some orchids, the gametophytes have sometimes been considered to be produced only rarely in nature. Boullard (1958*a*) succeeded in obtaining large numbers of the gametophytes of *Ophioglossum vulgatum* from natural sites, and made a special study of them as few others have been able to do. Later (1963*b*) he prepared a comparative account of gametophytes in Ophioglossaceae, showing that there is a considerable degree of variability in the extent to which they may become partially autotrophic. Prothalli of *Ophioglossum vulgatum*, *O. lusitanicum*, *O. pedunculosum*, and *Botrychium ternatum*, may vary in form and in the extent of chlorophyll development, although all are heavily mycotrophic. In this variability they not only resemble those of species of

TABLE XLVII

MYCORRHIZAL INFECTION IN THE PTERIDOPHYTA (Data from Boullard, 1957)

A = abundant, P = present, C = constant, V = variable in quantity and presence.

Family or order	Habit	Gametophyte Infection	Gametophyte Digestion	Sporophyte Infection	Sporophyte Digestion
Psilotaceae	Massive, colourless, underground	A.C.	Tolypophagy	A.C.	Tolypophagy
Lycopodiaceae	(1) Massive, colourless, underground	A.C.	Varied	V	Varied
	(2) Massive, partly green	A.C.	Varied	V	Varied
Selaginellaceae	Minute	—	—	V	Thamniscophagy
Equisetaceae	Laminate, green	—	—	Rare or absent	?
Ophioglossaceae	Massive, colourless, underground	A.C.	Thamniscophagy	A.C.	Thamniscophagy
Isoëtaceae	Minute	—	—	Very rare	?
Marattiaceae	Fleshy, green	A.C.	Thamniscophagy	A.C.	Thamniscophagy
Osmundaceae	Laminate, green	P.V.	Tolypophagy	P.C.	Tolypophagy
Schizaeaceae	Filamentous or laminate, green	P.V.	Tolypophagy	A.C.	Tolypophagy
Gleicheniaceae	Laminate, green	C	Tolypophagy	A.C.	Tolypophagy
Hymenophyllaceae	Filamentous or laminate, green	P	None	C or V	Tolypophagy or ptyophagy
Cyatheaceae	Laminate, green	—	—	C	Thamniscophagy
Polypodiaceae	Laminate, green	—	—	V	Variable
Hydropteridales	Minute	—	—	—	—

Lycopodium but also, in the immunity of the rhizome of the sporophyte to infection and in the reinfection of sporophyte roots from the soil, they are reminiscent of orchids although very different fungi are involved.

Attempts to isolate mycorrhizal endophytes from pteridophytes have yielded three types. First may be mentioned the slow-growing aseptate hyphae which often form vesicles on hyphae growing into the medium, but which have not been obtained in pure culture. These resemble *Endogone* in behaviour. Then there are the rapidly-growing septate mycelia belonging to *Rhizoctonia* or other septate genera, which are likely to be casual root-infecting fungi such as were discussed in connexion with the double infection hypothesis in the last chapter. Some of these latter are fairly constant in occurrence, and Boullard has mentioned the possibility that a species of *Chaetomium* may be a constant associate of *Selaginella kraussiana*. Thirdly, Hawker and her colleagues in Bristol isolated strains of *Pythium* from the roots of *Dryopteris* (Hawker *et al.*, 1957), and their report was followed by an investigation of the endophytes of numerous pteridophytes by Hepden (1960). Many of the results of these workers on the incidence, nature, and extent, of infection are like those of Boullard, but Hepden's isolation of *Pythium* is of great interest for that genus was not mentioned by Boullard as occurring in his isolation cultures. Hepden reports, giving drawings, that back-inoculation of *Pythium* into aseptic gametophytes of *Dryopteris filix-mas* and *Phyllitis scolopendrium*, resulted in an association that was histologically similar to vesicular-arbuscular mycorrhiza.

Gymnosperms

Endotrophic mycorrhiza involving aseptate phycomycetous mycelia is the common condition in the conifers. The Abietineae (Pinaceae) are exceptional in being entirely ectotrophic, though sporadic ectotrophic mycorrhiza may occur elsewhere—as in *Juniperus communis*, where *Cenococcum* infection sometimes occurs instead of its usual endotrophic mycorrhiza (Lihnell, 1939). We do not have any coherent monographic account of endotrophic mycorrhiza in gymnosperms, so a few examples will be presented here to illustrate a condition which is widespread.

Taxus baccata is a forest tree with the roots differentiated into pioneer and feeding roots, and exhibiting a differential distribution of infection of the *Paris* type.

Prat (1926) gave a particularly clear account of the structure, mode of growth, and infection, of the root system of *T. baccata*. This tree, like many other shade-tolerant species, has roots which exploit in particular the surface layers of soil, entering the humus and the litter zones where the feeding roots are formed in great quantity. There is marked differentiation between the pioneer and feeding roots. The long roots are of two kinds—pioneers of potentially unlimited growth, and their laterals, smaller in diameter, which have a markedly seasonal growth-period. The pioneers are the permanent axes of the root system and grow continually at the rate of about 10 to 30 cm. a year, carrying the root system far out into the humus layer around the tree. Their apices have a root-cap, and differentiation behind them is never fast enough to catch up with the growth-rate, so that a long zone of maturation is present. There is a well developed root-hair zone behind the apex, and proximally the root becomes secondarily thickened, sloughing off the cortex by periderm formation. These pioneers are rarely infected by endophytic hyphae, but if they are so infected it is only locally, yielding patches of tissue in which small arbuscules, soon to be digested, are formed.

The laterals of these pioneer roots pursue, by contrast, an indirect or tortuous path through the humus, ultimately branching to form very short sublaterals. Their root-hairs are fewer than those of the pioneers and are often deformed, and their apices do not grow continually at an even rate. The whole axis may be short-lived. Some die during the first winter after formation, while others persist for a few years. These more persistent shorter roots are penetrated by fungal hyphae which pass particularly through the root-hairs and may branch to occupy almost the whole cortical tissue. In the outer layers of the cortex they do not appear to be very active but may often be seen to be moribund or empty. In the middle cortex, coils as well as arbuscules and digestion stages are formed in profusion. The innermost one or few cortical layers are free from infection. Vesicles are both intercalary and terminal upon the main hyphae; they measure about 40 μ in diameter, are surrounded by a thick wall, and appear to be multinucleate. They are especially abundant in tissues which have been infected for a long time, being rare in newly-infected tissues. In second-year roots where phellogen has begun to separate the cortical tissues, vesicles are abundant. They appear to remain alive even in old tissue which has been abscissed.

Prat described the primary penetration of the roots by hyphae entering through the root-hairs. Swellings, resembling the appressoria of parasites, are often formed against the cell-wall, and from them narrower hyphae penetrate into the lumen.

The growth and branching of the feeding roots show many points of interest. Rudiments of branches are formed in the pericycle, and the young laterals grow out through the cortex. These are isolated from the living cortex through which they pass, by dead cortical cells which surround their path of growth. The fungal hyphae do not pass through these dead cells, so that the new root is at first uninfected, but later becomes infected from the soil.

The apices of those feeding roots which grow for more than one season grow intermittently. In autumn their growth becomes slow, and differentiation behind the dividing meristem nearly catches up with cell division. When growth is renewed in spring, a new growing-point is

Photo by G. Woods

(*a*)

Neottia nidus-avis. (*a*) Whole plant showing flowering shoot and the root-mass interwoven with beech roots.

PLATE 8

Photo by G. Woods

(*b*)

Neottia nidus-avis. (*b*) Details of the root system of *Neottia*, note the shoot bud on the upward-pointing root on the extreme right.

PLATE 8

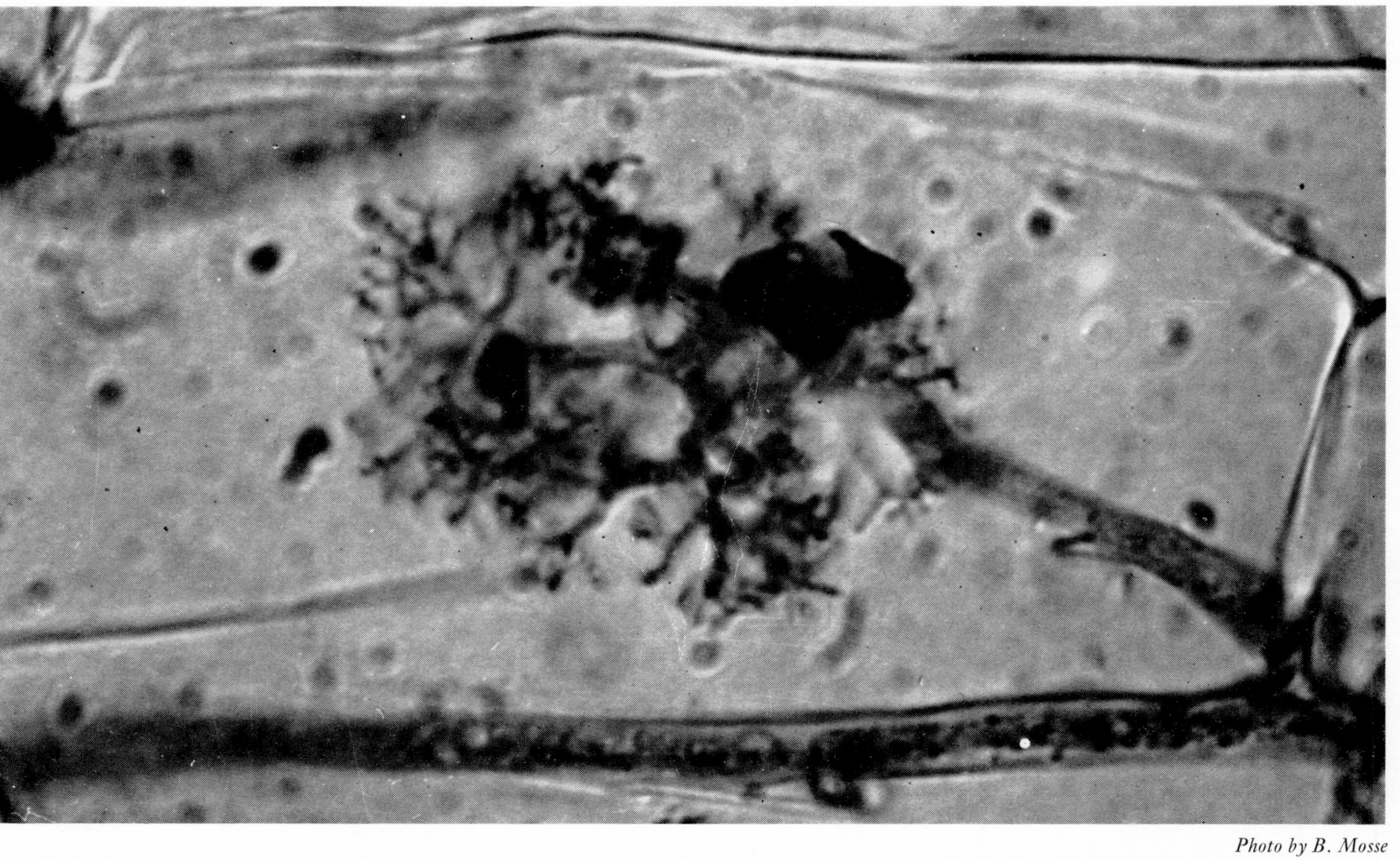

Photo by B. Mosse

Arbuscule in cell of onion.

Plate 9

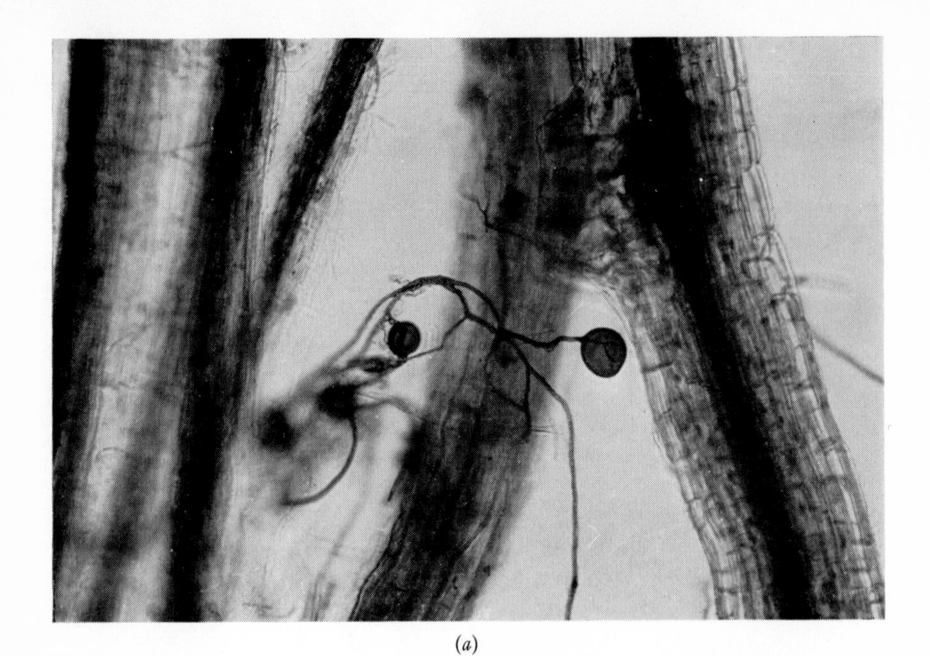

(*a*)

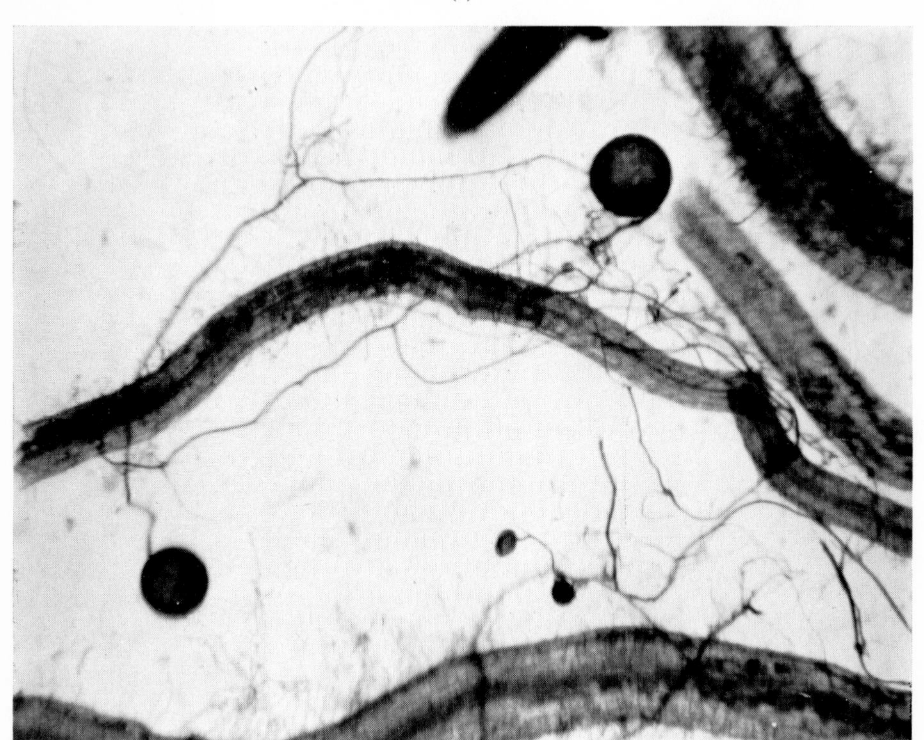

Photos by T. M. Nicolson

(*b*)

(*a*) Extramatrical mycelium of *Festuca rubra* var. *arenaria* with terminal vesicles and angular projections.
(*b*) Extramatrical mycelium of tomato showing chlamydospores and vesicles.

PLATE 10

formed endogenously near the apex, and the new growth breaks through the mature tissue in a manner analogous to the endogenous origin of branch roots. The young extension growths are also isolated from the cortex of their parent root by dead tissue, and are at first uninfected.

The fungus could be observed on the surfaces of roots and in the leaf-litter and humus around them. Here it produced, Prat observed, vesicles of a similar type to those within the tissues. The hyphae ramified irregularly, not forming any kind of aggregated strand, but the filaments, by virtue perhaps of a mucilaginous surface, were apt to appear quite large owing to the adherence to them of soil particles. Prat believed that the hyphae extended far into the soil.

Konoë (1957) described the similar localization of infection in the roots of members of the Taxodiaceae. He investigated *Cryptomeria*, *Sequoia*, *Sequoiadendron*, *Taxodium*, *Glyptostrobus*, *Cunninghamia*, *Taiwania*, *Sciadopitys*, and, most especially, *Metasequoia*. In the last-named there was a differentiation into long roots and short lateral systems which were usually racemosely branched. In the long roots few hyphae were present, but in the cortex of the short roots there was heavy infection. Hyphae were, however, absent from the exodermal layer, from the stelar tissue, and from all parts of the central cylinder. They were also absent from the meristematic and rapidly-growing region at the tips. From Konoë's description one would judge that the digestive pattern was tolypophagous, for although coils and vesicles are figured, no arbuscules are described.

The root system of *Juniperus* shows again many points of similarity. The young feeding roots are typically heavily infected. The infection mainly originates from the soil, not from the mother-root, since a layer of dead thickened cells seems to isolate the young root from the mother cortex as it passes through. Coils, vesicles, and arbuscules, are present, and thamniscophagous digestion seems to occur. In the previous chapter, D. Lihnell's work on the isolation of fungi from juniper was discussed, for it showed the presence of a multiplicity of rhizomycetes in addition to a phycomycetous mycorrhiza-former.

The root systems of *Agathis*, *Araucaria*, and the Podocarpaceae, provide examples of extreme heterorhizy. All members of the two families Araucariaceae and Podocarpaceae bear on their long roots lateral nodules which originate endogenously and are homologous with branches. These nodules, according to a recent paper by Baylis *et al.* (1963), are produced in aseptic conditions and are not the result of infection by micro-organisms.* They are nevertheless the main site of infection by aseptate mycelia, which bear vesicles and arbuscules and suffer thamniscophagous digestion. Mycelium is also present in the cells of the young long roots, but the main site of infection is the nodule. Each autumn the outer cortex of the nodule becomes thickened and lignified, but, in the spring,

* *See* Appendix §10.

renewed growth in the pericycle reforms the endodermis and cortex, which are then reinfected by hyphae from the soil—rarely from the cortex of the long root. The beaded appearance given by the long roots of these plants, which is due to alternating periods of growth and the seasonal growth of the nodules, is comparable with the phasic behaviour in *Taxus* described by Prat. It may, in a less acute form, be more widespread in plants with phycomycetous mycorrhizal infection (for instance, it has been described in *Acer* by Kessler [1966], and also in *Aesculus*).

Baylis *et al.* (1963) and Morrison & English (1967) have investigated the absorptive physiology of the mycorrhizal roots of *Agathis* and *Podocarpus*, and this will be discussed later.

Phycomycetous mycorrhizas have also been described in very many other gymnosperms including *Ginkgo*, *Cephalotaxus*, *Torreya*, *Cupressus*, *Chamaecyparis*, *Biota*, *Thuja*, *Thujopsis*, *Libocedrus*, etc.

Angiosperms

Phycomycetous mycorrhizas are widespread in the angiosperms. Some members of most families that are not infected in other ways, have been shown to be infected by aseptate mycorrhizal fungi. The number really investigated is small compared with the total number of species, but the mycotrophic habit is clearly exceedingly common. It is not yet fully known how physiologically and ecologically important this kind of association with fungal hyphae is, so that it is highly regrettable that studies such as those in the *Ecological Flora of the British Isles*, which purport to record the facts about mycorrhizal development, are so often plainly faulty. It is of great potential importance that many of the ecologically and economically important families of angiosperms such as Gramineae, Leguminosae, Palmae, and Rosaceae, to quote but a few, have a majority of or many mycotrophic members. In addition, individual crop plants such as *Hevea*, *Citrus*, cotton, tea, tobacco, oil palm, coconut palm, sugar-cane, ground-nut, avocado, apple, the cereal grasses, etc., are definitely known to be mycotrophic.

Of course it is possible to regard all these as infected by a kind of symptomless parasite which is ecologically, or in cultivation, of negligible importance: this is quite possible, because cultural conditions may be such that in rich or fertilized soils the effect of the rhizomycetes may be minimal (as it has been shown to be in ectotrophic mycorrhizas). However, the existence of whole families of saprophytic angiosperms (Burmanniaceae, Triuridaceae) with phycomycetous mycorrhiza (*see* Schmucker, 1959, for a review), should lead one to consider the ecological implications of this type of infection as worthy of study. Yet it is mainly with autotrophic angiosperms that recent work on the physiological effects of phycomycetous mycorrhiza has been performed, much of which demonstrates it to have an importance in nutrient absorption.

In this present section it is clearly impossible either to review the knowledge of phycomycetous mycorrhiza in the angiosperms in general or to consider details of more than a few examples. It may therefore be of interest first to refer to some of the ecological studies of a single author, in order to illustrate the commonness of the phycomycetous mycorrhizal condition. Boullard (1958, 1962, 1963; Boullard & Dominik, 1966) has recorded the presence of mycorrhizas in seashore and salt-marsh plants, in plants of *Pinus strobus* communities in North America, in shrubs of public parks, and in plants of semi-natural beech forests. A perusal of these papers will give a clear impression of the frequency, constancy, and types, of mycorrhizas—especially of phycomycetous mycorrhizas—in varied plants and habitats. The examples which follow are given as illustrations only, and are derived as much as possible from accounts which are readily available.

LEGUMINOSAE

Plants of the Leguminosae were investigated by F. R. Jones (1924). Although also symbiotic with *Rhizobium*, the plants showed a fungal infection of Gallaud's *Arum* type (*see* p. 245). The external hyphae were described as somewhat septate, but dead and functionless.

Jones observed crop plants growing in agricultural soil and also a few self-sown plants in other habitats. The fungus penetrates the surface cells of the root, not necessarily the root-hairs, sometimes forming appressoria as it does so. The hyphae increase in diameter within the tissues and spread up and down the root—mainly by means of hyphae which lie within the intracellular spaces, so that in transverse sections of the root they are cut across. In the outer cortical cells a few coils may be formed, but the main intracellular exploitation is in the inner cortex, where arbuscules are formed very profusely, and where digestive stages of those arbuscules are found in abundance. The very innermost cell layers of the cortex are free from hyphae.

The non-septate hyphae which enter the root are about 6 to 10 μ across, and they expand to 12 to 13 μ within the host. The thick hyphal walls often have small swellings or angular projections upon them, at the apices of which the walls are thinner. Occasionally, septa are present in the filaments within tissues, but these are formed more frequently in the extra-matrical hyphae. These two morphological features, the angular projections and occasional septa, are more common in phycomycetous endophytes than is usually realized (*see also* E. J. Butler, 1939), and a convincing explanation of their occurrence has been given by T. H. Nicolson in his work on the Gramineae which is described below. Jones believed that septa separated dead or moribund hyphae from living hyphae, and Nicolson agrees with this but has extended the explanation to the angular projections also.

The arbuscules, found only in the inner cortex, are variable in complexity, sometimes consisting of a few simple branches and sometimes of branch systems of great complexity. The vesicles occur in the outer cortex and are terminal in position on their hyphae. They are found in the intercellular cavities and are somewhat irregularly shaped, presumably expanding to fill the available space. Their dimensions are 75–150 $\mu \times$ 25–60 μ. Their walls are thick, and they are frequently separated from the hyphae which bear them by septa placed at some distance from the base of the vesicle.

Jones followed the progress of infection in plants of alfalfa and clover. In spring the roots which had survived the winter were usually heavily infected. As growth occurred, the new root was at first free from fungus but later became infected at its base. In contrast to Prat's description of *Taxus*, Jones reported that the hyphae from the infected mother-root entered the bases of the new roots of legumes and gradually exploited the new growth. When root-growth became slow, especially in times of drought, the fungal invasion extended up the roots as far as the meristem; but the faster the rate of growth, the more extensive was the region of uninfected tissue. Many of the large main roots of adult plants and the tap-roots of seedlings were, like the pioneer roots of *Taxus*, free from infection, being kept so by the onset of cork formation.

Jones made the important statement that the mycelium within the roots was not connected with the soil by living hyphae: 'No hyphae filled with protoplasm have been found connecting the mycelium in the root with that outside.' This, if it were true, would mean that the fungal hyphae could not act as organs absorbing material from the soil and translocating it to the host. Indeed, Jones was of the opinion that the fungal infection in this case was harmful and had the effect of shortening the active life of the roots.

Later work does not fully agree with Jones on these important points. Schrader (1958) observed the presence of external mycelium connected with the mycorrhizas of pea, and upon it vesicles were developed. Similarly in experimentally infected plants many hyphal connexions between the root and soil exist, at any rate in young mycorrhizas. Again, the descriptions of Schrader and of Asai (1944) associate the development of mycorrhiza with active growth of the host, and in Asai's experiments a causal relationship between infection and growth was believed to occur. Asai concluded that mycorrhizal infection was important for both the growth and development of bacterial nodules in many legumes, and his tables show that in many instances, growth stimulations were correlated with mycorrhizal development in both the presence and absence of nodules. He further concluded that mycorrhizal infection might stimulate the frequency of nodules. Lanowska (1966) also investigated the interaction of nodule activity and mycorrhizal infection. No mycorrhizas were observed in the immediate region of the bacterial nodules, and indeed they were not intensely developed during the period when nitrogen fixation appeared to be actively proceeding. Mycorrhizal infection increased in intensity late in the season, during the flowering and fruiting period of the host. The view was taken that nitrogen deficiency encouraged mycorrhizal development, since application of ammonium nitrate at a rate of 500 to 1000 mg. per 5 kg. of soil reduced infection.

Clearly these relationships of mycorrhizal intensity and nodule development in legumes merit further examination. The more general points of hyphal connexion with the soil, and the relative growth-rates of mycorrhizal and non-mycorrhizal plants, have been extensively examined in other places and will be considered later.

The frequency of mycorrhiza in the Leguminosae is shown by the fact that Asai observed it in 50 species from 30 genera—including annual and perennial herbs, lianes, shrubs, and trees. Jones observed it in species of 10 genera but found *Lupinus perennis*, *Cicer arietinium*, and a species of *Oxytropis*, to be free.

GRAMINEAE

The Gramineae are a family of great economic importance in which phycomycetous mycorrhizas are of frequent occurrence but are varied in minor points of structure. They are found both in the wild and the cultivated species. Boullard (1963*a*) gives a list of some 70 grass genera in which phycomycetous mycorrhiza has been observed in one or more of the species. In his list only about 30 species have been recorded as lacking mycorrhiza. *Glyceria* is the only genus in the list in which no species is reported by any observer to be mycotrophic. 'Pourtant, hormis les espèces des sables arides où des marais asphyxiques (pour les racines) elles sont en general richement mycorrhizées'.

Nicolson (1955, 1958, 1959, 1960, 1963) investigated the mycorrhiza of Gramineae in semi-natural and agricultural habitats in Britain. He observed that there were undoubtedly three hyphal phases: an external mycelium of considerable extent, living hyphal connexions with the root, and an active internal mycelium. The sequence of development of the hyphal organs, and the variation in their development in different species and even in different samples of the same species, are particularly interesting.

Nicolson describes the infection as characteristically of the *Paris* type, in which runner hyphae spread intracellularly from cell to cell in the outer cortex, and arbuscules were localized. In some species an *Arum* type of infection, with intercellular runner hyphae, occurred.

The hyphae, which enter the root either by root-hairs or by epidermal cells, penetrate the cell-walls by appressorium formation. Infection spreads intracellularly in all directions from the point of entry, the hyphae forming coils in the central cells. A single infection has been observed to spread over a length of 3 mm. in *Ammophila arenaria*. Septa in the hyphae are rare but, when present, cut off empty segments from active ones. In the inner cortex, arbuscules are formed either by direct penetration from the outer cortex or by runner-hyphae which pass along a cell layer and form arbuscules in each cell. These arbuscules are dichotomously branched systems of great complexity, and make the inner cortical zone very conspicuous in stained preparations. After the arbuscules have developed, vesicles are seen in the cortex—especially in the cell layers immediately surrounding the cells containing arbuscules (Fig. 21). These vesicles are terminal and usually intracellular. There is usually only one in a cell, but as many as five have been seen in a single cell. Their size is such as to fill the available space.

After the vesicles have been formed, digestion begins and the apices of the arbuscular hyphae swell to form oil-containing sporangioles. Their oil drops are released into the host-cell, where they coalesce to form large globules. As disintegration proceeds, the hyphal material shrinks and becomes structureless and the oil disappears from the cells. Nicolson has described many variations, such as the absence of hyphal coils, absence of arbuscules, presence of intercellular hyphae, and so on. These variations, which are not necessarily correlated with any particular species or habitat, seem to straddle many of the differences between types of phycomycetous infections that have been stressed by other observers.

The hyphae, which extend through the epidermal cells or through the root-hairs are connected with the external mycelium in the soil. This consists of two kinds of element (Fig. 23). The first kind is made up of the permanent hyphae of the mycelium which are thick-walled, stout, and aseptate, being 20 to 27 μ in diameter and giving rise to branches 7·5 to 10 μ thick. On the main hyphae and their primary branches, vesicles may be formed (Plate 10). These are both terminal and intercalary, and vary from 20 to 150 μ in diameter. Sometimes they are separated from the rest of the mycelium by a distant septum, in which case they are usually densely filled with contents. In many instances aggregates of vesicles, forming fairly compact bodies intertwined with hyphae, were observed.

The permanent hyphae give rise to lateral hyphae having thinner walls and smaller diameters. This second kind of external hypha is typically septate and about 3 to 5 μ in diameter. Such hyphae originate either as the ultimate branches of the main hyphae or, more abundantly, as lateral branches of them. They develop on thin-walled projections of the main hyphae and are at first non-septate, but later become septate. These fine hyphae are ephemeral elements of the hyphal system and are soon lost. The stages of their loss of activity are registered by the cutting off of empty portions by septa—until finally they fall away, leaving the characteristic angular projections that are so common on the main hyphae.

These observations explain the occurrence of the angular projections on both internal and external hyphae which have been described by Jones (1924) and E. J. Butler (1939), and are illustrated in our Fig. 22. Indeed

E. J. Butler observed that, inside the plant, fine hyphae were borne on 'angular or knob-like irregularities' on the walls of the main hyphae. The fine hyphae, he supposed, were the remains of the bases of arbuscules. From Nicolson's work it would appear that the differentiation of the external and internal mycelia is similar: both form main hyphae, lateral rhizoidal systems borne on angular projections, and, in addition, vesicles. The rhizoidal systems form the arbuscules within the tissue.

The observations of Winter (1951) on cultivated Gramineae are in interesting contrast with those of Nicolson in respect of the hyphal connexions with the soil. Winter concluded not only that intense infection

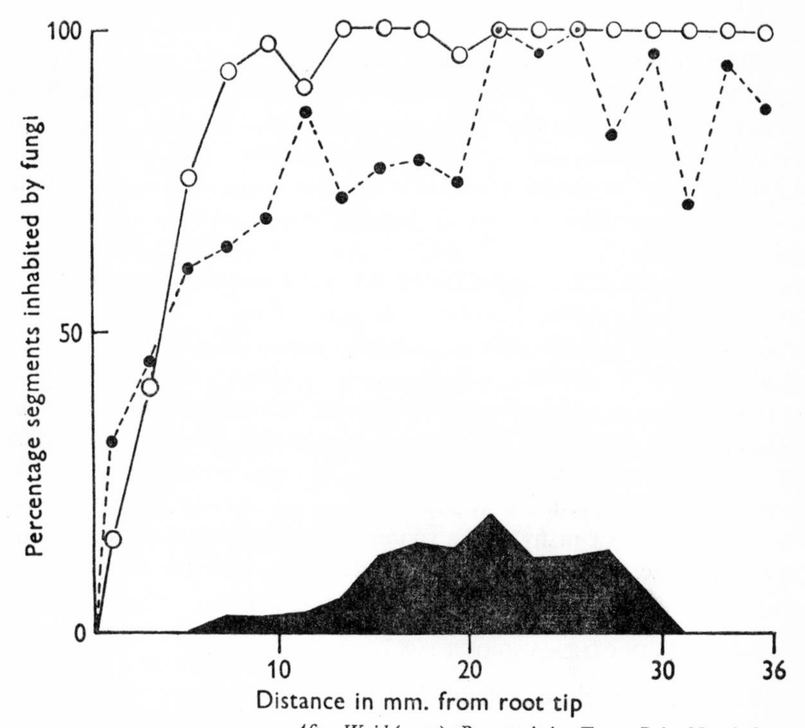

After Waid (1957). *By permission Trans. Brit. Mycol. Soc.*

FIG. 24.—Fungi on the apices of juvenile roots of Rye-grass. Surface population estimated by the direct (○——○) and culture (●——●) methods. Endophyte in the cortical cells, full black.

ran parallel with poor growth in field conditions, but also that there were very few connecting hyphae. He observed only 167 root-hairs penetrated by hyphae in oats as compared with 7253 uninfected hairs, and a much lower proportion in maize. There is, therefore, the same kind of variability recorded in the Gramineae as in the Leguminosae, the poorest development of extramatrical hyphae being observed in cultivated soils. This is a subject of considerable interest in regard to the effect of infection on the activity of the host, and will be discussed later.

A further point is made by Winter (1951), namely that septate and aseptate mycelia occur in oats but only septate mycelia in maize. This may merely reflect the extent of septation of old hyphae as described by Nicolson, but it may also be due to the occurrence of secondary infections by septate rhizomycetes.

The extent of mycorrhizal colonization of the roots of the grass *Lolium perenne* was studied by Waid (1957) in the course of an investigation of fungal colonization during maturation, senescence, and decay, of the root system. The mycorrhizal endophyte was an early colonizer and extended from the root-hair zone in juvenile roots for several centimetres proximally. Fig. 24 gives an interesting comparison between the colonization of the root surface by fungi and the colonization of the cortex by the phycomycetous endophyte. It is of interest that Waid observed that as the roots aged the central tissues became progressively colonized by dark sterile hyphae and sterile hyaline hyphae, which were followed in turn by *Fusarium* and other faster-growing sporing forms. The complexity of colonization of the functional cortices of roots is again apparent here, as it was in the other kinds of mycorrhiza. Nicolson (1959) also comments on this matter. He observed, 'In addition to the vesicular-arbuscular endophytes many other fungi of different types are common in or on grass roots'. Amongst these, he observed sterile hyphae, *Rhizoctonia*-like endophytes comparable with orchid mycorrhizal fungi, brown septate hyphae, chytrids, and Plasmodiophorales. He also observed fine hyphae enveloping and possibly entering the vesicles both within and without the tissues. These he viewed as a rudimentary hyphal sheath of protective function such as might not be unexpected if the endophytes were a species of *Endogone*. This condition was also observed by Godfrey (1957); but she showed clearly that many of the spores of *Endogone* were parasitized by septate hyphae, so that they became degenerate and inviable.

MYCORRHIZAS IN OTHER ANGIOSPERMS

The examples given are sufficient to indicate the general characteristics of phycomycete mycorrhiza. Plants of other families have been employed in detailed experimental work, and it is therefore expedient here to give a few references to papers on them.

Boullard (1953, 1953*a*) has investigated a number of plants belonging to the Araliaceae and Cornaceae. In them the short roots are described as being infected by aseptate hyphae which occupy all layers of the cortex. They spread inside the cells as single hyphae about 6 μ in diameter, to form coils of one or two turns in each cell. Vesicles about 70×30 μ, containing much oil, are found within the cells. The coils of hyphae undergo tolypophagous digestion similar to that occurring in some pteridophytes. In *Griselinia littoralis* (Cornaceae) however, Greenall (1963) has described arbuscules of a kind. Boullard believed, on the grounds of morphology and histology, that the endophyte of Cornaceae and Araliaceae differed from those of most other families. Moreover, Greenall (1963) and Baylis (1959) have experienced difficulty in obtaining infection of *Griselinia* with certain species of *Endogone*, *Pythium*, and '*Rhizophagus*', which have been successfully mycorrhizic with other plants.

Plants of the Rosaceae, especially apple and strawberry, have been used for experimentation. Otto (1962) has described the mycorrhizas of apple, and Mosse (1956) has described those of both apple and strawberry. Peuss (1958) describes tobacco infections and Waistie (1965) those of *Hevea*.

The Fungi of Phycomycetous Mycorrhizas

In the first edition of this book (1959) the fungi of phycomycetous mycorrhizas were jointly termed *Rhizophagus*, with the exception that the possibility was indicated that one or more species of *Pythium* might be involved in some of the associations. This terminology has been used also by others, e.g. Baylis (1962), but nevertheless it may lead to confusion and perhaps it is better dropped. There is, as has been made clear already, a great degree of similarity between kinds of phycomycetous mycorrhizas, yet at the same time there are marked variations in histology, in distribution of the endophyte, in digestion patterns, and in fungal organs. These may arise partly from variations in the host, partly from variations in the conditions of the habitat, and possibly also from differences in the genetic nature of the endophyte. For similar reasons the term vesicular-arbuscular mycorrhiza has not been used as a general term, but the name phycomycetous mycorrhiza has been substituted. It is somewhat misleading to use the former name if either vesicles or arbuscules are absent.

Many attempts have been made to isolate into pure culture the fungal endophytes, but many of the results obtained until recently had not been wholly successful or convincing; however, four widely different kinds of observation have repeatedly been made. Firstly, septate mycelia, such as *Rhizoctonia* and *Fusarium*, have been the most frequently isolated. Secondly, aseptate mycelia growing slowly from infected cells, but not capable of growth alone on artificial media, have been obtained by some of the most experienced workers with this kind of mycorrhizal material. Thirdly, a group of slow-growing aseptate mycelia have been isolated by special methods. Fourthly, species of *Pythium* with mycelia conforming in general structure to the endophytic mycelia, have been readily isolated in culture.

We need spend very little space in consideration of the work of those who claim that *Rhizoctonia* and other septate sterile mycelia cause the vesicular-arbuscular infection. The wide occurrence of such fungi on root surfaces and in root tissues has already been discussed. Although Demeter (1923), Jones (1924), McLuckie & Burges (1932), Neill (1944), and lately

Nicolson (1955), have all shown that the mycelium of the mycorrhizal fungus becomes somewhat septate in the outer cortex and in the soil, Nicolson has indicated clearly that this is associated with the loss of cell contents, particularly in the rhizoidal hyphae, and is not a property of the active mycelium. Hence it is not a reasonable expectation that the mycelium will produce truly septate hyphae in culture when it is non-septate in the tissues. We can discount *Rhizoctonia* as the causative genus and equally discount any other fungus with frequently septate active hyphae.

The striking point about the mycorrhizal fungus is that it may ramify about the roots in the soil, forming there large mycelial systems which are

TABLE XLVIII

ATTEMPTS, UP TO 1950, TO ISOLATE ENDOPHYTES HAVING ASEPTATE MYCELIA

Author	Plant Material	Notes
Bernard (1911)	*Solanum dulcamara*	3–4 mm. growth in 3–4 days.
Magrou (1935) Magrou (1937) Magrou (1936) Magrou (1939) Magrou (1946) Magrou & Magrou (1940)	Many, including: *Arum*, *Scilla*, *Veratrum*, *Solanum dulcamara*, *S. tuberosum*, *Orobus*	Greatest success obtained after an addition of vitamins and soil extract
Burgeff (1938)	*Psilotum*, *Marattia alata*	Good growth. The mycelia similar but not identical in the two hosts.
E. J. Butler (1939)	*Abutilon*	Slight growth from external mycelium
Lihnell (1939)	*Juniperus communis*	2–3 mm. in 12–13 days
Neill (1944)	*Citrus*	
Barrett (1947)	Pea	Growth on hemp-seed
M. Stahl (1949)	Thallose liverworts	Considerable growth in liquid culture

differentiated into runner-hyphae, short rhizoidal systems, and vesicles (Figs. 21 and 22). Moreover it readily dies back, and, when it does so, as Nicolson so clearly showed, it becomes septate. The fact that so many people have described the death of the outgoing hyphae suggests that the mycorrhizal fungus is dependent upon material of some kind, obtained from the roots, and which, as the roots age, becomes no longer available. This itself indicates that some strains or species of mycorrhizal fungi have specialized demands which must be met before they will grow in culture.

Table XLVIII gives a summary of the results of attempts up to 1950, to isolate endophytes having aseptate mycelia. Only the isolates of Barrett (1947) were maintained in culture after isolation. The remainder were not able to grow free from their host tissues, and attempts to obtain pure cultures of them by removal of hyphal tips, etc., were futile.

In 1936, Magrou reported briefly on a great advance which he had made in the isolation of mycorrhizal endophytes. He managed to encourage the endophyte of *Arum maculatum* to grow out of a root fragment across the medium to infect roots of *A. italicum*. In these, normal mycorrhizal infection was caused. This method of cross-inoculation was utilized by M. Stahl (1949) with thalli of Marchantiales. She placed gemmae of *Marchantia* or *Lunularia* species upon the surface of sterile sand which had been washed with rain-water. A mycorrhizal thallus was placed 3 to 4 cm. away on the sand surface. After from 6 to 8 weeks, the young plants developing from the gemmae often became infected in a normal manner. When a small clod of earth from a highly infected culture of *Marchantia* was substituted for the infected thallus, the young plant remained in every case uninfected. This seemed to indicate that the spreading of the mycelium from a living plant, or from living mycelium in plant tissue, was more active than the growth from mycelia in the soil.

By these means, M. Stahl obtained the following evidence of cross-inoculation.

Inoculum	Gemma	Result
Marchantia geminata	*M. geminata*	+
M. geminata	*M. multiradiata*	+
M. thermarum	*M. geminata*	+
Preissia commutata	*M. grisea*	+
P. commutata	*M. paleacea*	+
Androcryphia confluens	*M. grisea*	+
M. geminata	*Lunularia cruciata*	±
Lunularia cruciata	*L. cruciata*	±
Pellia epiphylla	*M. geminata*	−
Anthoceros sp.	*M. multiradiata*	−

These experiments of J. Magrou and M. Stahl opened the way to a method not only of studying the effect of infection on growth, but also of estimating the extent to which the histological pattern of infection in a given host was a property of that host rather than of the fungal strain.

In other experiments using this method of cross-inoculation, the specificity of many endophytes has been tested and the reaction of a given fungus to different hosts has been observed.

Endogone

The taxonomic affinity of the slow-growing phycomycetous endophytes has been a matter of considerable controversy, but a most important advance developed from the original observation by Peyronnel (1923), which he later repeated (*see* 1963) as did other observers, that hyphal connexions could be traced between mycorrhizas and the sporocarps of *Endogone* in the surrounding soil. E. J. Butler (1939) examined all the evidence then available and decided that '*Rhizophagus* [*sensu* Dangeard] endophytes may most satisfactorily be regarded as an imperfect genus of Endogonaceae'. Since then this view has been completely substantiated and there is now no reasonable doubt that most of the phycomycete mycorrhizas are caused by fungi of this affinity. The outstanding characteristic of the Endogonaceae (*see* Godfrey, 1957, 1957*a*, 1957*b*) is that they are zygomycetes which produce sporocarps containing either sexually produced zygospores or chlamydospores or both; but much is yet to be learned of their taxonomy, ecology, and physiology.

In 1953, Mosse confirmed the earlier reports by demonstrating that fructifications of a species of *Endogone* could be found attached by hyphae to the roots of strawberry. The fructifications were light-brown, spherical bodies consisting of loose hyphae containing 1–6(–32) spores each 60–160(–250) μ in diameter. The diameter of the fruit-body, about 1 mm., was small compared with that of many species of *Endogone*. In addition, thin-walled, probably immature, spores were present in the sporocarps. Godfrey likened this species to *Endogone macrocarpa*, and interpreted the thick-walled spores as chlamydospores rather than zygospores.

Following Mosse's work, Gerdemann (1955) extracted three types of spores in quantity from two soils in Illinois where clover and maize had been grown. The first, type A spores, were about 200 μ in diameter and attached to their parent hyphae without any swelling. They resembled the vesicles described on the internal and external mycelia of mycorrhizas. No infection was successfully obtained from these. The second, type B, were large, light-yellow spores 180–500 $\times$ 291–812 μ in dimensions—often spherical, ellipsoidal, or irregular in shape. They contained oil drops and were bounded by a double wall consisting of thin outer and thick inner coats. They were subtended on a hypha with a bulbous swelling and a lateral, antheridium-like hypha. These spores germinated in culture, and from them Gerdemann was able to infect host plants with typical phycomycetous mycorrhiza. The type C spores were also yellow and filled with oil drops. They occurred as clusters upon a single subtending hypha. They were 23–37 $\times$ 20–34 μ in dimensions, and spherical to clavate, with echinulate walls. They were borne on the same mycelia as type B spores, and were produced also on roots infected by mycelia derived from type B spores.

In a later paper, Gerdemann (1961) obtained yet another type of spore, about 250 μ in diameter, from the neighbourhood of maize roots. These spores were used to infect maize plants, in which typical vesicular-arbuscular mycorrhiza was then formed. This mycorrhiza produced thick-walled intra- and extra-radical vesicles, and also small sporocarps in the soil—similar to those found by Mosse attached to strawberry roots (*see* above).

In addition to the two unnamed species of *Endogone* confirmed by Gerdemann as mycorrhizal (one of which was probably conspecific with that obtained by Mosse), Dowding (1959) and Nicolson (1959) had suggested that *E. fasciculata* was probably mycorrhizal with grasses and other plants, and this was confirmed experimentally by Gerdemann (1965).

Besides the three species of *Endogone* thus proved to be mycorrhizal, it is now clear that there are many more. Gerdemann & Nicolson (1963) applied Gerdemann's method of wet sieving to Scottish soils and extracted six types of which four were mycorrhizal. Of these last, one was sporocarpic and the others chlamydosporic. It is apparent from these results that there is a constellation of species of *Endogone* which are probably widely distributed over the world, and unpublished work by Bowen and Mosse in Australia and Britain has further confirmed this. Taxonomic and ecological investigation of these mycorrhizal species of *Endogone* should prove of great interest, and it is being actively conducted by Mosse, Bowen, Gerdemann, and Nicolson. Nicolson, in an article to be published in *Science Progress*, has made a provisional statement of the findings so far, but since much other work by Bowen and Mosse is in preparation, the matter must be left to the future for full consideration.*

ENDOPHYTES ISOLATED BY J. T. BARRETT

Dr. Barrett (1947, 1958, 1961) reported the successful isolation of several strains of rhizomycete from the mycorrhizas of a number of species of plants. In his paper of 1961 he gave a short account of his method, which utilized pieces of boiled hemp embryo as bait. The mycorrhizal root was submerged in water, and pieces of embryo from hemp-seed were placed upon extraradical mycelium and vesicles. A white mycelial mass bearing vesicles developed on the submerged hemp-seeds. The bait was transferred, after washing to remove contaminants, to boiled tap-water containing new hemp-seed bait. After some weeks a considerable mycelium, contaminated with *Mortierella*, grew over the bait. The mycelium was transferred to hemp-seed agar. Barrett reported (1961) that he succeeded in isolating a similar fungus from 14 out of about 50 trials, and purified 11 of the isolates. He stated that contamination by *Pythium* gives much trouble in the isolation procedure, and that its species are difficult to separate from the slow-growing endophyte. The 11 successfully purified isolates were obtained from garden pea, carrot, celery, strawberry, sweet potato, wild oats, iris, tomato, sweet corn, *Vinca minor*, and apple. Mosse has informed me that Gerdemann has successfully used this isolation method and obtained an essentially similar rhizomycete.

Barrett grew his isolates on standard malt agar with or without plant decoction. The cultural behaviour and appearance enabled Barrett to divide his strains into two groups differing in appearance, frequency of vesicles, rate of growth, etc. In experimental syntheses, mycorrhizas were established on maize and a number of other plants, and he concluded that the position of the main hyphae, localization of arbuscules, and other features which separate the *Paris* from the *Arum* type of infection (cf. pp. 245–6), are a property of the host and not of the fungus.

* *See* Appendix §9.

Other observers—Mosse, Greenall, and Gerdemann—have attempted experimental synthesis of mycorrhizas with Barrett's isolates or with similar fungi. Gerdemann,* according to Mosse (1963), failed to obtain successful synthesis, as did Greenall (1963) using both Barrett's isolate and *Pythium* with *Griselinia*; however, Mosse (1961) obtained mycorrhizas with apple—but not with clover, onion, strawberry, or hemp.

It seems possible, on general grounds of similarity of growth and vesicle formation, that this group of rhizomycetes belongs taxonomically to the Endogonaceae. Barrett and others have used the generic name *Rhizophagus* for it. However, G. T. S. Baylis and other New Zealand workers have used the name *Rhizophagus* to embrace a wider range of rhizomycetes in which species of *Endogone* seem to be included. At the moment, to avoid confusion, it would seem advisable, pending further work, to use neither *Rhizophagus* nor *Endogone* for these strains obtained by Barrett.

Pythium

Since the late nineteenth century it has been repeatedly suggested that phycomycetous mycorrhizas might be caused by species of *Pythium*, and this possibility has been reconsidered by L. E. Hawker and her colleagues in a series of experimental papers. Harrison (1955) elaborated an isolation technique whereby the fungi could be grown out from the cells of sections of root into culture media, to give rise to colonies of *Pythium*. Hawker *et al.* (1957) further described the method. Roots were scrubbed free from soil and shaken in sterile water with sterile sand as abrasive. The longitudinal sections of the roots were mounted on a cover-glass over a ring-cell in water agar containing 0·2 per cent of aureomycin. Tip isolations of the hyphae, seen to be growing out of cells of the root, were transferred to fresh medium. In this way *Pythium* was isolated from cells of the roots of *Allium ursinum*, *Endymion non-scriptus*, and *Dryopteris filix-mas*. Later, Hepden (1960) isolated *Pythium* from *Phyllitis scolopendrium* and Carré & Harrison (1961) obtained a similar fungus from *Conocephalum conicum*. Back-inoculation of these strains of *Pythium* into sterile host tissues is described by all the workers to produce typical vesicular-arbuscular infection. The plants tested include *Allium cepa*, *A. schoenoprasum*, *A. ursinum*, *Phyllitis scolopendrium*, *Conocephalum conicum*, *Lactuca* sp. and apple. Figures are given only by Hawker *et al.* (1957) and Hawker & Ham (1957) of sections of infected tissue showing runner hyphae, vesicles, and, in one case, arbuscules.

The most successful back-inoculation experiments were carried out with onion as the test plant, but, as Ham (1962) showed, an approach to the typical histological pattern of vesicular-arbuscular mycorrhiza was only obtained under restricted conditions. With apple, as Hawker & Ham (1957) point out, vesicles form before arbuscules, and this in itself is not usual. *Pythium* seems to form mycorrhiza most readily in media in the

* *See* Appendix §9.

absence of glucose and nitrate, for in the presence of these nutrients the host may be parasitized. The degree of parasitism depends on both the strain of fungus and its vigour. The vigour of the host was also involved. Adequate lighting and good drainage of soil reduce parasitic attack, and the mycorrhizal condition develops only in plants that are growing well.

Mosse (1963) has tested *Pythium* isolates, obtained from the Bristol workers, in inoculation experiments with apple and onion. Apple roots were not penetrated in her experiments, whereas those of onion might be parasitized or might show typical vesicular-arbuscular penetration.

It is of interest in assessing the situation to note that there is evidence for the production of mycorrhizas by *Pythium* only in narrowly defined conditions with a few hosts. There can be no doubt that the wide range of phycomycete mycorrhizal infections are not caused by fungi of this kind.

XIV

THE ECOLOGY AND PHYSIOLOGY OF PHYCOMYCETOUS MYCORRHIZAS

RESEARCH into the functional significance of phycomycetous mycorrhizas has been beset with difficulties, not the least of these being an inability to grow the important fungi in culture. As a result, much of the past experimentation has been imperfect and many of the hypotheses and conclusions reached have not always been well-founded; nor have they paved the way for further advances. In recent years, thanks to the efforts of a few research workers, the subject has escaped from a morass of uncritical speculation and ignorance. This chapter will consider a selection of the researches which help to indicate the part played by this kind of infection, but it should be noted that worthwhile results have as yet been obtained from the study of only a mere handful of examples. Although these belong to many plant families, they are not really an adequate sample from which to generalize with great confidence.

ECOLOGICAL VARIATION OF MYCORRHIZAL INFECTION

As mycorrhizal infection by phycomycetous endophytes is so widespread, it is not surprising that there is considerable variation in the extent of infection, which may be dependent on the kind of higher plant involved, its state of growth, and its habitat conditions. In some kinds of plant, infection seems to be always light and sporadic. Instances of this have already been given. For example, amongst the pteridophytes, *Equisetum* is only slightly infected, whereas by contrast the Ophioglossales are intensely colonized by mycorrhizal endophytes. In Nicolson's work on sand-dune grasses (1959, 1960), different grass species showed very different intensities of infection, even in the same habitats.

A few studies have been made, analogous to those with ectotrophic mycorrhiza, on the variation in degree of infection with habitat conditions and with the magnitude of specific environmental factors. Otto (1962) made an intensive investigation of mycorrhizal development of apple seedlings over a three-year period in various cultivated soils. His estimates of mycorrhizal infection are given both as frequency of infection, that is as percentage of roots infected, and as intensity of infection, estimated on a scale of 0 to 5 based upon the relative numbers of cells containing fungus as seen in microscope sections. Both kinds of estimate were subjected to statistical tests of significance. The frequency of infection varied from site to site in the first two years, but showed an increase with time on all sites. By the end of the third year, all had reached a frequency of 100 per cent. The intensity at first varied in a similar way to frequency, and increased with time; but the final intensity was related to habitat conditions and the vegetative vigour of the host plant. Various adverse conditions of phosphate supply, water regime, and so on, which reduced growth, also reduced the intensity of mycorrhizal development. Otto also described a seasonal change in intensity. An increase occurred through spring and summer to a maximum in autumn, and there was a subsequent decline to a low value in winter which was associated with a stagnation in fungal activity.

It is noteworthy that Otto recorded that he observed no quantitative relationship of yield to mycorrhizal development, the implication being that the mycorrhizal fungi do not intervene in activities which affect the growth of the host. Indeed this is explicitly stated in his papers of 1962. In view of the experience obtained with ectotrophic mycorrhizas in this regard, his conclusion should be viewed with care. It will be recalled that the great advance in their study by Hatch (1937) was due to his analysis of results of this kind in conjunction with other results showing the reverse. For the record we may quote Winter (1951), 'Starke Verpilzung geht überaus häufig mit schlechter Entwicklung der Pflanzen parallel'. But by contrast again others (e.g. Schrader, 1958) state that heavy infection and considerable development of internal and external vesicles are correlated with strong root and shoot growth.

Variation in the degree of infection in a natural succession was studied by Nicolson (1959, 1960) on sand-dunes. He found that both frequency and extent of infection were related to the successional stage. In the stable phases of dune succession, where organic matter increased to significant amounts in the rooting region, the infection rating was high. In the unstable and building phases, where the roots persisted for shorter periods as functional entities, and where sand that was low in organic matter accumulated, infection was less. However, in the most mature fixed zones infection once again decreased, and this may have been related to the closing of the community and to consequent changes of habitat factors—including an increase in the general microbial activity in the soil.

The effects of specific factors on infection have also been examined in the laboratory. From an early date, infection has been regarded as dependent upon an adequate supply of light to the host. Schrader (1958) stated that the infection of pea plants under experimental conditions is dependent upon light supply and therefore upon the carbohydrate status of the host. Peuss (1958) described an experiment in which tobacco plants were grown in full light, 50 per cent of full light, and with the photosynthetic area reduced by the removal of leaves. Both treatments greatly reduced the degree of infection as compared with the control. The shading treatment reduced the infection rating from 85 to about 30.

Boullard (1957*a*, 1959, 1960) has shown, by altering the period for which his experimental plants were illuminated each day, that the degree of infection is related to light supply in *Aster tripolium*, *Hedera helix*, and *Bupleurum falcatum*. By contrast, Baylis (1967) observed no effect on mycorrhization of *Coprosma* by reducing the light to 30 per cent of daylight. The same caveat applies here that applies to similar work with ectotrophic mycorrhizas. It is not improbable that these effects of light, where they exist, act by way of changes in the constitution, growth, or excretions of the roots, and are not solely or directly due to carbohydrate supply.

Good quantitative data are absent on the effect of manurial treatments on mycorrhizal development. Winter (1951) stated that there is no relation between the nutrient status or humus content of the soil and mycorrhizal development. However, as Mosse (1963) pointed out, there have been many reports that organic or farmyard manures alter the condition of infection and increase the extent of arbuscule production, and that mineral fertilizers reduce infection. Recently, in the course of work on nutrient absorption by mycorrhizal plants, Baylis (1967) observed that small phosphate additions to soil greatly reduce or eliminate mycorrhiza formation both in heated and unheated soil. Daft & Nicolson (1966) showed that the percentage of infection in roots of tomato was inversely related to phosphate supply added as bone-meal. This is indeed a general view, that a high nutrient state of the soil reduces infection (*see* Lanowska, 1966); but there is a great need for more fact-finding experimentation on the effects of external factors on mycorrhizal development.

The Endophytes in the Soil

The papers by E. J. Butler and Dowding which were quoted in the last chapter, emphasize sufficiently the extensive distribution of Endogonaceae in the soil and of the mycorrhizal species in particular. Careful removal of plants from the soil shows that there is usually considerable development of mycelium around the root, extending for one or more centimetres from its surface, and although there are firm statements to the contrary (Winter, 1951), this seems to be the condition during the active vegetative period of most plants. Nicolson's work with grasses (1955–60) showed that the extra-radical mycelium not only bears spores and sporocarps but may be dimorphic in structure, being divided into thick-walled permanent hyphae and finer ephemeral absorbing hyphae. Its extent varies, and in some conditions may be small. There is evidence of considerable exploitation or colonization of humic or organic particles, which has raised the question of absorption of organic substances external to the root. Mosse (1959*a*) made an experimental study of apple and strawberry mycorrhizas, and showed that the extramatrical development was dependent on season and increased to reach a high value late in the year. In experimental systems where fungus was allowed to grow out from infected roots on to filter paper, the dimorphic state described by Nicolson (cf. p. 260) was observed.

The extent of connexion between the external and the internal mycelia, is clearly important to the functioning of the fungus in nutrient absorption. Mosse (1959*a*) has estimated the frequency of hyphal connexions (entry points) in a variety of conditions. Their numbers varied through the season, increasing from spring to autumn; they also varied with the nutrient status of the soil. Values of 2·6 to 21·1 per mm. of root length were recorded for strawberry, and 4·6 to 10·7 per mm. for apple. Other estimates have been made on other plants—Lihnell (1939) gives 4 to 16 per mm. for *Juniperus*, while McLuckie & Burges (1932) give 2·5 to 4·1 per mm. for *Eriostemon crowei*. The exact significance of these figures and others like them cannot yet be assessed, because too little is known of the quantitative ability of fungal hyphae to translocate nutrients to estimate whether such a frequency of connexions is, or is not, sufficient for their supposed role in the absorption and supply of nutrients.

Apart from the localization of mycelium in the environs of the root, little is known of its distribution in the soil in an active vegetative condition. There seems no doubt that it has some powers of spreading from newly-decaying plant remains, but nothing is known of its longevity or persistence. The study of the distribution of spores and sporocarps in the soil has been pursued to some degree. Nicolson (1967) points out that seasonal, annual, and local, variations in their frequency and in the frequency of different types, may be detected. On the whole he has observed higher frequency of *Endogone* spores in cultivated soils than in natural and semi-natural communities. In sand-dunes for instance he encountered few, and here *Endogone fasciculata*, whose spores are not readily recoverable by wet sieving, was the main mycorrhiza-former. Similarly, Waistie (1965) observed one type of sporocarp with associated chlamydospores to be most prevalent in soils of rubber plantations in Malaya, but three other types of spore were occasionally encountered.

In highly cultivated soils, on the other hand, more varied and numerous populations of spores were encountered. Nicolson (1967) gives a table of data collected from a number of soils bearing different crops. In each, a number of kinds of spore was encountered though usually one type predominated. Nicolson stressed the difficulties involved in this kind of work. Not only is the technique of extraction laborious and subject to error, but it is selective for those spores which are easily extracted by sieving. Again, the variations of distribution of spores within the soil are such that the differences between replicates make it necessary to sieve large numbers of samples to obtain credible results. Even allowing for these sources of error, seasonal variation and differences in population associated with different crops may be detectable.

The longevity of the spores as viable propagules, and their almost universal distribution in soil, ensures that most plants everywhere have a chance of being infected. A particularly graphic example is provided by an experiment by Gerdemann (1964). In an attempt to obtain uninfected plants without soil-sterilization, he grew maize on soil obtained from three feet below the original surface underneath a five-year-old house. The soil had grown alfalfa as its last crop. After sampling, it was kept dry in a greenhouse for one and a half years. Hence this soil had not borne vegetation for over six years, yet plants of maize became infected in it. It will be readily appreciated that this kind of possibility makes the provision of uninfected control plants in experiments difficult.

Germination of Spores and the Inoculation of Host Plants

As Godfrey (1957*b*) showed, the spores of Endogonaceae are very difficult to germinate under artificial conditions. She herself obtained some slight success with only *Endogone microcarpa* and *E. macrocarpa*, the spores of which germinated sporadically in a manner that appeared to be much dependent on their age and state and little upon applied external conditions. Gerdemann (1955) obtained 40 per cent germination of his type B spores in distilled water after sterilizing them in sodium hypochlorite. Germination took place at 20° C., and one or more germ-tubes were put out. Small, short-lived mycelial colonies were obtained on hemp-seed and on a variety of agar media. Mosse (1956) observed that, in the spring, a spontaneous spore-germination could be obtained on filter paper, and that spores with vacuolated contents would germinate whereas those with opaque contents would not. If selected spores, after surface sterilization, were mounted on cellophane disks over plates of soil, considerable germination took place. The necessary conditions included adequate moisture and the presence of active micro-organisms in the soil. Germination was believed to be stimulated by substances produced by soil micro-organisms. In this way up to 80 per cent germination could be expected. The extent of hyphal growth from germinating spores varied with conditions, and certain substances, such as tartaric acid or unknown compounds in sonically disintegrated roots, improved growth.

Mosse (1962) tested the ability of germinating spores to infect host-plants under sterile conditions. She found that if sterile sporelings on a disk of cellophane were used as inocula for sterile host-plants under aseptic conditions, no mycorrhizal infection occurred. She observed, however, that in the presence of a bacterial contaminant, mycorrhizal infection was readily produced both in solid and in liquid cultures. If sterile roots were treated with mild pectolytic or cellulolytic enzymes, or with culture filtrates of a species of *Pseudomonas*, some degree of infection also resulted. The stimulation of mycorrhiza formation by species of *Pseudomonas* was evident with many kinds of host plant, including both dicotyledons and monocotyledons, but of greater importance was the possibility that, by using pure enzymes, a pure two-membered culture might be set up.

Once established in a root system, the fungus not only produced extraradical mycelium on the agar medium but this could be used as an inoculum for other hosts planted in that agar after the first plant was removed. In this way, aseptic two-membered cultures without bacteria could be created in reasonable quantity.

Various other methods of inoculation have been used for experimental purposes. Selected spores or sporocarps, obtained from the soil, may be washed or surface-sterilized before being added to the substrate on which the host is grown. A variant is to wash the spores with water before use, and to add the wash-water to a control soil without any spores (Gerdemann, 1964). In this way, control soils with the same contaminating organisms but without the mycorrhizal fungus are obtained. Gerdemann (1965) passed the isolated selected spores through a funnel containing a filter pad of cotton-wool. He then added sterile soil to the funnel and grew the host plant upon it. In this way, stocks of plants inoculated from selected sources of spores could be maintained for future use. Daft & Nicolson (1966) maintained stocks of inoculated plants in a greenhouse as a source of spores and sporocarps for inocula, but kept parallel uninoculated (but otherwise similar) pots as control sources of contaminating soil organisms—so that experimental controls could be supplied with a microbial population lacking the mycorrhizal fungus.

The inoculum most commonly used by early experimenters was a small quantity of soil, containing spores or sporocarps, that was added to the sterile soil on which test plants were to be grown. Sterilized soil was added in similar quantity to the control samples. Many experimenters have used fragments of mycorrhizal root as inocula, and treated the controls with pieces of sterilized or non-mycorrhizal root. Peuss (1958) used a grafting or implantation method. Pieces of living mycorrhiza were introduced into the tissues of the roots of test plants, and similar pieces of non-mycorrhizal roots into the controls. By this elegant means, mycorrhizal plants were secured in soil or water culture.

A method of a different sort has been used in many experiments by Baylis (*see* 1961*a*, 1967) upon *Griselinia* and other plants. Seedlings are germinated from sterilized seed, and one sample is transplanted to sterile soil while another is transplanted to soil beneath an established *Griselinia*. After some months, both sets are brought together under similar experimental conditions.

Methods of these kinds have been used in work on the specificity of the fungi and upon the effect of infection upon the host.

Specificity of the Endophytes

As was mentioned earlier, the experiments of J. Magrou and M. Stahl, which tested cross-inoculation between *Arum italicum* and *A. maculatum*, and between various species and genera of liverworts, gave an impression of lack of specificity of endophytes to particular hosts. This work also suggested that the different histological patterns in infected organs, and different distributions of fungal hyphae, were often influenced more by the host than by the fungus. J. T. Barrett's work with his isolates also led to similar conclusions. It is, therefore, of great interest that similar results were obtained by Koch (1961), who showed that the endophyte of *Atropa belladonna* might form mycorrhiza with over forty species of diverse families.

Using spores of type B, collected from soil bearing either red clover or maize, Gerdemann (1955) tested their ability to form mycorrhizas with other plants. Those from soil bearing red clover infected maize, strawberry, and sweet clover as well as red clover, but did not infect oats. Those from soils bearing maize only reinfected maize in Gerdemann's experiments. Similarly, the spores used by Mosse as a source of inocula (*see* 1962, 1963) were by no means specific to the plant of origin but exhibited a wide host-range.

By contrast, some degree of specificity in infection has sometimes been described. Tolle (1958) observed, when she tested cross-inoculations between oat, barley, wheat, and rye, that the fungi from oats and barley only cross-inoculated roots of the same species, whereas those of rye and wheat cross-inoculated both rye and wheat.

The general impression gained from this rather inadequate information on specificity is that the various rhizomycetes are relatively unspecific —a condition also encountered amongst mycorrhizal fungi of ectotrophs, of orchids, and amongst the nodule bacteria of Leguminosae. Just as in other cases, so here, the ecological impression from work on the rhizomycetes in the soil suggests the existence of preferred associations, and the collection of different kinds of spore from different cultivated sites, reported by Nicolson (1967), is evidence of this. But it must be appreciated that non-specificity in infection does not mean that all infective rhizomycetes will have similar effects on the growth of a given host. This is clearly not so among strains of *Rhizobium*, nor among the Basidiomycetes of ectotrophic mycorrhiza, and it probably is not so with phycomycetous endophytes.

The Effect of Infection on the Host Plant

We have already noted that, in some cases, a correlation has been observed between intense infection and thrifty growth of the host, whilst in others the reverse has been recorded. Such observations made under field conditions, are almost valueless in assessing the effects of infection on the growth of the host. Only those experiments in which mycorrhizal and non-mycorrhizal plants are directly compared under uniform conditions, are of real value.

In his study of mycorrhiza in Leguminosae, Asai (1944) grew a large number of species in autoclaved soil, and compared them with plants raised in similar conditions except that the soil had been reinoculated with various kinds of unsterile soil. Those on inoculated soil developed mycorrhizas. He recorded, with many estimates, that the mycorrhizal plants grew faster than the non-mycorrhizal ones. Moreover when nodules were developed, the nodulated plants were not greatly larger than the nodule-free plants unless they were also mycorrhizic. This dependence of the efficacy of the nodules on coincident mycorrhizal infection has not been a general experience, and hence Asai's work has not been widely considered. At all events it is better at this stage to collate the experimental work on simpler systems, where two organisms only are in close association and bacterial nodules are absent.

Table XLIX give a brief summary of some of the more recent experiments on the effect of infection upon the growth of the host plants. In these examples, several plants of very different kinds were used, and a variety of experimental methods were employed. Some made use of steamed or autoclaved soil, some used sand culture, and one employed a subsoil which was deficient of mycorrhizal fungi and required no sterilization. Inoculation was brought about by infected roots, implantation, sporocarps, or from adult plants. In all these cases but one, considerable stimulation of growth as indicated by dry-weight increment, usually in statistically significant amounts, was observed. It is noteworthy that the one clear exception was Holevas's experiment (1966) with strawberries on phosphorus-enriched soil, where no greater shoot-growth was observed in mycorrhizal plants than in uninfected ones. This table does not exhaust the list of experiments recording this kind of effect—other examples will be quoted later; nor does it include cases where stimulation was not observed. The latter are by no means lacking in the literature but figures of weights are rarely given. For instance, Waistie (1965) states that a pot-test with *Hevea* plants in Malaya gave no significant differences between plants from sterilized and unsterilized soils after 18 months' growth. Similarly, the account of Dr. Mosse's recent work in the Rothamsted Experimental Station report for 1966 states that no marked differences were observed between mycorrhizal and non-mycorrhizal plants of tomato in three separate experiments. It is essential, therefore, to consider more completely the evidence available from inoculation experiments, and to try to determine whether any destructive criticism may be levelled against the experimental methods and whether any explanation can be found for the variable results.

One criticism which has been made of many experiments performed with soil, is that heat-sterilization produces toxic effects which may be eliminated by fungal growth. The stimulations of mycorrhizal plants are therefore sometimes ascribed to such an effect, which might be brought about by the mycorrhizal endophyte. This criticism is not generally valid.

TABLE XLIX

SOME EXAMPLES OF THE EFFECT OF MYCORRHIZAL INFECTION OBTAINED BY INOCULATION WITH *Endogone* ON THE DRY-WEIGHT (gm.) OF THE HOST-PLANT

Observer	Plant	Organ	Uninfected	Mycorrhizal	Mode of infection	Note
Mosse, 1957	Apple	Stem	0·8	1·3		
		Root	2·1	2·5	sporocarps	autoclaved soil
		Total	2·9	3·8		
Winter & Meloh,		Stem	1·70	2·18		
1958	Maize	Root	0·667	0·815	infected	sand culture
		Total	2·367	2·995	roots	
Peuss, 1958	Tobacco	Stem	1·22	2·01		unsterilized subsoil and
		Root	0·43	0·71	implantation	sand
		Total	1·65	2·72	method	
Baylis, 1959	*Griselinia*	Stem	—	—	beneath	controls in autoclaved
		Root	—	—	infected	soil during infection
		Total	5·14	12·48	adult	period. Growth-period in autoclaved soil
Baylis, 1959	*Griselinia*	Stem	—	—	beneath	as above but growth
		Root	—	—	infected	period in autoclaved
		Total	5·87	7·85	plant	sand/soil mixture
Baylis, 1959	*Griselinia*	Stem	—	—	beneath	as above but sterilized
		Root	—	—	infected	with electric soil steri-
		Total	2·13–1·66	6·10	plant	lizer at 100°C.
Baylis *et al.*, 1963	*Podocarpus*	Stem	—	—	beneath	sterilized with electric
	totara	Root	—	—	infected	soil sterilizer at 100°C.
		Total	0·38	2·63	plant	
Gerdemann, 1964	Maize	Stem	2·7	10·4	sporocarps	steam-sterilized soil,
		Root	1·0	2·9	or spores	wash-water form sporo-
		Total	3·7	13·3		carps added to soil
Gerdemann, 1965	Maize	Stem			sporocarps	steam-sterilized soil.
		Root			of *Endogone*	wash-water from sporo-
		Total	6·1	14·4	*fasiculata*	carps added to controls
Holevas, 1966	Strawberry	Stem	—	—	sporocarps	soil/sand mixture auto-
		Root	—	—		claved. Low P content
		Total	6·691	16·043		
Holevas, 1966	Strawberry	Stem	—	—	sporocarps	soil/sand mixture auto-
		Root	—	—		claved. High P content
		Total	18·832	17·041		
Daft & Nicolson, 1966	Tomato	Stem			spores	sand. Low P
		Root				
		Total	0·088	0·393		
Daft & Nicolson, 1966	Tomato	Stem			spores	sand. Very high P
		Root				
		Total	0·702	0·737		

Not only is it customary, as stated above, for the washings of spores or roots used as inocula to be added to the controls, or for precautions to be taken to ensure that the controls are equipped with soil-fungi, but many of the experiments do not involve soil-sterilization at all. They may use sand, pure solutions, water cultures, etc.

Baylis (1967) has made a valuable study of this problem of sterilization, Plants of *Griselinia littoralis*, *Coprosma robusta*, and *Pittosporium tenuifolium*, were grown in a mycorrhizal condition in normal soils and in a non-mycorrhizal condition in steamed soils. In every case a greater dry-weight increment was recorded from the mycorrhizal plants. The same species were therefore tested in a soil from beneath old trees of *Melicytus ramiflorus* in which early growth of seedlings and mycorrhiza formation were erratic. In each case the resulting seedlings, after six months or a year of growth, were classified into three categories: mycorrhizal, partly mycorrhizal, and non-mycorrhizal. The results showed that those seedlings that had become wholly or partly mycorrhizal were significantly greater in dry-weight than their uninfected counterparts. This experiment is clearly reminiscent of similar experiments with ectotrophic mycorrhiza, and is subject to the same criticisms in that perhaps the large seedlings were more susceptible to mycorrhizal infection; but this criticism is as unlikely to be just here as it was with ectotrophic mycorrhizas. Baylis went on to show that steaming the soil might indeed have a stimulating effect on the growth of mycorrhizal plants, for he grew *Griselinia* and *Coprosma* as mycorrhizal plants in steamed soil and observed a 30 per cent increase of growth in dry-weight as compared with mycorrhizal controls in unsteamed soils. In a further experiment it was shown that heating the soil had a stimulating effect on all seedlings, mycorrhizal or not, but in every case an additional stimulation associated with mycorrhizal infection was observed. These results definitely dispose of the criticism which states that the alleviation of a toxicity of steamed or heated soil is a general explanation of the effects of mycorrhizal inoculation in experiments.

A second source of variability in response may be due to the nutrient status of the soil. If the effects of mycorrhizal infection arise from an increased absorption of nutrients from the soil, the differences between mycorrhizal and non-mycorrhizal seedlings might be expected to be least in soils of high nutrient status. The experiments of Daft & Nicolson (1966) are especially relevant to this point. They grew tomato plants in a sterile mixture of dune sand and horticultural sand in the presence or absence of an inoculum of spores of a selected endophyte. The sand-cultures were treated with six different levels of bone-meal. Dry-weight, percentage infection of the roots, and phosphate absorption, were estimated. In the three lowest phosphate levels the infection rating was about 80 to 90 per cent, and highly significant increases of dry-weights were attained by the infected plants. In the two highest phosphate levels the dry-weight differences were not significant, although the infection rating was still 60–70 per cent. In the remaining treatment, medium phosphate concentration, infection (63–73 per cent) gave a significant dry-weight increase in one replicate but not in the other. In every treatment, however, phosphate absorption was greatest into the infected plants. Essentially similar results as regards both dry-weight and infection ratings were obtained by Baylis

(1967) for *Griselinia* and *Coprosma*. It should be noted that in these cases, as in the work of Holevas, uninfected and mycorrhizal plants grew equally well in soils of high nutrient status; but this was not the experience of Peuss (1958) with tobacco. She stated that even with large applications of fertilizers to nutrient-deficient soils, the mycorrhiza-free plants did not grow as well as the mycorrhizal ones.

A further source of variability may lie in the kind of the fungus used for inoculation. Although the fungi seem to be relatively unspecific with respect to their ability to infect host-plants, this does not show them to be unspecific in their physiological effects. Daft & Nicolson (1966) tested this possibility experimentally, using three different endophytes—one being sporocarpic and two non-sporocarpic. In an experiment with tobacco, they measured leaf-areas and infection rating. Endophyte No. 3 (one of the non-sporocarpic species), which gave the highest infection rating, gave an initial stimulation of leaf-area; but over a period of 83 days the final score was equal to that of the controls, whilst the other species gave some increase of leaf-area. Using tomato in full nutrient solution in sand-culture once again, endophyte No. 3 was associated with a dry-weight of the host that was very little above the control, whereas endophyte No. 1 (sporocarpic) gave a dry-weight which was significantly greater than that of the control. A different cultivar of tomato gave different results again, for all the endophytes increased its dry-weight increment in phosphate-deficient conditions. The differences perceived in these experiments of Daft & Nicolson, although small, are in some cases statistically significant, and indicate that this partial explanation of the variability of the effect of infection requires further analysis. The experiments of Meloh (1963) lead to similar conclusions. Using maize, she showed a stimulating effect on root and shoot dry-weight at each sampling time up to 14 weeks of growth when mycorrhizas derived from spore inocula were used. On the other hand, sporocarpic mycorrhiza had no such stimulating effect.

Yet another source of variability may arise from the fact that the mycorrhizal fungus derives material from its host. Therefore, if conditions occurred in which this drain on the host exceeded or balanced any stimulating effect due to the fungus, a decreased or equal growth-rate might occur in infected plants. An example of this is to be found in experiments of Baylis (1967), where decreased dry-weight was observed in infected plants of *Coprosma* supplied with high or very high levels of phosphate. Similarly, in the experiments of Winter & Peuss-Schönbeck (1963) a study was made of the growth of tobacco plants at various infection intensities. At a rating of 19 per cent infection the dry-weight of the shoots of infected plants was somewhat lower than that of the uninfected controls, whereas at infection intensities of 45–84 per cent their dry-weight was significantly higher.

If we accept these various sources of variability of behaviour, it is evident that we must accept the probability that many experimental results will show little effect of infection upon the growth of the host. Moreover, they strengthen the evidence in favour of a positive effect of infection in certain conditions and emphasize the fact that phycomycete mycorrhiza must be studied in respect of its physiological role in ecologically probable conditions.

Explanation of the Effects of Infection on Growth

The results of infection experiments indicate that there is a greater production of plant material in some circumstances by the mycorrhizal host than by the non-mycorrhizal host. The necessary conditions for this to occur were outlined in a previous paper (Harley, 1950). Infection must bring about one of the following effects:

1. Assimilation of gaseous nitrogen.
2. Conduction of material from the soil to the host.
3. Secretion of substances into the soil which affect the availability of nutrients.
4. A change in the nature or area of the absorbing surface.
5. Production of vitamins or growth stimulators by the fungus within the tissues.

Each of these possibilities has been canvassed from time to time, and there is a little evidence for some of them. Mosse (1962) produced evidence in favour of the fourth—that the mycorrhizal roots of *Trifolium* in culture were longer than the uninfected ones, and produced more first-order laterals, although the general macroscopic appearance of the infected seedlings was not greatly different from that of the uninfected controls. Morrison & English (1967) have recorded slight nitrogen fixation in the root region of *Agathis australis*, but this is not associated with the presence or absence of the mycorrhizal fungus.* In general, however, the evidence is clearly in favour of an intervention of mycorrhizal infection in nutrient absorption from the soil.

Mosse (1957) showed that in mycorrhizal apple plants, greater amounts of potassium, calcium, magnesium, and iron, were absorbed per unit of dry-weight, and less manganese, than in their non-mycorrhizal counterparts. Gerdemann (1964), using maize, showed that larger amounts of phosphate per unit of dry-weight were present in mycorrhizal plants than in the uninfected controls, but that the amounts of potassium and magnesium were significantly less, and there were also significant differences in the absorption of minor elements. Again Baylis (1967), using *Griselinia littoralis*, *Podocarpus totara*, *Coprosma robusta*, *Pittosporium tenuifolium*, and *Myrsine australis*, showed that mycorrhizal plants contained much greater quantities of phosphorus, potassium, and calcium, per unit of weight, but less nitrogen, than their uninfected controls. In almost all the experiments where phosphate content has been analysed, it has been

* *See* Appendix §10.

observed that the most striking difference between mycorrhizal and non-mycorrhizal plants lies in their phosphate content. As a consequence, much recent experimentation has concentrated on phosphate uptake. In the investigations of Daft & Nicolson (1966), tomatoes, grown at all the experimental levels of bone-meal application, absorbed more phosphate when in the mycorrhizal condition, irrespective of whether mycorrhizal infection was correlated with dry-weight increase or not. However with strawberry, Holevas (1966) observed that mycorrhizal plants at the high phosphate level were similar to uninfected plants both in phosphate content and dry-weight, although, at the lower level, mycorrhizal plants were much greater in both respects. The experiments of Baylis (1967) are especially convincing, and he concludes, with some confidence, that 'the phycomycetous endophytes assist uptake of phosphate from soils far below the minimum agricultural standard of fertility. They can occur at higher levels of available phosphate and their presence may be detrimental to the growth of their host, but plants that are growing rapidly because phosphate is readily available and no other factor is limiting, are often free from infection'.*

Following these results on phosphate uptake there has been, as with ectotrophic mycorrhiza, an examination of short-term phosphate absorption in experiments using isotopic phosphate. Gray & Gerdemann (1967) have shown that mycorrhizal plants of both *Liriodendron tulipifera* and *Liquidambar styraciflua* absorb phosphate from radioactive solution much faster on a dry-weight basis than do uninfected plants. Working with *Agathis australis*, Morrison & English (1967) have obtained results which have a bearing on this enhanced phosphorus uptake. They found that the uptake of phosphorus was markedly stimulated by mycorrhizal development and was temperature-sensitive, being diminished by low temperature and almost abolished at 1° C. Hence it is probable that this is a metabolically-dependent phosphate uptake, as might be expected.

G. D. Bowen, in experiments yet to be published, co-operated with B. Mosse in an examination of short-term phosphate uptake, using clover and onion. The mycorrhizal roots absorbed phosphate twice as fast as the uninfected controls from 5×10^{-6} M. phosphate solution over a few hours. The intervention of the fungus in this process was indicated by the fact that, in the short periods of the experiments, the mycelial parts of the root became the most radioactive, and more accumulation occurred in the fungal structures than elsewhere.

The study of phycomycetous endotrophic mycorrhizas is therefore seen to have emerged as a reputable pursuit. There are many outstanding problems; but it is now abundantly clear that mycorrhiza is an ecologically significant factor which can no longer be ignored or passed over in a superficial manner.

* *See* Appendix §11.

BIBLIOGRAPHY

ADDOMS, R. M. (1937). Nutritional studies on Loblolly Pine. *Plant Physiol.*, 12, 199–205.

—— & MOUNCE, F. C. (1931). Notes on the nutrient requirements and histology of the cranberry (*V. macrocarpon*) with special reference to mycorrhiza. *Plant Physiol.*, 6, 563–68.

—— & —— (1932). Further notes on the nutrient requirements and histology of the cranberry (*V. macrocarpon*) with special reference to the sources of nitrogen. *Plant Physiol.*, 7, 643–56.

ALDRICH-BLAKE, R. N. *See* BLAKE, R. N. ALDRICH-

ALLEN, P. J. (1954). Physiological aspects of fungus diseases of plants. *Annu. Rev. Pl. Physiol.*, 3, 225–48.

ALLINGTON, W. B. & CHAMBERLAIN, D. W. (1949). Trends in the population of pathogenic bacteria within leaf tissues of susceptible and immune plants. *Phytopathology*, 39, 656–60.

ANONYMOUS (1931). Establishing pines. Preliminary observations on the effects of soil inoculation. *Rhodesia Agric. J.*, 28, 185–7.

AP REES, T. *See* REES, T. AP

ARDITTI, J. (1966). The effects of niacin, ribose, and nicotinamide co-enzyme on germinating orchid seeds and young seedlings. *Amer. Orch. Soc. Bull.*, 35, 892–8.

—— (1967). Factors affecting the germination of orchid seeds. *Bot. Rev.*, 33, 1–97.

ARK, P. A. (1937). Little leaf or rosette of fruit trees VII. *Proc. Amer. Soc. Hort. Sci.*, 1936, 216–21.

ASAI, T. (1944). Über die Mykorrhizabildung der Leguminosen-Pflanzen. *Jap. J. Bot.*, 13, 463–85.

BAILLON, G. & BAILLON, M. (1924). The non-symbiotic germination of orchid seeds in Belgium. *Orchid Rev.*, 32, 305–8.

BAIN, H. F. (1937). Production of synthetic mycorrhiza in the cultivated cranberry. *J. Agric. Res.*, 55, 811–35.

BAKER, K. F. & SNYDER, W. C. (Eds.) (1965). *Ecology of Soil-borne Plant Pathogens.* University of California Press, Berkeley & Los Angeles, 571 pp.

BARBER, D. A. (1966). Effect of micro-organisms on nutrient absorption by plants. *Nature (London)*, 212, 638–40.

—— (1967). Effect of micro-organisms on the absorption of inorganic nutrients by intact plants. I. Apparatus and culture technique. *J. Exp. Bot.*, 18, 163–9.

—— & LOUGHMAN, B. C. (1967). The effect of micro-organisms on the absorption of inorganic nutrients by intact plants II. Uptake and utilization of phosphate by barley plants grown under sterile and non-sterile conditions. *J. Exp. Bot.*, 18, 170–6.

Barnes, R. L. & Naylor, A. W. (1959). Effect of various nitrogen sources on the growth of isolated roots of *Pinus serotina*. *Physiol. Plant.*, 12, 82–9.
Barrett, J. T. (1947). Observations on the root endophyte, *Rhizophagus*, in culture. *Phytopathology*, 37, 359–60.
—— (1958). Synthesis of mycorrhiza with pure cultures of *Rhizophagus*. *Phytopathology*, 48, 391.
—— (1961). Isolation, culture and host relation of the phycomycetoid vesicular-arbuscular mycorrhizal endophyte, *Rhizophagus*. *Recent Advances in Botany, University of Toronto Press*, 2, 1725.
Barrows, F. L. (1936). Propagation of *Epigaea repens* I. *Contr. Boyce Thompson Inst.*, 8, 81–97.
—— (1941). Propagation of *Epigaea repens* II. *Contr. Boyce Thompson Inst.*, 11, 431–48.
Baylis, G. T. S. (1959). Effect of vesicular-arbuscular mycorrhiza on the growth of *Griselinia littoralis* (Cornaceae). *New Phytol.*, 58, 274–80.
—— (1961). The significance of mycorrhizas and root nodules in New Zealand vegetation. *Proc. Roy. Soc. N.Z.*, 89, 45–50.
—— (1961*a*). The effect of vesicular-arbuscular mycorrhizas on the growth of *Griselinia littoralis* (Cornaceae). *Recent Advances in Botany*, 2, 1732–5.
—— (1962). *Rhizophagus*. The Catholic Symbiont. *Aust. J. Sci.*, 25, 195–200.
—— (1967). Experiments on the ecological significance of phycomycetous mycorrhizas. *New Phytol.*, 66, 231–43.
——, McNabb, R. F. R. & Morrison, T. M. (1963). The mycorrhizal nodules of podocarps. *Trans. Brit. Mycol. Soc.*, 46, 378–84.
Beau, C. (1920). Sur le rôle trophique des endophytes d'Orchidées. *C. R. Acad. Sci., Paris*, 171, 675–7.
Beevers, H. (1961). *Respiratory Metabolism in Plants*. Row Peterson, New York, 232 pp.
Bergemann, J. (1955). Die Mykorrhiza-Ausbildung einiger Koniferen-Arten in verschiedenen Böden. *Z. Welftw.*, 18, 184–202.
—— (1956). Das Mykorrhiza-Problem in der Forstwirtschaft. *Allg. Forstz.*, 23/24, 297–304.
Bernard, N. (1899–11). References to Bernard's work are quoted in full by M. C. Rayner (1927). The following papers have been specifically mentioned in the text.
—— (1905). Nouvelles espèces d'endophyte d'Orchidées. *C. R. Acad. Sci., Paris*, 140, 1272–3.
—— (1909). L'évolution dans la symbiose. *Ann. Sci. Nat. Bot.*, 9, 1–96.
—— (1911). Les mycorrhizes des *Solanum*. *Ann. Sci. Nat. Bot.*, 14, 235–58.
—— (1911*a*). Sur la fonction fungicide des bulbes d'Ophrydées. *Ann. Sci. Nat. Bot.*, 14, 223–34.
Bierhorst, D. W. (1953). Structure and development of the gametophyte of *Psilotum nudum*. *Amer. J. Bot.*, 40, 649–58.
Bisby, G. R., James, N. & Timonin, M. I. (1933). Fungi isolated from Manitoba soils by the plate method. *Canad. J. Res.*, 8, 253–75.
Björkman, E. (1940). [In Swedish, English summary.] Mycorrhiza in pine and spruce seedlings grown under varied radiation intensities in rich soils with and without nitrate added. *Medd. Skogsförsöksanst*, 32, 23–74.

BJÖRKMAN, E (1941). [In Swedish, German summary.] Die Ausbildung und Frequenz der Mykorrhiza in mit Asche gedüngten und ungedüngten Teilen von entwässertem Moor. *Medd. Skogsförsöksanst*, 32, 255–96.

—— (1942). Uber die Bedingungen der Mykorrhizabildung bei Kiefer und Fichte. *Symb. Bot. Upsaliens.*, 6(2), 1–191.

—— (1944). The effect of strangulation on the formation of mycorrhiza in pine. *Svensk Bot. Tidskr.*, 38, 1–14.

—— (1949). The ecological significance of the ectotrophic mycorrhizal association in forest trees. *Svensk Bot. Tidskr.*, 43, 223–62.

—— (1953). [In Swedish, English summary.] Factors arresting early growth of the spruce after plantation in northern Sweden. *Norrlands Skogsv. Förb. Tidskr.*, 2, 285–316.

—— (1956). Uber die Natur der Mykorrhizabildung unter besonderer Berüchsichtigung der Waldbaüme und die Anwendung in der forstlichen Praxis. *Forstwiss. Zbl.*, 75, 257–512.

—— (1960). *Monotropa hypopitys* L. an epiparasite on tree roots. *Physiol. Plantarum*, 13, 308–29.

—— (1961). The influence of ectotrophic mycorrhiza on the development of forest tree plants after planting. *Proc. XII I.U.F.R.O. Congress, Wien*, seperatum, 24, 1–3.

BLAIR, I. D. (1943). Behaviour of the fungus *Rhizoctonia solani* Kühn in the soil. *Ann. Appl. Biol.*, 30, 118–27.

BLAKE, R. N. ALDRICH- (1930). The root system of the Corsican Pine. *Oxford For. Mem.*, 12, 1–64.

BOND, G. & SCOTT, G. D. (1955). An examination of some symbiotic systems for fixation of nitrogen. *Ann. Bot.* (*London*), 19, 67–77.

BOSE, S. R. (1947). Hereditary (seed-borne) symbiosis in *Casuarina equisetifolia. Nature* (*London*), 159, 512–14.

BOULLARD, B. (1951). Champignons endophytes de quelques fougères indigènes et observations relatives à *Ophioglossum vulgatum* L. *Le Botaniste*, 35, 257–80.

—— (1953). Les champignons endophytes des Araliacées. *Bull. Soc. Bot. Fr.*, 100, 75–77.

—— (1953*a*). Les champignons endophytes des Cornacées. *Bull. Soc. Bot. Fr.*, 100, 150–2.

—— (1954). Nouvelles observations relatives à l'infection mycorhizienne experimentale des plantules d'*Hedera helix* L. *Bull. Soc. Bot. Fr.*, 101, 361–3.

—— (1956). Progrès recents dans l'étude des mycorhizes endotrophes. *Bull. Soc. Bot. Fr.*, 103, 75–90.

—— (1957). La mycotrophie chez les ptéridophytes, sa fréquence, ses charactères, sa signification. *Le Botaniste*, 45, 1–185.

—— (1957*a*). Premières observations concernant l'influence du photopériodisme sur la formation des mycorrhizes. *Mem. Soc. Sci. Nature et Math. Cherbourg*, 48, 1–12.

—— (1958). Les mycorrhizes des espèces de contact marin et de contact salin. *Rev. Mycol.*, 23, 282–317.

—— (1958*a*). Découverte de gamétophyte de *l'Ophioglossum vulgatum* L. en Normandie. *Bull. Soc. Linn. Normandie*, Série 9, 9, 112–18.

BOULLARD, B. (1959). Relations entre la photopériode et l'abondance des mycorrhizes chez l'*Aster tripolium* L. (Composées). *Bull. Soc. Bot. Fr.*, **106**, 131–4.

—— (1960). La lumière et les mycorrhizes. *Ann. Biol.*, 36, 231–48.

—— (1960*a*). Un facteur à ne pas négliger pour mener à bien les opérations de boisement des sols incultes: le facteur biologique. *Bull. Soc. Linn. de Normandie*, Série 10, 1, 47–76.

—— (1960*b*). La culture des plantes forestières en sachets. *Bull. Soc. Linn. de Normandie*, Série 10, 1, 221–4.

—— (1961). Influence du photopériodisme sur le mycorrhization de jeunes conifères. *Bull. Soc. Linn. Normandie*, Série 10, **2**, 30–46.

—— (1962). Mycorrhizes endotrophes chez diverses plantes collectée dans les peuplements nord-américaines de pin Weymouth (*P. strobus* L.). *Bull. Soc. Linn. Normandie*, Série 10, 3, 30–43.

—— (1963). Les arbrisseaux de nos parcs et les mycorrhizes. *Bull. Soc. Hist. Nat. Doubs.*, 65, 31–42.

—— (1963*a*). Les mycorrhizes des Graminées. *Journ. d'Agric. et Bot. Appliquée*, **10**, 411–37.

—— (1963*b*). Le gamétophyte des Ophioglossacées. *Bull. Soc. Linn. de Normandie*, Série 10, 4, 81–97.

—— & DOMINIK, T. (1960). Recherches comparatives entre le mycotrophisme du *Fagetum carpaticum* de Babia Góra et celui d'autres Fageta précédement étudiés. *Zesz. Nauk. W. S. R. Szczecin*, 3, 3–20.

—— & —— (1966). Les associations mycorrhiziennes dans les hêtraies françaises II. *Bull. Mus. Hist. Nat. Marseille*, **26**, 5–19.

BOURSNELL, J. G. (1950). The symbiotic seed-borne fungus in the Cistaceae. *Ann. Bot.* (*London*), 14, 217–43.

BOWEN, G. D. (1963). The natural occurrence of mycorrhizal fungi for *Pinus radiata* in South Australian soils. *C.S.I.R.O. Divisional Report*, 6/63, 11 pp.

—— (1964). The uptake of phosphate by mycorrhizas of *Pinus radiata*. Paper read to 3rd Australian Conference in Soil Science, Canberra.

—— & ROVIRA, A. D. (1961). The effects of micro-organisms on plant growth, I. Development of roots and root-hairs in sand and agar. *Plant and Soil*, 15, 166–88.

—— & —— (1966). Microbial factors in short-term phosphate uptake studies with plant roots. *Nature* (*London*), **211**, 665–6.

—— & THEODORU, C. (1967). Studies on phosphate uptake by mycorrhizas. *Proc. 14th I.U.F.R.O. Congress, Munich* 1967, **24**, 116–38.

BRIAN, P. W., HEMMING, H. G. & MCGOWAN, J. C. (1945). Origin of a toxicity to mycorrhiza in Wareham Heath. *Nature* (*London*), 155, 637–8.

BRIERLEY, J. K. (1955). Seasonal fluctuations of the concentrations of oxygen and carbon dioxide in the litter layer of beech woods with reference to salt uptake by excised mycorrhizal roots of the beech. *J. Ecol.*, 43, 404–8.

BRISCOE, C. B. (1959). Early results of mycorrhizal inoculation in pine in Puerto Rico. *Caribbean Forestry*, **20**, 73–77.

BROOK, P. J. (1952). Mycorrhiza of *Pernettya macrostigma*. *New Phytol.*, 51, 388–97.

BROWN, R. (1946). Biological stimulation in germination. *Nature* (*London*), 157, 64–9.

BROWN, W. (1922). Studies in the physiology of parasitism VIII. On the exosmosis of nutrient substances from the host into the infection drop. *Ann. Bot. (London)*, 36, 101–19.

—— (1936). The physiology of host-parasite relations. *Bot. Rev.*, 2, 236–81.

BULLER, A. H. R. (1933). *Researches on Fungi* V. Longmans, Green, London, 416 pp.

BULTEL, G. (1926). Les Orchidées germinées sans champignons sont des plants normales. *Rev. Hort., Paris*, 98, 125.

BURGEFF, H. (1909). *Die Wurzelpilze der Orchideen, ihre Kultur und ihr Leben in der Pflanze*. G. Fischer, Jena, 220 pp.

—— (1911). *Die Anzucht tropischer Orchideen aus Samen*. G. Fischer, Jena, 89 pp.

—— (1932). *Saprophytismus und Symbiose*. G. Fischer, Jena, 249 pp.

—— (1934). Pflanzliche Avitaminose und ihre Behebung durch Vitaminzufuhr. *Ber. Dtsch. Bot. Ges.*, 52, 384–90.

—— (1936). *Samenkeimung der Orchideen*. G. Fischer, Jena, 312 pp.

—— (1938). Mycorrhiza. In *Manual of Pteridology*. Chronica Botanica, The Hague, pp. 159–91.

—— (1961). *Mikrobiologie des Hochmores*. Gustav Fischer Verlag, Stuttgart, 197 pp.

BURGES, N. A. (1936). On the significance of mycorrhiza. *New Phytol.*, 35, 117–31.

—— (1939). Soil fungi and root infection. *Broteria*, 8, 1–18.

—— (1939*a*). The defensive mechanism in orchid mycorrhiza. *New Phytol.*, 38, 273–83.

—— (1963). Some problems in soil microbiology. *Trans. Brit. Mycol. Soc.*, 46, 1–14.

—— (1965). Soil microflora—its nature and biology. In *Ecology of Soilborne Plant Pathogens*, Ed. K. F. Baker & W. C. Snyder. University of California Press, Berkeley & Los Angeles, pp. 21–32.

—— & FENTON, E. (1953). The effect of carbon dioxide on the growth of soil fungi. *Trans. Brit. Mycol. Soc.*, 36, 104–8.

BUTLER, E. J. (1939). The occurrences and systematic position of the vesicular arbuscular type of mycorrhizal fungi. *Trans. Brit. Mycol. Soc.*, 22, 274–301.

BUTLER, F. C. (1953). Saprophytic behaviour of some cereal root-rot fungi I-III. *Ann. Appl. Biol.*, 40, 284–97, 298–304, 305–11.

CAMPBELL, E. (1963). *Gastrodia minor* Petrie, an epiparasite of Manuka. *Trans. Roy. Soc. N. Zealand*, 2, 73–81.

CAPPELLETTI, C. (1935). Sulla fruttificazione basidiofora dell' *Hypochnus catonii* (Burgeff). *Nuovo G. Bot. Ital.*, 42, 265–6.

CARRÉ, C. G. & HARRISON, R. W. (1961). Studies on the vesicular-arbuscular endophytes III. An endophyte of *Conocephalum conicum* (L.) Dum. identified with a strain of *Pythium*. *Trans. Brit. Mycol. Soc.*, 44, 565–72.

CARRODUS, B. B. (1965). Absorption and assimilation of nitrogen compounds by mycorrhizas. *D. Phil. Thesis, Oxford University* (typescript).

—— (1966). Absorption of nitrogen by mycorrhizal roots of beech I. Factors affecting the assimilation of nitrogen. *New Phytol.*, 65, 358–71.

—— (1967). Absorption of nitrogen by mycorrhizal roots of beech II. Ammonia and nitrate as sources of nitrogen. *New Phytol.*, 66, 1–4.

CARRODUS, B. B. & HARLEY, J. L. (1968). Note on the incorporation of acetate and the T.C.A. cycle in mycorrhizal roots of beech. *New Phytol.*, **67**, 557–60.

CARTWRIGHT, K. ST. G. & FINDLAY, W. P. K. (1946). *Decay of Timber and its Prevention.* London, H.M.S.O., 294 pp.

CATONI, G. (1929). La fruttificazione basidiofora di un endofita dell' Orchidee. *Boll. Staz. Pat. Veg. Roma*, **9**, 66–74.

CHARLES, A. (1954). The respiratory fluctuations of starving detached leaves. *New Phytol*, **53**, 81–91.

CHESTERS, C. G. C. (1948). A contribution to the study of fungi in the soil. *Trans. Brit. Mycol. Soc.*, **30**, 100–17.

CHILVERS, G. A. & PRYOR, L. D. (1965). The structure of eucalypt mycorrhizas. *Austral. J. Bot.*, **13**, 245–59.

CHRISTOPH, H. (1921). Untersuchungen über die mycotrophen Verhältnisse der 'Ericales' und die Keimung von Pyrolaceen. *Beih. Bot. Zbl.*, **38**, 115–17.

CLARK, F. B. (1964). Micro-organisms and soil structure affect Yellow Poplar growth. *U.S. For. Serv. Res. Paper*, C.S. **9**, 11 pp.

CLEMENT, E. (1924). Germintaion of *Odontoglossum* and other seed without fungal aid. *Orchid Rev.*, **32**, 233–8.

—— (1924*a*). The non-symbiotic germination of orchid seeds. *Orchid Rev.*, **32**, 359–65.

—— (1926). The non-symbiotic and symbiotic germination of orchid seeds. *Orchid Rev.*, **34**, 165–9.

CLODE, J. J. E. (1956). As micorrizas na nigraçao do fosforo, estudo com O ^{32}P. *Publ. Serv. Flor. Aquic. Portugal*, **23**, 167–205.

CLOWES, F. A. L. (1949). The morphology and anatomy of the roots associated with ectotrophic mycorrhiza. *D. Phil., Thesis, Oxford University* (typescript).

—— (1950). Root apical meristems of *Fagus sylvatica. New Phytol.*, **49**, 249–68.

—— (1951). The structure of mycorrhizal roots of *Fagus sylvatica. New Phytol.*, **50**, 1–16.

—— (1954). The root-cap of ectotrophic mycorrhizas. *New Phytol.*, **53**, 525–9.

CORDENOY, H. J. DE (1904). Sur une fonction spéciale des mycorrhizes des racines laterales de la vanille. *C.R. Acad. Sci., Paris*, **138**, 391–3.

COUNCILMAN, W. T. (1923). The root-system of *Epigaea repens* and its relation to the fungi of the humus. *Proc. Nat. Acad. Sci., Washington*, **9**, 279–85.

CROMER, D. A. (1935). The significance of the mycorrhiza of *Pinus radiata. Bull. For. Bur. Aust.*, **16**, 19 pp.

CROSSETT, R. N. & LOUGHMAN, B. C. (1966). The absorption and translocation of phosphorus by seedlings of *Hordeum vulgare* L. *New Phytol.*, **65**, 459–68.

CRUICKSHANK, I. A. M. (1963). Phytoalexins. *Annu. Rev. Phytopath.*, **1**, 351–74.

—— (1965). Pisatin studies: the relation of phytoalexins to disease reactions in plants. In *Ecology of Soilborne Plant Pathogens*, Ed. K. F. Baker, & W. C. Snyder. University of California Press, Berkeley & Los Angeles, pp. 325–36.

CURTIS, J. T. (1939). The relation of the specificity of orchid mycorrhiza fungi to the problem of symbiosis. *Amer. J. Bot.*, **26**, 390–8.

DAFT, M. J. & NICOLSON, T. H. (1966). Effect of *Endogone* mycorrhiza on plant growth. *New Phytol.*, **65**, 343–50.

DALE, J., MCCOMB, A. L. & LOOMIS, W. F. (1955). Chlorosis, mycorrhizae and the growth of pines on a high lime soil. *Forest Sci.*, **1**, 148–57.

DANIELS, J. (1963). Saprophytic and parasitic activities of some isolates of *Corticium solani*. *Trans. Brit. Mycol. Soc.*, 46, 485–502.

DE CORDENOY, H. J. *See* CORDENOY, H. J. DE

DEMETER, K. (1923). Ueber Plasmoptysenmykorrhiza. *Flora (Jena)*, 116, 405–56.

DENIS, M. (1919). Sur quelques thalles d'*Aneura* dépourvus de chlorophylle. *C.R. Acad. Sci., Paris*, 168, 64–66.

DE SILVA, R. L. *See* SILVA, R. L. DE

DIMBLEBY, G. W. (1953). Natural regeneration of pine and birch on the heather moors of north-east Yorkshire. *Forestry*, 26, 41–52.

DOAK, K. D. (1928). The mycorrhizal fungus of *Vaccinium*. *Phytopathology*, 18, 148.

—— (1955). Mineral nutrition of mycorrhizal association of Bur Oak. *Lloydia*, 18, 101–8.

DOMINIK, T. (1955). [In Polish, German summary.] Vorschlag einer neuer Klassifikation der ektotrophen Mykorrhizen auf morphologisch-anatomischen Merkmalen begrundet. *Roczn. Nauk. Rol.*, 14, 223–45.

—— & BOULLARD, B. (1961). Les associates mycorrhiziennes dans les hêtraies Françaises. *Prace Inst. Badwcz. Lesn., Warszawa*, No. 207, 3–30.

DOWDING, E. S. (1959). Ecology of *Endogone*. *Trans. Brit. Mycol. Soc.*, 42, 449–57.

DOWNIE, D. G. (1940). On the germination and growth of *Goodyera repens*. *Trans. Bot. Soc. Edinb.*, 33, 36–51.

—— (1941). Notes on the germination of British orchids. *Trans. Bot. Soc. Edinb.*, 33, 94–103.

—— (1943). Notes on the germination of *Corallorhiza innata*. *Trans. Bot. Soc. Edinb.*, 33, 380–92.

—— (1943*a*). The source of the symbiont of *Goodyera repens*. *Trans. Bot. Soc. Edinb.*, 33, 383–90.

—— (1957). *Corticium solani*, as an orchid endophyte. *Nature* (*London*), 179, 160.

DUGGAR, B. M. & DAVIS, A. R. (1916). Studies in the physiology of fungi I. Nitrogen fixation. *Ann. Mo. Bot. Gdn.*, 3, 413–37.

FASSI, B. (1965). Micorrize ectotrofiche di *Pinus strobus* L. prodotte da un' endogone (*Endogone lactiflua* Berk.). *Allionia*, 11, 7–15.

FAVARGER, C. (1953). Sur la germination des gentianes. *Phyton*, 4, 275–89.

FERDINANSEN, C. & WINGE, O. (1925). *Cenococcum* Fr. A monographic study. *Kgl. Veterin. Landbohöjsk. Aarskr.*, pp. 332–82.

FINN, R. F. (1942). Mycorrhizal inoculation of soil of low fertility. *Black Rock For. Pap.*, 1, 116–17.

FORTIN, J. A. (1966). Synthesis of mycorrhiza on explants of the root hypocotyl of *Pinus sylvestris* L. *Canadian Journ. Bot.*, 44, 1087–92.

FRANCKE, H. L. (1934). Beiträge zue Kenntnis der Mykorrhiza von *Monotropa hypopytis* L. Analyse und Synthese der Symbiose. *Flora* (*Jena*), 129, 1–5.

FRANK, A. B. (1885). Ueber die auf Wurzelsymbiose beruhende Ernährung gewisser Bäume durch Unterirdische Pilze. *Ber. Dtsch. bot. Ges.*, 3, 128–45.

—— (1894). Die bedeutung der Mykorrhizapilze für die gemeine Kiefer. *Forstwiss. Zbl.*, 16, 185–90.

FRASER, L. (1931). An investigation of *Lobelia gibbosa* and *L. dentata*. I. Mycorrhiza, latex system and general biology. *Proc. Linn. Soc. N.S.W.*, 56, 497–525.

FRAYMOUTH, J. (1956). Haustoria of the Peronosporales. *Trans. Brit. Mycol. Soc.*, **39**, 79–107.
FREISLEBEN, R. (1933). Über experimentelle Mykorrhiza-Bildung bei den Ericaceen. *Ber. Dtsch. Bot. Ges.*, **51**, 351–6.
—— (1934). Zur Frage der Mykotrophie der Gattung *Vaccinium* L. *Jb. Wiss. Bot.*, **80**, 421–56.
—— (1936). Weitere Untersuchungen über die Mykotrophie der Ericaceen. *Jb. Wiss. Bot.*, **82**, 413–59.
FRIES, N. (1941). Über die Sporenkeimung bei einigen Gasteromyceten und mykorrhizabildenen Hymenomyceten. *Arch. Mikrobiol.*, **12**, 266–84.
—— (1943). Untersuchungen über Sporenkeimung und Mycelentwicklung bodenbewohnender Hymenomyceten. *Symb. Bot. Upsaliens.*, 6(4), 1–81.
GALLAUD, I. (1905). Études sur les mycorrhizes endotrophes. *Rev. Gén. Bot.*, **17**, 5–48, 66–83, 123–36, 223–39, 313–25, 425–33, 479–500.
GAMS, W. (1963). *Mycelium radicis atrovirens* in forest soils. Isolation from soil micro-habitats and identification. In *Soil Organisms*, Ed. J. Doeksen & J. van der Drift, North Holland Publ. Co., Amsterdam, pp. 176–82.
GARDE, R. V. LA (1929). Non-symbiotic germination of orchids. *Ann. Mo. Bot. Gdn.*, **16**, 499–514.
GARRETT, S. D. (1950). Ecology of root-inhabitating fungi. *Biol. Rev.*, **25**, 220–54.
—— (1951). Ecological groups of soil fungi: a survey of substrate relationships. *New Phytol.*, **50**, 149–66.
—— (1952). The soil fungi as a microcosm for ecologists. *Sci. Progr.*, **40**, 436–50.
—— (1956). *Biology of Root-infecting Fungi.* Cambridge University Press, 293 pp. [This book gives references to the literature on association of fungi and roots, and to Dr. Garrett's previous work.]
—— (1962). Decomposition of cellulose in soil by *Rhizoctonia solani* Kühn. *Trans. Brit. Mycol. Soc.*, **45**, 115–20.
GAST, P. R. (1937). Studies in the development of conifers in raw humus III. *Medd. Skogsförsöksanst*, **29**, 587–82.
GÄUMANN, E. (1950). *Principles of Plant Infection.* Crosby Lockwood, London, 543 pp.
——, BRAUN, R. & BAZZIGHER, G. (1950). Über induzierte Abwehrreaktionen bei Orchideen. *Phytopath. Z.*, **17**, 36–62.
—— & HOHL, H. R. (1960). Weitere Untersuchungen über die chemischen Abwehrreaktionen der Orchideen. *Phytopath. Z.*, **38**, 93–104.
—— & JAAG, O. (1945). Über induzierte Abwehrreaktionen bei Orchideen. *Experimentia*, **1**, 21–22.
—— & KERN, H. (1959). Über die chemische Abwehrreaktionen bei Orchideen. *Phytopath. Z.*, **35**, 1–26.
—— & —— (1959*a*). Über die Isolierung und chemischer Nachweis des Orchinols. *Phytopath. Z.*, **35**, 347–56.
——, MULLER, E., NUESCH, J. & RIMPAU, R. H. (1961). Über die Wurzelpilze von *Loroglossum hircinum* (L.) Rich. *Phytopath. Z.*, **41**, 89–96.
——, NUESCH, J. & RIMPAU, R. H. (1960). Weitere Untersuchungen über die chemisher Abwehrreaktionen der Orchideen. *Phytopath. Z.*, 38, 274–308.
GERDEMANN, J. W. (1955). Relation of a large soil-borne spore to phycomycetous mycorrhizal infections. *Mycologia*, **47**, 619–32.

GERDEMANN, J. W. (1961). A species of *Endogone* from corn causing vesicular-arbuscular mycorrhiza. *Mycologia*, 53, 254–61.

—— (1964). The effect of mycorrhizas on the growth of maize. *Mycologia*, 56, 342–9.

—— (1965). Vesicular-arbuscular mycorrhizae formed on maize and tulip tree by *Endogone fasciculata*. *Mycologia*, 57, 562–75.

—— & NICOLSON, T. M. (1963). Spores of mycorrhizal *Endogone* species extracted from soil by wet sieving. *Trans. Brit. Mycol. Soc.*, 46, 235–44.

GERRETSEN, F. C. (1937). Manganese deficiency of oats and its relation to soil bacteria. *Ann. Bot.* (*London*), New Series 1, 208–30.

—— (1948). The influence of micro-organisms on phosphate intake by the plant. *Plant and Soil*, 1, 51–81.

GOBL, F. (1965). Mykorrhiza Untersuchungen in einem subalpinen Fichtewald. *Beiträge Subalpinen Waldforchung*, 66, 173–95.

GODFREY, R. M. (1957). Studies on British species of *Endogone* I. Taxonomy and morphology. *Trans. Brit. Mycol. Soc.*, 40, 117–35.

—— (1957*a*). Studies on British species of *Endogone* II. Fungal parasites. *Trans. Brit. Mycol. Soc.*, 40, 136–44.

—— (1957*b*). Studies on British species of *Endogone* III. Germination of spores. *Trans. Brit. Mycol. Soc.*, 40, 203–10.

GORDON, H. D. (1937). Mycorrhiza in *Rhododendron*. *Ann. Bot.* (*London*) New Series **I**, 593–613.

GOSS, R. W. (1960). Mycorrhizae of ponderosa pine in Nebraska grassland soils. *Univ. Nebraska Coll. of Agric. Res. Sta. Bull.*, 192, 47 pp.

GOSSELIN, R. (1944). Studies on *Polystictus circinatus* and its relation to butt-rot of spruce. *Farlowia*, 1, 525–68.

GRAY, L. E. & GERDEMANN, J. W. (1967). Influence of vesicular-arbuscular mycorrhizas on the uptake of phosphorus-32 by *Liriodendron tulipifera* and *Liquidambar styraciflua*. *Nature* (*London*), 213, 106–7.

GRAY, P. H. H. & MCMASTER, N. B. (1933). A microbiological study of podsol soil profiles. *Canad. J. Res.*, 8, 375–89.

—— & TAYLOR, C. B. (1935). A microbiological study of podsol soil profiles. II. Laurentian soils. *Canad. J. Res.*, 13, 251–62.

GREENALL, J. M. (1963). The mycorrhizal endophytes of *Griselinia littoralis* (Cornaceae). *N.Z. Journ. Bot.*, 1, 389–400.

GRIFFITHS, D. A. (1965). The mycorrhizas of some conifers grown in Malaya. *Malayan Forester*, 28, 118–21.

GÜNNEWIG, J. (1933). Beiträge zur Kenntnis des Loliumpilzes. *Beitr. Biol. Pfl.*, 20, 227–54.

HAASOUNA, M. C. & WAREING, P. F. (1964). Possible role of rhizosphere bacteria in the nutrition of Ammophila arenaria. *Nature* (*London*), 202 467.

HACSKAYLO, E. (1951). A study of the roots of *Pinus virginiana* in relation to certain Hymenomycetes suspected of being mycorrhizal. *J. Wash. Acad. Sci.*, 41, 399–400.

—— (1953). Pure culture syntheses of pine mycorrhizae in terralite. *Mycologia*, 45, 971–5.

—— (1961). Factors influencing mycorrhizal development in some cultural practices. *Recent Advances in Botany*, 2, pp. 1744–8.

HACSKAYLO, E. & PALMER, J. G. (1955). Hymenomycetous species forming mycorrhiza with *Pinus virginiana*. *Mycologia*, **47**, 145–57.

—— & SNOW, A. G. (1959). Relation of soil nutrients and light to prevalence of mycorrhizae. *U.S.D.A. Station Paper No.* 125, *North Eastern Forest Service*, 13 pp.

——, —— & VOZZO, J. A. (1965). Effect of temperature on growth and respiration of ectotrophic mycorrhizal fungi. *Mycologia*, **57**, 748–56.

HALKET, A. C. (1930). The rootlets of *Amyelon radicans* Will.; their anatomy, apices and endophytic fungus. *Ann. Bot.* (*London*), **44**, 865–904.

HAM, A. M. (1962). Studies on vesicular-arbuscular endophytes IV. Inoculation of species of *Allium* and *Lactuca sativa* with *Pythium* isolates. *Trans. Brit. Mycol. Soc.*, **45**, 179–89.

HAMADA, M. (1940). Studien über die Mykorrhiza von *Galeola septentrionalis* Reichb. Ein neuer Fall der Mykorrhiza-Bildung durch intraradikale Rhizomorpha *Jap. J. Bot.*, **10**, 151–212.

—— (1940*a*). Physiologische-morphologische Studien über *Armillaria mellea* (Vahl) Quel. mit besonderer Rücksicht auf die Oxalsäurebildung. *Jap. J. Bot.*, **10**, 387–464.

—— & NAKAMURA, S. I. (1963). Wurzelsymbiose von *Galeola altissima* Reichb. f., einer chlorophyll-freien Orchidee, mit dem holzstörenden Pilze *Hymenochaete crocicreas* Berk. et Br. *Sci. Rep. Tôhoku University* (4 Series), **29**, 227–38.

HANDLEY, W. R. C. (1963). Mycorrhizal associations and *Calluna* heathland afforestation. *Forestry Comm. Bull.*, **36**, H.M.S.O. London, 70 pp.

—— & SANDERS, C. J. (1962). The concentration of easily soluble reducing substances in roots and the formation of ectotrophic mycorrhizal associations. A re-examination of Björkman's hypothesis. *Plant and Soil*, **16**, 42–61.

HANNAY, J. W. & BUTCHER, D. N. (1961). An ageing process in excised roots of groundsel (*Senecio vulgaris* L.). *New Phytol.*, **60**, 9–20.

HARLEY, J. L. (1937). Ecological observations on the mycorrhiza of beech. *J. Ecol.*, **25**, 421–3.

—— (1940). A study of the root system of the beech in woodland soils with especial reference to mycorrhizal infection. *J. Ecol.*, **28**, 107–17.

—— (1948). Mycorrhiza and soil ecology. *Biol. Rev.*, **23**, 127–58.

—— (1950). Recent progress in the study of endotrophic mycorrhiza. *New Phytol.*, **49**, 213–47.

—— (1952). Associations between micro-organisms and higher plants (Mycorrhiza). *Annu. Rev. Microbiol.*, **6**, 367–86.

—— (1964). Incorporation of carbon dioxide into excised beech mycorrhizas in the presence and absence of ammonia. *New Phytol.*, **63**, 203–8.

—— & BEEVERS, H. (1963). Acetate utilization by maize roots. *Pl. Physiol.*, **38**, 117–23.

—— & BRIERLEY, J. K. (1954). Uptake of phosphate by excised mycorrhizal roots of the beech VI. *New Phytol.*, **53**, 240–52.

—— & —— (1955). Uptake of phosphate by excised mycorrhizal roots of the beech VII. *New Phytol.*, **54**, 296–301.

——, —— & MCCREADY, C. C. (1954). Uptake of phosphate by excised mycorrhizal roots of the beech V. *New Phytol.*, **53**, 92–98.

HARLEY, J. L. & JENNINGS, D. H. (1958). The effect of sugars on the respiratory responses of beech mycorrhiza to salts. *Proc. Roy. Soc. B.*, 148, 403–18.

—— & LOUGHMAN, B. C. (1963). The uptake of phosphate by excised mycorrhizal roots of the beech IX. *New Phytol.*, 62, 350–9.

—— & —— (1966). Phosphohexoisomerase in beech mycorrhiza. *New Phytol.*, 65, 157–60.

—— & MCCREADY, C. C. (1950). Uptake of phosphate by excised mycorrhizal roots of the beech I. *New Phytol.*, 49, 388–97.

—— & —— (1952). Uptake of phosphate by excised mycorrhizal roots of the beech II. *New Phytol.*, 51, 56–64.

—— & —— (1952*a*). Uptake of phosphate by excised mycorrhizal roots of the Beech III. *New Phytol.*, 51, 342–8.

—— & —— (1953). A note on the effect of sodium azide on the respiration of beech mycorrhiza. *New Phytol.*, 51, 342–4.

——, —— & BRIERLEY, J. K. (1953). Uptake of phosphate by excised mycorrhizal roots of the Beech IV. *New Phytol.*, 52, 124–32.

——, —— & —— (1958). Uptake of phosphate by excised mycorrhizal roots of the beech VIII. *New Phytol.*, 57, 353–62.

——, ——, —— & JENNINGS, D. H. (1956). The salt respiration of excised beech mycorrhizas II. *New Phytol.*, 55, 1–28.

——, —— & GEDDES, J. A. (1954). The salt respiration of excised beech mycorrhizas I. *New Phytol.*, 53, 427–44.

—— & REES, T. AP (1959). Cytochrome oxidase in mycorrhizal and uninfected roots of *Fagus sylvatica*. *New Phytol.*, 58, 364–86.

—— & SMITH, D. C. (1956). Sugar absorption and surface carbohydrase activity of *Peltigera polydactyla*. *Ann. Bot.* (*London*), New Series, 20, 513–43.

—— & WAID, J. S. (1955). A method of studying active mycelia on living roots and other surfaces in the soil. *Trans. Brit. Mycol. Soc.*, 38, 104–18.

—— & —— (1955*a*). The effect of light upon the roots of beech and its surface population. *Plant and Soil*, 7, 96–112.

—— & WILSON, J. M. (1959). The absorption of potassium by beech mycorrhizas. *New Phytol.*, 58, 281–98.

—— & —— (1963) Die Wechselwirkung der Kationen während der Aufnahme durch Buchenmykorrhiza. In *Mykorrhiza*, Ed. W. Rawald & H. Lyr. G. Fischer, Jena, pp. 261–72.

HARPER, J. L. (1950). An investigation of the interaction of soil micro-organisms with especial reference to the study of bacterial populations of plant roots. *D.Phil.*, *Thesis*, *Oxford University* (typescript).

HARRISON, R. W. (1955). A method of isolating vesicular-arbuscular endophytes from roots. *Nature* (*London*), 175, 432.

HARVAIS, G. & HADLEY, G. (1967). The relation between host and endophyte in orchid mycorrhiza. *New Phytol.*, 66, 205–15.

HASSLEBAUM, G. (1931). Cytologische and physiologische Studien in ericoiden endotrophen Mycorrhiza von *Empetrum nigrum*. *Bot. Arch.*, 31, 386–440.

HAASAUNA, M. C. & WAREING, P. F. (1964). Possible role of rhizosphere bacteria in the nitrogen nutrition of *ammophila anemia*. *Nature* (*London*), 202, 467.

HATCH, A. B. (1934). A jet-black mycelium forming ectotrophic mycorrhizae. *Svensk Bot. Tidskr.*, 28, 369–83.

HATCH, A. B. (1937). The physical basis of mycotrophy in the genus *Pinus*. *Black Rock For. Bull.*, 6, 168 pp.

—— & DOAK, K. D. (1933). Mycorrhizal and other features of the root system of *Pinus*. *J. Arnold Arbor.*, 14, 85–99.

—— & HATCH, C. T. (1933). Some Hymenomycetes forming ectotrophic mycorrhizae with *Pinus strobus* L. *J. Arnold Arbor.*, 14, 324–34.

HAUSEN, S. S. VON (1948). On the role of growth substances in higher plants. *Physiol. Plant.*, 1, 85–94.

HAWKER, L. E. (1954). British hypogeous fungi. *Phil. Trans.*, 237, 429–545.

—— (1962). Studies on vesicular-arbuscular endophytes V. A review of the evidence relating to identity of the causal fungi. *Trans. Brit. Mycol. Soc.*, 45, 190–9.

—— & HAM, A. M. (1957). Vesicular-arbuscular mycorrhizas in apple seedlings. *Nature* (*London*). 180, 995.

——, HARRISON, R. W., NICHOLS, V. O. & HAM, A. M. (1957). Studies on vesicular-arbuscular endophytes I. *Trans. Brit. Mycol. Soc.*, 40, 375–90.

HENDERSON, M. W. (1919). A comparative study of the structure and saprophytism of the Pyrolaceae and Monotropaceae with reference to their derivation from the Ericaceae. *Contr. Bot. Lab. Univ. Pa.*, 5, 42–51.

HENRY, A. W. (1931). Occurrence and sporulation of *Helminthosporium sativum* in the soil. *Canad. J. Res.*, 5, 407–13.

HEPDEN, P. M. (1960). Studies on vesicular-arbuscular endophytes II. Endophytes in the pteridophyta with special reference to leptosporangiate ferns. *Trans. Brit. Mycol. Soc.*, 43, 559–70.

HILDEBRAND, A. A. (1934). Recent observations on the strawberry root-rot in the Niagara peninsula. *Canad. J. Res.*, 11, 18–31.

—— & KOCK, L. W. (1936). A microscopical study of infection of the roots of strawberry and tobacco seedlings by micro-organisms of the soil. *Canad. J. Res.*, 14, 11–26.

—— & WEST, P. M. (1941). Strawberry root-rot in relation to microbiological changes induced in root-rot soil by the incorporation of certain cover crops. *Canad. J. Res.*, 19, 183–98.

HILTNER, L. (1904). Über neuer Erfahrungen und Probleme auf dem Gebeit der Bodenbacteriologie und unter besonderer Beruchsichtigung der Gründüngung und Brache. *Arb. Dtsch. Landw. Ges.*, 98, 59–78.

HOAGLAND, D. R., CHANDLER, W. H. & STOUT, P. R. (1937). Little leaf or rosette of fruit trees VI. *Proc. Amer. Soc. Hort. Sci.*, 1936, 210–12.

HOLEVAS, C. D. (1966). The effect of vesicular-arbuscular mycorrhiza on the uptake of soil phosphorus by strawberry (*Fragaria* sp. var. Cambridge Favourite). *J. Hort. Sci.*, 41, 57–64.

HOLLÄNDER, S., *in* BURGEFF, H. (1936).

HOW, J. E. (1940). The mycorrhizal relations of larch I. A study of *Boletus elegans* (Schum.) in pure culture. *Ann. Bot.* (*London*), New Series 4, 135–50.

—— (1941). The mycorrhizal relations of larch II. The role of the larch root in the nutrition of *Boletus elegans* (Schum). *Ann. Bot.* (*London*), New Series, 5 121–3.

HUBERMAN, M. A. (1940). Normal growth and development of southern pine seedlings in the nursery. *Ecology*, 21, 323–34.

IRMISCH, T. (1853). *Beiträge zur Biologie und Morphologie der Orchideen.* Leipzig, 82 pp.

JAHN, E. (1934). Die peritrophe Mykorrhiza. *Ber. Dtsch. Bot. Ges.*, **52**, 463.

JAMES, W. O. & WARD, M. M. (1957). The dieca effect in the respiration of barley. *Proc. Roy. Soc. London*, **B147**, 309–15.

JENNINGS, D. H. (1956). The respiration of beech mycorrhiza. *D.Phil. Thesis, Oxford University* (typescript).

—— (1958). Changes in the internal carbohydrates of beech mycorrhizas during treatment with azide. *New Phytol.*, **57**, 254–5.

—— (1964). Changes in the size of orthophosphate pools in mycorrhizal roots of beech with reference to absorption of the ion from the external medium. *New Phytol.*, **63**, 181–93.

—— (1964*a*). The effect of cations on the absorption of phosphate by beech mycorrhizal roots. *New Phytol.*, **63**, 348–57.

JONES, F. R. (1924). A mycorrhizal fungus in the roots of legumes and some other plants. *J. Agric. Res.*, **29**, 459–70.

JONES, W. NEILSON. *See* NEILSON JONES, W.

KAMIENSKI, F. (1881). Die Vegetationsorganen der *Monotropa hypopytis* L. *Bot. Ztg.*, **39**, 458–61.

KATZNELSON, H. (1940). Survival of azobacter in soil. *Soil Sci.*, **49**, 21–35.

—— (1940*a*). Survival of micro-organisms introduced into soil. *Soil Sci.*, **49**, 283–93.

—— (1946). The 'rhizosphere effect' of mangels on certain groups of soil micro-organisms. *Soil Sci.*, **62**, 443–54.

—— (1965). Nature and importance of the rhizosphere. In *Ecology of Soilborne Plant Pathogens*, Ed. K. F. Baker & W. C. Snyder. University of California Press, Berkeley & Los Angeles, pp. 187–209.

—— & RICHARDSON, L. T. (1948). Rhizosphere studies and associated micro-biological phenomena in relation to strawberry root-rot. *Sci. Agric.*, **28**, 293–308.

——, ROUATT, J. W. & PAYNE, T. M. B. (1955). The liberation of amino-acids and reducing compounds by plant roots. *Plant and Soil*, **7**, 35–48.

——, —— & PETERSON, F. A. (1962). The rhizosphere effect of mycorrhizal and non-mycorrhizal roots of yellow birch seedlings. *Canad. J. Bot.*, **40**, 377–82.

KELLER, H. G. (1950). Die Verwertbarkeit verschiedener Kohlehydrate und Dicarbonsäuren durch *Cenococcum graniforme. Schweiz. Z. All. Path.*, **13**, 565–9.

—— (1952). *Untersuchungen über das Wachstum von* Cenococcum graniforme *(Sow.) Ferd. u. Winge, auf verschieden Kohlenstoffquellen.* Thesis, Juris-Verlag, Zürich, 123 pp.

KELLEY, A. P. (1950). *Mycotrophy in Plants.* Chronica Botanica, Waltham, Mass., 206 pp.

KESSELL, S. L. (1927). Soil organisms: the dependence of certain pine species on a biological soil factor. *Emp. For. J.*, **6**, 70–74.

KESSLER, K. J. (1966). Growth and development of mycorrhizae on sugar maple (*Acer saccharum* Marsh). *Canad. J. Bot.*, **44**, 1413–25.

KEYWORTH, W. G. (1947). Verticillium wilt of the hop (*Humulus lupulus*) II. The selection of wilt-resistant varieties. *J. Pomol.*, **23**, 99–108.

KIDSTON, R. & LANG, W. H. (1921). On the Old Red Sandstone plants showing structure from the Rhynie Chert bed, Aberdeenshire, pt. V. *Trans. Roy. Soc. Edinb.*, 52, 855–902.

KLJUSNIK, P. I. (1952). [In Russian.] Oak mycorrhiza. *Lesn. Khoz.*, 5, 63. From *Rev. Appl. Myc.*, 35, 315.

KNUDSON, L. (1922). Non-symbiotic germination of orchid seeds. *Bot. Gaz.*, 75, 1–25.

—— (1924). Further observations on non-symbiotic germination of orchid seeds. *Bot. Gaz.*, 77, 212–19.

—— (1925). Physiological study of the germination of orchid seeds. *Bot. Gaz.*, 79, 345–79.

—— (1927). Symbiosis and asymbiosis relative to orchids. *New Phytol.*, 26, 328–36.

—— (1929). Seed germination and growth of *Calluna vulgaris*. *New Phytol.*, 28, 369–85.

—— (1930). Flower production by orchid grown non-symbiotically. *New Phytol.*, 32, 192–9.

—— (1933). Non-symbiotic development of seedlings of *Calluna vulgaris*. *New Phytol.*, 32, 127–55.

KOCH, H. (1961). Untersuchungen über die Mykorrhiza der Kulturpflanzen unter besonderer Berücksichtigung von *Althaea officinalis* L., *Atropa belladonna* L., *Helianthus annua* L., und *Solanum lycopersicum* L. *Gartenbauwiss.*, 26, 5.

KONOË, R. (1957). Über das Vorkommonen der Wurzelpilze bei *Metasequoia* und den nächst verwandten Pflanzen. *J. Inst. Polytechnics. Osaka City Univ.*, 8, 179–84.

KRAMER, P. J. & HODGSON, R. H. (1954). Differences between mycorrhizal and non-mycorrhizal roots of Loblolly Pine. *VIIIth Int. Bot. Cong., Paris*, 13, 133–4.

—— & WILBUR, K. M. (1949). Absorption of radioactive phosphorus by mycorrhizal roots of pine. *Science*, 110, 8–9.

KRASILNIKOV, N. A. (1961). *Soil Micro-organisms and Higher Plants*. Israel programme for Scientific Translations, 474 pp.

——, KRISS, A. E. & LITVINOV, M. A. (1936). Influence of root excretions on the development of *Azotobacter* and of other soil microbes I. *Mikrobiology U.S.S.R.*, 5, 87–98.

——, —— & —— (1936*a*). Ibid. II. *Mikrobiology U.S.S.R.*, 5, 270–82.

KUBIKOVA, J. (1965). Mycoflora synusias on the roots of woody plants. *Rozpravy Česk. Akad. Věd.*, 75, 54 pp.

KUSANO, S. (1911). *Gastrodia elata* and its symbiotic association with *Armillaria mellea*. *J. Agric. Tokio*, 4, 1–66.

LA GARDE, R. V. *See* GARDE, R. V. LA

LAIHO, O. (1965). Further studies on the ectendotrophic mycorrhizas *Acta For. Fenn.*, 79, 3, 35 pp.

—— & MIKOLA, P. (1964). Studies on the effect of some eradicants on mycorrhizal development in forest nurseries. *Acta For. Fenn.*, 77, 33 pp.

LAING, E. V. (1923). Tree roots: their action and development. *Trans. R. Scot. Arb. Soc.*, 37, 6–21.

LANOWSKA, J. (1966). [In Polish, English summary.] Influence of different sources of nitrogen on the development of mycorrhiza in *Pisum sativum. Pamietnik Pulowski*, 21, 365–86.

LEACH, R. (1937). Observations on the parasitism and control of *Armillaria mellea. Proc. Roy. Soc. B.*, 121, 561–73.

—— (1939). Biological control and ecology of *Armillaria mellea* (Vahl) *Fr. Trans. Brit. Mycol. Soc.*, 23, 320–9.

LEVISOHN, I. (1952). Forking in pine roots. *Nature* (*London*), 169, 175.

—— (1953). Growth response to mycorrhizal mycelia in the absence of mycorrhizal association. *Nature* (*London*), 172, 316–17.

—— (1954). Aberrant root infections of pine and spruce seedlings. *New Phytol.*, 53, 284–90.

—— (1955). Isolation of ectotrophic mycorrhizal mycelia from rhizomorphs. *Nature* (*London*), 176, 519.

—— (1956). Growth stimulation of forest tree seedlings by the activity of free-living mycorrhizal mycelia. *Forestry*, 29, 53–59.

—— (1956*a*). Effect of mulching on mycorrhizal fungi. *Emp. For. Rev.*, 35, 281–6.

—— (1957). Antagonistic effects of *Alternaria tenuis* on certain root-fungi of forest trees. *Nature* (*London*), 179, 1143–4.

—— (1957*a*). Differential effects of root infecting mycelia on young trees in different environments. *Emp. For. Rev.*, 36, 281–6.

—— (1959). Strain differentiation in a root-infecting fungus. *Nature (London)*, 183, 1065–6.

—— (1960). Physiological and ecological factors influencing the effects of mycorrhizal inoculation. *New Phytol.*, 59, 42–47.

LEWIS, D. H. (1963). Uptake and utilization of substances by beech mycorrhiza. *D.Phil. Thesis*, *Oxford University*, 153 pp. (typescript).

—— & HARLEY, J. L. (1965). Carbohydrate physiology of mycorrhizal roots of beech. I. Identity of endogenous sugars and utilization of exogenous sugars. *New Phytol.*, 64, 224–37.

—— & —— (1965*a*). *Idem.* II. Utilization of exogenous sugars by uninfected and mycorrhizal roots. *New Phytol.*, 64, 238–55.

—— & —— (1965*b*). *Idem.* III. Movement of sugars between host and fungus. *New Phytol.*, 64, 256–69.

LEWIS, F. J. (1924). An endotrophic fungus in the Coniferae. *Nature* (*London*), 114, 860.

LIHNELL, D. (1939). Untersuchungen über die Mykorrhizen und die Wurzelpilze von *Juniperus communis. Symb. Bot. Upsaliens.*, 3(3), 1–141.

—— (1942). *Cenococcum graniforme* als Mykorrhizabildner von Waldbäumen. *Symb. Bot. Upsaliens.*, 5(2), 1–18.

—— (1942*a*). Keimungsversuche mit Pyrolasamen. *Symb. Bot. Upsaliens.*, 6(3), 1–37.

LINDEBERG, G. (1944). Über die physiologie lignin-abbauender Boden-Hymenomyceten. *Symb. Bot. Upsaliens.*, 8(2), 1–183.

—— (1948). On the occurrence of polyphenol oxidases in soil-inhabiting Basidiomycetes. *Physiol. Plant*, 1, 196–205.

LINDQUIST, B. (1932). Den sydskandinaviska kulturgromomskogens reproduktionsförhällanden. *Svenska Skogsv. fören. Tidskr.*, 30, 17–38 (with German summary).

LINDQUIST, B. (1939). Die Fichtenmykorrhiza im Lichte der modernen Wuchstofforschung. *Bot. Notiser*, 1939, 315–56.
LINNEMANN, G. (1955). Untersuchungen über die Mykorrhiza von *Pseudotsuga taxifolia* Britt. *Zbl. Bakt.*, 108, 398–410.
—— (1959). Beobachtungen über die Mykorrhiza-Typen bei *Pseudotsuga. Allg. Forstzeitschrifte*, 51, 1–4.
—— (1960). Rassenunterscheide bei *Pseudotsuga taxifolia* hinsichlich der Mykorrhiza. *Allg. Forst. und Jagdzeit*, 131 (2), 41–47.
—— (1964). Mykorrhiza und Düngung. *Allg. Forst. und Jagdzeit*, 9, 228–33.
LIPS, S. H. & BIALE, J. B. (1966). Stimulations of oxygen uptake by electron transfer inhibitors. *Plant Physiol.*, 41, 797–802.
LOBANOW, N. W. (1960). *Mykotrophie der Holzpflanzen.* VEB Deutscher Verlag der Wissenschaften, Berlin, 352 pp.
LOUGHMAN, B. C. (1966). The mechanism of absorption and utilization of phosphate by barley plants in relation to transport to the shoot. *New Phytol.*, 65, 388–97.
—— & MARTIN, R. P. (1957). Methods and equipment for the study of incorporation of phosphorus by intact barley plants in experiments of short duration. *J. Exp. Bot.*, 8, 272–9.
—— & RUSSELL, R. S. (1957). The absorption and utilization of phosphate by young barley plants IV. *J. Exp. Bot.*, 8, 280–93.
LUCAS, R. L. (1960). Transport of phosphorus by fungal mycelium. *Nature (London)*, 188, 763.
LÜCK, R. (1940). Zur Biologie der heimischen Pirola-arten. *Schr. Phys.-ökon. Ges. Königsb.*, 71, 300–34.
—— (1941). Zur Keimung der heimischen Pirola-arten. *Flora (Jena)*, 135, 1–5.
LUNDEBERG, G. (1960). The relationship between pine seedlings (*Pinus sylvestris* L.) and soil fungi. *Svensk Bot. Tidskr.*, 54, 346–59.
—— (1963). Die beziehungen zwischen Kiefensamlingen und Bodenpilzen—einige Untersuchungen mit eines neuen Wassersteril-kultur Methode. In *Mykorrhiza*, Ed. W. Rawald & H. Lyr. International Mykorrhiza Symposium, Weimar, 1960, pp. 185–9.
LYR, H. (1963). Über die Abnahme der Mykorrhiza—und Knöllchenfrequenz mit zunehmender Bodentiefe. In *Mykorrhiza*, Ed. W. Rawald & H. Lyr. International Mykorrhiza Symposium, Weimar, 1960, pp. 303–13.
—— (1963*a*). Zur Frage des Streuabbaus durch ectotrophe Mykorrhizapilze. In *Mykorrhiza*, Ed. W. Rawald & H. Lyr. International Mykorrhiza Symposium, Weimar, 1960, pp. 123–145.
MCCOMB, A. L. (1938). The relation between mycorrhizae and the development and nutrient absorption of pine seedlings in a prairie nursery. *J. For.*, 36, 1148–54.
—— (1943). Mycorrhizae and phosphorus nutrition of pine seedlings. *Bull. Ia. Exp. Sta.*, 314, 582–612.
—— & GRIFFITH, J. E. (1946). Growth stimulation and phosphorus absorption of mycorrhizal and non-mycorrhizal White Pine and Douglas Fir seedlings in relation to fertilizer treatments. *Plant Physiol.*, 21, 11–17.
MCCREADY, C. C. (1952). Studies in the physiology of some ectotrophic mycorrhizas. *D.Phil. Thesis, Oxford University* (typescript).

MacDougal, D. T. & Dufrenoy, J. (1944). Mycorrhizal symbiosis in *Aplectrum*, *Corallorhiza* and *Pinus*. *Pl. Physiol.*, **19**, 440–65.

—— & —— (1946). Criteria of nutritive relations of fungi and seed plants in mycorrhizae. *Pl. Physiol.*, **21**, 1–10.

MacLennan, D. H., Beevers, H. & Harley, J. L. (1963). Compartmentation of acids in plant tissues. *Biochem. J.*, **89**, 316–327.

McLennan, E. (1928). Growth of fungi in soil. *Ann. Appl. Biol.*, **15**, 95–109.

—— (1935). Non-symbiotic development of seedlings of *Epacris impressa* Labill. *New Phytol.*, **34**, 55–63.

McLuckie, J. N. & Burges, A. (1932). Mycotrophism in the Rutaceae I. The mycorrhiza of *Eriostemon Crowei* F. v. M. *Proc. Linn. Soc., N.S.W.*, **57**, 291–312.

McNabb, R. F. R. (1961). Mycorrhiza in the New Zealand Ericales. *Aust. J. Bot.*, **9**, 57–61.

Magrou, J. (1924). A propos du pouvoir fungicide des tubercules d'Ophrydées. *Ann. Sci. Nat. Bot.*, **6**, 265–70.

—— (1935). Essais de culture des champignons de mycorrhizes. *C.R. Acad. Sci., Paris*, **201**, 1038–9.

—— (1936). Culture et inoculation du champignon symbiotique de l'*Arum maculatum*. *C.R. Acad. Sci., Paris*, **203**, 887–8.

—— (1937). Sur la culture des champignons de mycorrhizes. *Ann. Sci. Nat. Bot.*, **19**, 359–70.

—— (1939). Nouveaux essais de culture des champignons de mycorrhizes. *C.R. Acad. Sci., Paris*, **208**, 923–5.

—— (1946). Sur la culture de quelques champignons de mycorrhizes à arbuscules et vésicules. *Rev. Gén. Bot.*, **53**, 49–77.

—— & Magrou, M. (1940). Essai de culture du champignon symbiotique de la pomme de terre. *C. R. Acad. Sci., Paris*, **211**, 234–6.

Malmborg, S. von (1933). Ein saprophytisches Lebermoos, *Cryptothallus* nov. gen. *Ann. Bryol., Hague*, **6**, 122–3.

—— (1934). Weiteres über die Gattung *Cryptothallus*. *Ann. Bryol., Hague*, **7**, 108–10.

Mariat, F. (1952). Recherches sur la physiologie des embryons d'Orchidées. *Rev. Gén. Bot.*, **59**, 324–77.

Martin, T. L., Anderson, D. A. & Goates, R. (1942). The influence of chemical composition on organic matter and development of mold flora in soil. *Soil Sci.*, **54**, 297–302.

Marx, D. H. & Davey, C. B. (1967). Ectotrophic mycorrhizae as determinants of pathogenic root infections. *Nature (London)*, **213**, 1139.

—— & Zak, B. (1965). Effect of pH on mycorrhizal formation of slash pine in aseptic culture. *Forest Science*, **11**, 66–75.

Masui, K. (1926). A study of mycorrhiza of *Abies firma*, S. & Z., with special reference to its mycorrhizal fungus *Cantharellus floccosus* Schw. *Mem. Coll. Sci. Kyoto Imp. Univ., B.* **2**, 16–84.

—— (1926*a*). On the renewed growth of the mycorrhizal root. *Mem. Coll. Sci. Kyoto Imp. Univ., B.* **2**, 85–92.

Melin, E. (1917). Studier över de norrländska myrmarkernas vegetation med särskildhänsyn till deras skogsvegetation efter torrläggning. *Akad. Avh. Uppsala*, 1–426.

MELIN, E. (1921). Till kännedomen om mykorrhizasvamparnas spridningssätt hos Ericaceerna. *Bot. Notiser*, 1921, 283–6.

—— (1923). Experimentelle Untersuchungen über die Konstitution und Ökologie der Mykorrhizen von *Pinus sylvestris* und *Picea abies*. *Mykologische Untersuch.*, 2, 73–331.

—— (1923*a*). Experimentelle untersuchungen über Birken-und Espenmykorrhiza und ihre Pilzen-symbionten. *Svensk Bot. Tidskr.*, 17, 479–520.

—— (1924). Zur Kenntnis der Mykorrhizenpilze von *Pinus montana*. *Bot. Notiser*, 1924, 69–92.

—— (1925). *Untersuchungen über die Bedeutung der Baummykorrhiza. Eine ökologische physiologische Studie.* G. Fischer, Jena, 152 pp.

—— (1927). [In Swedish, German summary.] Studier över Barrträdsplantans utreckling i råhumus. *Medd. Skogsförsöksanst, Stockh.*, 23, 433–84.

—— (1936). Methoden der experimentellen Untersuchung mykotropher Pflanzen. *Handb. Biol. ArbMeth.*, *Abt.* II, 4, 1015–18.

—— (1946). Die Einfluss von Waldstreuextrakten auf das Wachstum von Bodenpilzen mit besonderer Berückssichtigung der Wurzelpilze von Baümen. *Symb. Bot. Upsaliens.*, 8(3), 1–116.

—— (1948). Recent advances in the study of tree mycorrhiza. *Trans. Brit. Mycol. Soc.*, 30, 92–99.

—— (1953). Physiology of mycorrhizal relations in plants. *Annu. Rev. Pl. Physiol.*, 4, 325–46.

—— (1954). Growth factor requirements of mycorrhizal fungi of forest trees. *Svensk Bot. Tidskr.*, 48, 86–94.

—— (1959). Mykorrhiza. *In* W. Ruhland's *Handbuch der Pflanzenphysiologie*, XI, 605–38.

—— (1963). Some effects of forest tree roots on mycorrhizal Basidiomycetes. In *Symbiotic Associations*, Ed. P. S. Nutman & B. Mosse, Cambridge University Press, pp. 125–45.

—— & DAS, V. S. R. (1954). The influence of root-metabolites on the growth of tree mycorrhizal fungi. *Physiol. Plant.*, 7, 851–8.

—— & LINDEBERG, G. (1939). Über die Einfluss von Aneurin und Biotin auf das Wachstum einiger Mykorrhizapilze. *Bot. Notiser*, 1939, 241–5.

—— & NILSSON, H. (1950). Transfer of radioactive phosphorus to pine seedlings by means of mycorrhizal hyphae. *Physiol. Plant.*, 3, 88–92.

—— & —— (1952). Transfer of labelled nitrogen from an ammonium source to pine seedlings through mycorrhizal mycelium. *Svensk Bot. Tidskr.*, 46, 281–5.

—— & —— (1953). Transfer of labelled nitrogen from glutamic acid to pine seedlings through the mycelium of *Boletus variegatus* (Sw). Fr. *Nature (London)*, 171, 134.

—— & —— (1953*a*). Transport of labelled phosphorus to pine seedlings through the mycelium of *Cortinarius glaucopus* (Schaeff. ex Fr.) Fr. *Svensk Bot. Tidskr.*, 48, 555–8.

—— & —— (1955). Ca^{45} used as an indicator of transport of cations to pine seedlings by means of mycorrhizal mycelia. *Svensk Bot. Tidskr.*, 49, 119–22.

—— & —— (1957). Transport of C^{14} labelled photosynthate to the fungal associate of pine mycorrhiza. *Svensk. Bot. Tidskr.*, 51, 166–86.

—— (1958). Translocation of nutritive elements through mycorrhizal mycelia to pine seedlings. *Bot. Not.*, 111, 251–6.

MELIN, E. & NILSSON, H. & HACSKAYLO, E. (1958). Translocation of cations to seedlings of *Pinus virginiana* through mycorrhizal mycelia. *Bot. Gaz.*, 119, 241–6.

—— & NORKRANS, B. (1942). Uber die Einfluss der Pyramidin und der Thiazolekomponente des Aneurins auf das Wachstum von Wurzelpilze. *Svensk Bot. Tidskr.*, 36, 271–86.

—— & —— (1948). Amino-acids and the growth of *Lactarius deliciosus* (L.) Fr. *Physiol. Plant.*, 1, 174–84.

—— & NYMAN, B. (1940). Weitere Untersuchungen über die Wirkung von Aneurin und Biotin auf das Wachstum von Wurzelpilze. *Arch. Mikrobiol.*, 11, 318–28.

—— & —— (1941). Uber das Wuchstoffbedürfnis von *Boletus granulatus* (L.). *Arch. Mikrobiol.*, 12, 254–9.

MELOH, V. A. (1963). Untersuchungen zur Biologie der endotrophen Mycorrhiza bei *Zea mais* L. und *Avena sativa* L. *Archiv. Mikrobiol.*, 46, 369–81.

MEYER, F. H. (1962). Die Buchen und Fichten Mykorrhiza in verschiedenen Bodentypen, ihre Beeinflussung durch Mineraldüngung sowie für die Mykorrhizabildung wichtige Faktoren. *Mitt. d. Bundefor. Aust. Forst. u. Holz.*, 54, 73 pp.

—— (1964). Neue Erkenntnisse über des Zusammenleben von Pilz und Baum. *Die Umschau Wiss. u. Technik.*, 11, 325–28.

—— (1964*a*). The rôle of the fungus *Cenococcum graniforme* (Sow.) Ferd. et Winge, in the formation of mor. In *Soil Microbiology*, Ed. A. Jongerius. Elsevier Publishing Company, Amsterdam, pp. 23–31.

—— (1966). Mycorrhiza and other plant symbiosis. In *Symbiosis*, Ed. S. M. Henry. Academic Press, New York & London, Vol. 1, pp. 171–255.

MIKOLA, P. (1948). On the black mycorrhiza of birch. *Arch. Soc. Zool.-Bot. Fenn. Vanamo*, 1, 81–85.

—— (1948*a*). On the physiology and ecology of *Cenococcum graniforme*. *Commun. Inst. For. Finl.*, 36, 1–104.

—— (1953). An experiment on the invasion of mycorrhizal fungi into prairie soil. *Karstenia*, 2, 33–4.

—— (1955). Artificial inoculation of mycorrhizal fungi. *Arch. Soc. Zool.-Bot. Fenn. Vanamo*, 9, Suppl. 197–201.

—— (1965). Studies on the ectendotrophic mycorrhiza of pine. *Acta. For. Fenn.*, 79, 56 pp.

—— & LAIHO, O. (1962). Mycorrhizal relations in the raw humus layer of northern spruce plants. *Commun. Inst. For. Finl.*, 55, 1–13.

MILLARD, V. A. & TAYLOR, C. B. (1927). Antagonism of micro-organisms as the controlling factor in the inhibition of scab by green manuring. *Ann. Appl. Biol.*, 14, 202–15.

MITCHELL, H. L. (1934). Pot culture tests of soil fertility. *Black Rock For. Bull.*, 5, 1–138.

—— (1939). The growth and nutrition of White Pine (*Pinus strobus* L.) seedlings in cultures of varying nitrogen, phosphorus, potassium and calcium. *Black Rock For. Bull.*, 9, 1–135.

——, FINN, R. F. & ROSENDAHL, R. O. (1937). The relation between mycorrhizae and the growth and nutrient absorption of coniferous seedlings in nursery beds. *Black Rock For. Pap.*, 1, 58–73.

MODESS, O. (1941). Zur Kenntnis der Mykorrhizabildner von Kiefer und Fichte. *Symb. Bot. Upsaliens.*, 5(1), 1–147.

MÖLLER, C. (1947). Mycorrhizae and nitrogen assimilation. *Forstl. Forsöksv. Danm.*, **19**, 105–208.

MOLLIARD, M. (1937). Sur la biologie du *Calluna vulgaris* L. Signification des mycorrhizes chez les Ericacées. *Ann. Sci. Nat. Bot.*, **19**, 401–8.

MONSON, A. M. & SUDIA, T. W. (1963). Translocation in *Rhizoctonia solani*. *Bot. Gaz.*, **124**, 440–3.

MONTFORT, C. (1940). Beziehungen zwischen morphologischen und physiologischen Reduktionserscheinungen im Bereicht der Licht-Ernährung bei saprophytischen Orchideen. *Ber. Dtsch. Bot. Ges.*, **58**, 41–9.

—— & KUSTERS, E. (1941). Saprophytismus und Photosynthese I. *Bot. Arch.*, **40**, 571–633.

MORRISON, T. M. (1954). Uptake of phosphorus-32 by mycorrhizal plants. *Nature* (*London*), **174**, 606.

—— (1957). Mycorrhiza and phosphorus uptake. *Nature* (*London*), **179**, 907.

—— (1957*a*). Host-endophyte relationship in mycorrhiza of *Pernettya marostigma*. *New Phytol.*, **56**, 247–57.

—— (1962). Absorption of phosphorus from soils by mycorrhizal plants. *New Phytol.*, **61**, 10–20.

—— (1962*a*). Uptake of sulphur by mycorrhizal plants. *New Phytol.*, **61**, 21–7.

—— & ENGLISH, D. A. (1967). The significance of mycorrhizal nodules of *Agathis australis*. *New Phytol.*, **66**, 245–50.

MOSER, M. (1956). Die Bedeutung der Mykorrhiza für Aufforstungen in Hochlagen. *Forstw. Cbl.*, **75**, 8–18.

—— (1958). Die kunstliche Mykorrhiza impfung an Forstpflanzen I. *Forstw. Cbl.*, **77**, 1–64.

—— (1958*a*) Die kunstliche Mykorrhiza impfung an Forstpflanzen II. *Forstw. Cbl.*, **77**, 254–320.

—— (1958*b*). Der Einfluss tiefer temperaturen auf das Wachstum und die Lebenstätigkeit höherer Pilze mit spezieller Berücksichtigung von Mykorrhizapilzen. *Sydowia Ann. Myc.*, Ser. 11, **12**, 386–99.

—— (1959). Beiträge zur kenntnis der Wuchsstoff-beziehungen im Bereich ectotrophen Mykorrhizen. *Arch. Mikrobiol.*, **34**, 251–69.

—— (1959*a*). Die kunstliche Mykorrhiza impfung an Forstpflanzen III. *Forstw. Cbl.*, **78**, 193–202.

—— (1963). Forderung der Mykorrhizabildung in der Forstliche Praxis. *Mitt. Forst. Bundes.-Versuchsanstalt*, *Mariabrunn*, **60**, 693–720.

MOSSE, B. (1953). Fructifications associated with mycorrhizal strawberry roots. *Nature* (*London*), **171**, 974.

—— (1956). Fructifications of an *Endogone* species causing endotrophic mycorrhiza in fruit plants. *Ann. Bot.* (*London*), New Series, **20**, 349–62.

—— (1957). Growth and chemical composition of mycorrhizal and non-mycorrhizal apples. *Nature* (*London*), **179**, 922–4.

—— (1959). The regular germination of resting spores and some observations on the growth requirements of an *Endogone* sp. causing vesicular-arbuscular mycorrhiza. *Trans. Brit. Mycol. Soc.*, **42**, 273–86.

—— (1959*a*). Observations on the extramatrical mycelium of a vesicular-arbuscular endophyte. *Trans. Brit. Mycol. Soc.*, **42**, 439–48.

MOSSE, B. (1961). Experimental techniques for obtaining pure inoculum of *Endogone* sp. and some observations on the vesicular-arbuscular infestations caused by it and other fungi. *Recent Advances in Botany*, vol. 2, pp. 1728–32.

—— (1962). The establishment of vesicular-arbuscular mycorrhiza under aseptic conditions. *J. Gen. Microbiol.*, **27**, 509–20.

—— (1963). Vesicular-arbuscular mycorrhiza: an extreme form of fungal adaptation. In *Symbiotic Associations*, Ed. P. S. Nutman & B. Mosse, Cambridge University Press, pp. 146–70.

—— (1966). Rothamsted Experimental Station Report, p. 79.

NAKAMURA, S. I. (1962). Zur Samenkeimung einer Chlorophyllfreien Erdorchidee *Galeola septentrionalis* Reichb. f., *Zeit. für Bot.*, **50**, 487–97.

NEILL, J. C. (1940). The endophyte of Rye-grass (*Lolium perenne*). *N.Z. J. Sci. Tech.*, **21**, 280–91.

—— (1942). The endophytes of *Lolium* and *Festuca*. *N.Z. J. Sci. Tech.*, **23**, 185–93.

—— (1944). *Rhizophagus* in Citrus. *N.Z. J. Sci. Tech.*, **25**, 191–201.

NEILSON JONES, W. & SMITH, M. L. (1928). On the fixation of atmospheric nitrogen by *Phoma radicis callunae* including a new method of investigating nitrogen fixation by micro-organisms. *Brit. J. Exp. Biol.*, **6**, 167–89.

NELSON, C. D. (1964). The production and translocation of photosynthate-C^{14} in conifers. In *The Formation of Wood in Forest Trees*, Ed. M. H. Zimmermann. Maria Moors Cabot Foundation for Botanical Research, New York, pp. 243–59.

NĚMEC, B. (1899). Die Mykorrhiza der Lebermoose. *Ber. Dtsch. Bot. Ges.*, **17**, 311–17.

—— (1904). Über die Mykorrhiza bei *Calypogæa trichomanes*. *Beih. Bot. Zbl.*, **16**, 253–68.

NEUMANN, G. (1934). Über die Mykorrhiza in der Gattung *Gentiana*. *Zbl. Bakt.*, II, **89**, 433–58.

NEUSCH, J. (1963). Defence reactions in orchid bulbs. In *Symbiotic Associations*, Ed. P. S. Nutman & B. Mosse. Cambridge University Press, pp. 335–43.

NICHOLAS, G. (1932). Associations des bryophytes avec d'autres organismes. In *Manual of Bryology*, Chronica Botanica, The Hague, pp. 109–28.

NICOLSON, T. H. (1955). The mycotrophic habit in grasses. *Thesis*, *Nottingham University* (typescript.)

—— (1958). Vesicular-arbuscular mycorrhiza in the Gramineae. *Nature* (*London*), **181**, 718.

—— (1959). Mycorrhiza in the Gramineae I. Vesicular-arbuscular endophytes with special reference to the external phase. *Trans. Brit. Mycol. Soc.*, **42**, 421–38.

—— (1960). Mycorrhiza in the Gramineae II. Development in different habitats, particularly sand-dunes. *Trans. Brit. Mycol. Soc.*, **43**, 132–45.

—— (1963). Vesicular-arbuscular Mykorrhiza bie den Gramineen—Morphologische und ökologische aspekte. In *Mykorrhiza*, Ed. W. Rawald & H. Lyr. Internat. Mykorrhizasymposium, Weimar, 1960. G. Fischer, Jena, pp. 57–66.

—— (1967). Vesicular-arbuscular mycorrhiza—a universal plant symbiosis. *Sci. Prog.* (*Oxford*), **55**, 561–81.

NIEWIECZERZALOWNA, B. (1932). Recherches morphologiques sur la mycorrhize des Orchidées indigenes. *C.R. Soc. Sci. Varsovie*, 25, 86–115.

NOBÉCOURT, P. (1923). Sur la production d'anticorps par les tubercules des Ophrydées. *C. R. Acad. Sci., Paris*, 177, 1055–7.

NOGGLE, C. R. & WYND, F. L. (1943). Effect of vitamins on germination and growth of orchids. *Bot. Gaz.*, 104, 455–9.

NORKRANS, B. (1949). Some mycorrhiza-forming *Tricholoma* species. *Svensk Bot. Tidskr.*, 43, 485–90.

—— (1950). Studies in growth and cellulolytic enzymes of *Tricholoma*. *Symb. Bot. Upsaliens.*, 11(1), 1–126.

—— (1953). The effect of glutamic acid, aspartic acid and related compounds on the growth of certain *Tricholoma* species. *Physiol. Plant.*, 6, 584–93.

NUTMAN, P. S. (1958). The physiology of nodule formation. In *Nutrition of Legumes*, Ed. E. G. Hallsworth. Butterworths Scientific Publications, London, pp. 87–107.

OLIVEROS, S. (1932). Effect of soil inoculation on the growth of Benguet pine (quoted from Hatch, 1937).

ORLOV, A. Y. (1957). Observations on absorbing roots of spruce (*Picea excelsa* Link) in natural conditions. *Botanicheskii Zhurnal S.S.S.R.*, 42, 1072–181.

—— (1960). Further observations on the growth of absorbing roots of *Picea excelsa* Link. *Botanicheskii Zhurnal, S.S.S.R.*, 45, 888–96.

OSBORN, E. M. & HARPER, J. L. (1951). Antibiotic production by growing plants of *Leptosyne maritima*. *Nature (London)*, 167, 685–6

OSBORN, T. G. B. (1909). Lateral roots of *Amyelon radicans* and their mycorrhiza. *Ann. Bot. (London)*, 23, 603–10.

OTTO, G. (1959). Beitrag zur Frage der funktionellen Bedeutung der Vesikel der endotrophen Mycorrhiza an Samlingen von *Malus communis* L. *Arch. Mikrobiol.*, 32, 373–92.

—— (1962). Das Auftreten und die Entwicklung der endotrophen Mykorrhiza an ein-bis dreijährigen Appelsämlingen auf verschiedenen Standorten, I and II. *Zentbl. Bakt. Parasitkde.*, 115, 404–38, 525–44.

PARKINSON, D. & COUPS, E. (1963). Microbiological activity in a podsol. In *Soil Organisms*, Ed. J. Doeksen & J. van der Drift. N. Holland Publ. Co., Amsterdam, pp. 167–75.

PEARSON, R. & PARKINSON, D. (1961). The sites of excretion of ninhydrin-positive substances by broad bean seedlings. *Plant and Soil*, 13, 391–6.

PENNINGSFELD, F. (1950). Über Atmungsuntersuchungen an einigen Bodenprofilen. *Z. PflErnähr., Düng.*, 50, 135–64.

PÉREMBOLON, M. & HADLEY, G. (1965). Production of pectic enzymes by pathogenic and symbiotic *Rhizoctonia* strains. *New Phytol.*, 64, 144–51.

PEUSS, H. (1958). Untersuchungen zur Ökologie und Bedeutung der Tabakmycorrhiza. *Archiv. für Mikrobiologie*, 29, 112–42.

PEYRONEL, B. (1921). Nouveaux cas de rapports mycorrhiziques entre Phanérogames et Basidiomycetes. *Bull. Soc. Mycol. Fr.*, 37, 143–8.

—— (1922). Sulla normale presenza di micorize nel grano e in altre piante coltivate e spontanee. *Boll. Staz. Pat. Veg. Roma*, 3, 43–50.

—— (1922*a*). Nuovi casi di rapporti micorizici tra Basidiomiceti e Fanerogame arboree. *Boll. Soc. Bot. Ital.*, 1, 3–14.

PEYRONEL, B. (1923). Fructification de l'endophyte à arbuscules et a vésicules des mycorrhizes endotrophes. *Bull. Soc. Mycol. Fr.*, **39**, 119–26.

— (1924). Prime recherche sulle micorize endotrofiche e sulla microflora radicola normale delle fanergame. *Riv. Biol.*, **5**, 463–85; **6**, 17–53.

— (1963). Mykorrhizenstruktur und mykorrhizagene Pilze. In *Mykorrhiza*, Ed. W. Rawald & H. Lyr. Internat. Mykorrhizasymposium, Weimar, 1960, G. Fischer, Jena, pp. 15–25.

— & FASSI, B. (1956–57). Micorrize ectotrofiche in una Cesalpiniacea del Congo Belga. *Atti. Accad. Sci, Torino*, **91**, 1–8.

PIJL, L. VAN DER (1934). Die Mykorrhiza von *Burmannia* und *Epirrhizanthes* und die Fortflanzung ihres Endophyten. *Rec. Trav. Bot. Néerl.*, **31**, 761–79.

PRAT, H. (1926). Étude des mycorrhizes du *Taxus baccata*. *Ann. Sci. Nat. Bot.*, **8**, 141–62.

PRESTON, R. D. (1952). *The Molecular Architecture of Plant Cell Walls*. Chapman & Hall, London, 311 pp.

PRESTON, R. J. (1943). Anatomical studies of the roots of juvenile lodgepole pine. *Bot. Gaz.*, **104**, 443–8.

PRYOR, L. D. (1956). Ectotrophic mycorrhiza in renantherous species of *Eucalyptus*. *Nature* (*London*), **177**, 587–8.

QUENDOW, K. G. (1930). Beiträge zur Frage der Aufnahme gelöster Kohlenstoffverbindungen durch Orchideen und andere Pflanzen. *Bot. Arch.*, **30**, 51–108.

RAMSBOTTOM, J. (1922). Orchid mycorrhiza. *Trans. Brit. Mycol. Soc.*, **8**, 28–66.

RAWALD, W. (1963). Untersuchungen zur Stickstoffernährung der höheren Pilze. In *Mykorrhiza*, Ed. W. Rawald & H. Lyr. Internat. Mykorrhizasymposium, Weimar, 1960, G. Fischer, Jena, pp. 67–83.

RAYNER, M. C. (1915). Obligate symbiosis in *Calluna vulgaris*. *Ann. Bot.* (*London*), **29**, 97–133.

— (1925). The nutrition of mycorrhiza plants: *Calluna vulgaris*. *Brit. J. Exp. Biol.*, **2**, 265–91.

— (1927). Mycorrhiza. *New Phytol.*, reprint 15, 246 pp. (Refer to this for Dr. Rayner's earlier papers.)

— (1929). Biology of fungus infection in the genus *Vaccinium*. *Ann. Bot.* (*London*), **43**, 55–70.

— (1929*a*). Seedling development in *Calluna vulgaris*. *New Phytol.*, **28**, 377–85.

— (1938). Use of soil or humus inocula in nurseries and plantations. *Emp. For. J.*, **17**, 236–43.

— & LEVISOHN, I. (1940). Production of synthetic mycorrhiza in the cultivated cranberry. *Nature* (*London*), **145**, 461.

— & — (1941). The mycorrhizal habit in relation to forestry IV. Studies on mycorrhizal response in *Pinus* and other conifers. *Forestry*, **15**, 1–36.

— & NEILSON JONES, W. (1944). *Problems in Tree Nutrition*. Faber & Faber, London, 184 pp. (Refer to this for papers on tree mycorrhiza by Dr. Rayner and collaborators.)

— & SMITH, M. L. (1929). *Phoma radicis callunae*. A physiological study. *New Phytol.*, **28**, 261–90.

RAYNER, R. W. (1948). Latent infection in *Coffea arabica* L. *Nature* (*London*), **161**, 245–6.

REDMOND, D. R. (1957). Observations on rootlet development in Yellow Birch. *Forestry Chronicle* 33, 208–212.

REES, T. AP (1957). The effect of fungal infection on the respiratory metabolism of plant tissues. *D.Phil. Thesis*, *Oxford University* (typescript).

REITSMA, J. (1932). Studien über *Armillaria mellea* (Vahl) Quel. *Phytopath. Z.*, 4, 461–522.

RENNER, O. (1938). Über blasse saprophytische *Cephalanthera alba* und *Epipactis latifolia*. *Flora* (*Jena*), 32, 225–36.

RENNIE, P. J. (1955). The uptake of nutrients by mature forest growth. *Plant and Soil*, 7, 49–95.

RHEINHOLD, E. L. (1949). Studies on the uptake of sugars and sugar derivatives by root storage tissue. *Ph.D. Thesis*, *Cambridge University* (typescript).

RICHARDS, B. N. (1961). Soil pH and mycorrhiza development in *Pinus*. *Nature* (*London*), **190**, 105–6.

—— (1965). Mycorrhiza development of loblolly pine seedlings in relation to soil reaction and supply of nitrate. *Plant and Soil*, **22**, 187–99.

—— & VOIGT, G. K. (1964). Role of mycorrhiza in nitrogen fixation. *Nature* (*London*), **201**, 310–11.

—— & WILSON, G. L. (1963). Nutrient supply and mycorrhiza development in Caribbean pine. *For. Sci.*, **9**, 405–12.

RICHARDSON, D. H. S., SMITH, D. C. & LEWIS, D. H. (1967). Carbohydrate movement between the symbionts of lichens. *Nature (London)*, **214**, 879–82.

RISHBETH, J. (1950). Observations on the biology of *Fomes annosus* with particular reference to East Anglian pine plantations I. *Ann. Bot.* (*London*), New Series, 14, 365–85.

—— (1951). Observations on the biology of *Fomes annosus* with particular reference to East Anglian pine plantations II. *Ann. Bot.* (*London*), New Series 15, 1–21.

—— (1951*a*). Observations on the biology of *Fomes annosus* with particular reference to East Anglian pine plantations III. *Ann. Bot.* (*London*), New Series, **15**, 38, 221–46.

RIVETT, M. (1924). The root tubercles in *Arbutus unedo*. *Ann. Bot.* (*London*), **38**, 661–77.

ROBERTSON, N. F. (1954). Studies on the mycorrhiza of *Pinus sylvestris* I. *New Phytol.*, **53**, 253–83.

ROBINSON, R. K. (1963). A study of some aspects of the biology of certain root-infecting fungi. *D.Phil. Thesis*, *Oxford University* (typescript), 115 pp.

ROELOFFS, J. W. (1930). Over Kunstmatige verjonging van *Pinus merkussii*, Jungh et de Vr. en *P. Khasya* Royle. *Tectona*, **23**, 874–907.

ROMELL, L. G. (1938). A trenching experiment in spruce forest and its bearing on problems of mycotrophy. *Svensk Bot. Tidskr.*, **32**, 89–99.

—— (1939) The ecological problem of mycotrophy. *Ecology*, **20**, 163–7.

—— (1939*a*). Barrskogens marksvampar och deras roll i skogens liv. *Svenska Skogsv. fören. Tidskr.*, **37**, 348–75.

ROSENDAHL, R. O. & WILDE, S. A. (1942). Occurrence of ectotrophic mycorrhizal fungi in soils of cut-over areas and sand dunes. *Bull. Ecol. Soc. Amer.*, **23**, 73–74.

ROTHSTEIN, A. (1954). The enzymology of the cell surface. *Protoplasmatologia*, **2**, 1–86.

ROUTIEN, J. B. & DAWSON, R. F. (1943). Some interrelationships of growth, salt absorption, respiration and mycorrhizal development in *Pinus echinata*. *Amer. J. Bot.*, **30**, 440–51.

ROVIRA, A. D. (1956). Plant root excretions in relation to the rhizosphere effect. *Plant and Soil*, **7**, 209–17.

—— (1959). Root excretions in relation to the rhizosphere effect IV. Influence of plant species, age of plant, light, temperature, and calcium nutrition on exudation. *Plant and Soil*, **11**, 53–64.

—— (1965). Plant root exudates and their effects upon micro-organisms. In *Ecology of Soilborne Plant Pathogens*, Ed. K. F. Baker & W. C. Snyder. University of California Press, Berkeley & Los Angeles, pp. 170–86.

—— & BOWEN, G. D. (1966). Phosphate incorporation by sterile and non-sterile plant roots. *Aust. J. Biol. Sci.* **19**, 1167–9.

RUBENCHIK, L. I. (1960). *Azotobacter and its Uses in Agriculture*. Translated by A. Artman. Nat. Sci. Foundation, Washington, D.C., and Department of Agriculture, by the Israel Program of Scientific Translations, 1963, 278 pp.

RUINEN, J. (1953). Epiphytosis. A second view on epiphytism. *Ann. Bogor.*, **1**, 101–57.

RUSSELL, E. W. (1950). [Sir E. J. Russell's] *Soil Conditions and Plant Growth*, 8th edn. Longmans, Green, London, xvi+635 pp.

RUSSELL, R. S. & SHORROCKS, V. M. (1959). The relation between transpiration and the absorption of inorganic ions by intact plants. *J. Exp. Bot.*, **10**, 301–16.

SAMPSON, K. (1933). The systemic infection of grasses by *Epichloë typhina* (Pers.) Tul. *Trans. Brit. Mycol. Soc.*, **18**, 30–47.

—— (1935). The presence and absence of an endophytic fungus in *Lolium temulentum* and *L. perenne*. *Trans. Brit. Mycol. Soc.*, **19**, 337–43.

—— (1938). Further observations on the systemic infection of *Lolium*. *Trans. Brit. Mycol. Soc.*, **21**, 84–96.

—— (1939). Additional notes on the systemic infection of *Lolium*. *Trans. Brit. Mycol. Soc.*, **23**, 316–19.

SANFORD, G. B. (1926). Some factors affecting the pathogenicity of *Actinomyces scabies*. *Phytopathology*, **16**, 525–47.

—— & BROADFOOT, W. C. (1931). Studies on the effects of other soil-inhabiting micro-organisms on the virulence of *Ophiobolus graminis*. *Sci. Agric.*, **11**, 512–28.

SANTORO, T. & CASIDA, L. E. (1962). Growth inhibition of mycorrhizal fungi by gibberellin. *Mycologia*, **54**, 70–71.

SANTOS, N. F. (1941). Elementos para o Estado des micorrizas ectendotroficas do *Pinus pinaster* Sol. *Publ. Div. Geral. Serv. Florestais Aquicolas*, **8**, 65–95.

SARAUW, G. F. L. (1893). Über die Mykorrhizen unserer Waldbäume. *Bot. Zbl.*, **53**, 343–5.

SCHAFFERSTEIN, G. (1938). Untersuchungen über die Avitaminose der Orchideen Keimlinge. *Jb. Wiss. Bot.*, **86**, 720–52.

—— (1941). Die Avitaminose der Orchideen Keimlinge. *Jb. Wiss. Bot.*, **90**, 141–98.

SCHELLING, C. L. (1950). Die Verwertbarkeit verschiedener Kohlenstoffquellen durch *Mycelium radicis atrovirens* Melin. *Schweiz. Z. Allg. Path.*, **13**, 570–4.

SCHMUCKER, T. H. (1959). Saprophytismus bei Kormophyten. *Encyclopaedia Plant Phys.*, 11, 386–428.

SCHRADER, R. (1958). Untersuchungen zur Biologie der Erbsenmycorrhiza. *Arkiv für Mikrobiol.*, 32, 81–114.

SCHÜTTE, K. H. (1956). Translocation in the fungi. *New Phytol.*, 55, 164–82.

SEALY, J. R. (1949). *Arbutus unedo*. *J. Ecol.*, 37, 365–88.

—— & WEBB, D. A. (1950). Biological flora of the British Isles. No 1311, *Arbutus* L. *J. Ecol.*, 38, 223–36.

SEWARD, A. C. (1898). *Fossil Plants*, vol. I, p. 207. Cambridge University Press, 452 pp.

SHIRLEY, H. L. (1945). Light as an ecological factor and its measurement. *Bot. Rev.*, 9, 497–532.

SHIROYA, T., LISTER, G. R., SLANKIS, V., KROTKOV, G. & NELSON, C. D. (1962). Translocation of the products of photosynthesis to roots of pine seedlings. *Canad. J. Bot.*, 40, 1125–36.

SILOW, R. A. (1933). A systemic disease of Red Clover, caused by *Botrytis anthophila* (Bond.). *Trans. Brit. Mycol. Soc.*, 18, 239–48.

SILVA, R. L. DE & WOOD, R. K. S. (1964). Infection of plants by *Corticium solani* and *C. praticola*—effect of plant exudates. *Trans. Brit. Mycol. Soc.*, 47, 15–24.

SIMMONDS, P. M. (1947). The influence of antibiosis in the pathogenicity of *Helminthosporium sativum*. *Sci. Agric.*, 27, 625–32.

—— & LEDINGHAM, R. J. (1937). A study of the fungus flora of wheat roots. *Sci. Agric.*, 18, 49–59.

SINGH, K. G. (1964). Fungi associated with roots and rhizosphere of Ericaceae. *Ph.D. Thesis, Durham University* (typescript).

—— (1965). Comparison of techniques for isolation of root infecting fungi *Nature* (*London*), 206, 1169–70.

—— (1966). Ectotrophic mycorrhiza in equatorial rain forests. *Malayan Forestry*, 29, 13–18.

SLANKIS, V. (1948). Einfluss von Exudaten von *Boletus variegatus* auf die dichotomische Verzweigung isolierter Kiefernwurzeln. *Physiol. Plant.*, 1, 390–400.

—— (1948*a*). Verschiedene Zuckerarten als Kohlenhydratquelle für isolierter Wurzeln von *Pinus sylvestris*. *Physiol. Plant.*, 1, 278–89.

—— (1949). Wirkung von β-Indolylessigsäure auf die dichotomische Verzweigung isolierter Wurzeln von *Pinus sylvestris* L. *Svensk Bot. Tidskr.*, 43, 603–7.

—— (1951). Über den Einfluss von β-Indolylessigsäure und andere Wirksstoffen auf das Wachstum von Kiefernwurzeln. *Symb. Bot. Upsaliens.*, 11, 1–63.

—— (1958). The role of auxin and other exudates in mycorrhizal symbiosis of forest trees. *In* K. V. Thimann, Ed., *Physiology of Forest Trees*. Ronald Press, New York, pp. 427–43.

—— (1961). On the factors determining the establishment of ectotrophic mycorrhiza of forest trees. *Recent Advances in Botany IXth International Bot. Congress*, 1960, pp. 1738–42.

—— (1963). Der gegenwärtiger Stand unseres Wissens von der Bildung der ektotrophen Mykorrhizen bei Waldbaumen. In W. Rawald & H. Lyr *Mykorrhiza*, Ed., International Mykorrhizasymposium, Weimar, 1960, G. Fischer, Jena, pp. 175–83.

SMITH, G. & KERSTEN, H. (1942). The relation between xylem thickenings in primary roots of *Vicia faba* seedlings and elongation, as shown by soft X-ray irradiation. *Bull. Torrey Bot. Cl.*, 69, 221–34.

SMITH, S. E. (1966). Physiology and ecology of *Orchis* mycorrhizal fungi with reference to seedling nutrition. *New Phytol.*, 65, 488–99.

—— (1967). Carbohydrate translocation in orchid mycorrhizas. *New Phytol.*, 66, 371–8.

SPOEHR, H. A. (1942). The culture of albino maize. *Plant Physiol.*, 17, 397–410.

SPRÄU, F. (1937). Beiträge zur Mykorrhizenfrage. Die Fructification eins aus *Orchis mascula* isolierten Wurzelpilzes *Corticium masculi* (nov. spec.). *Jb. Wiss. Bot.*, 85, 151–68.

STAHL, E. (1900). Der Sinn der Mykorrhizenbildung. *Jb. Wiss. Bot.*, 34, 534–668.

STAHL, M. (1949). Die Mykorrhiza der Lebermoose mit besonderer Beruck-sichtigung der thallosen Formen. *Planta*, 37, 103–48.

STALDER, L. & SCHULTZ, F. (1957). Untersuchungen über die kausalen Zusammenhange des Erikawurzelsterbens. *Phytopath. Zeit.*, 30, 117–48.

STARKEY, R. L. (1929). Some influences of the development of higher plants upon micro-organisms in the soil I. *Soil Sci.*, 27, 319–34.

—— (1929*a*). Some influences of the development of higher plants upon the micro-organisms in the soil II. *Soil Sci.*, 27, 355–78.

—— (1929*b*). Some influences of the development of higher plants upon the micro-organisms in the soil III. *Soil Sci.*, 27, 433–44.

—— (1931). Some influences of the development of higher plants upon the micro-organisms in the soil IV. *Soil Sci.*, 32, 367–92.

—— (1931*a*). Some influences of the development of higher plants upon the micro-organisms in the soil V. *Soil Sci.*, 32, 395–404.

—— (1938). Some influences of the development of higher plants upon the micro-organisms in the soil VI. *Soil Sci.*, 45, 207–49.

STENLID, G. (1947). Exudation from excised pea roots as influenced by inorganic ions. *Ann. Roy. Agric. Coll. Sweden*, 14, 301.

STEVENSON, G. (1959). Fixation of nitrogen by non-nodulated seed plants. *Ann. Bot.* (*London*), New Series 23, 622–35.

STOATE, T. N. (1950). Nutrition of the pine. *Bull. For. Bur. Aust.*, 30, 1–61.

STONE, E. L. (1950). Some effects of mycorrhizae on the phosphorus nutrition of Monterey pine seedlings. *Proc. Soil. Sci. Soc. Amer.*, 1949, 14, 340–5.

STOVER, R. H. & WAITE, F. (1953). An improved method of isolating *Fusarium* spp. from plant tissue. *Phytopathology*, 43, 700.

STREET, H. E. & LOWE, J. S. (1950). The carbohydrate nutrition of tomato roots II. The mechanism of absorption by excised roots. *Ann. Bot.* (*London*), New Series, 14, 30–41.

—— & ROBERTS, E. H. (1952). Factors controlling meristematic activity in excised roots I. *Physiol. Plantarum*, 5, 498–509.

SUMMERHAYES, V. S. (1951). *Wild Orchids of Britain*, Collins, London, 365 pp.

TERNETZ, C. (1907). Über die Assimilation des atmosphärischen Stickstoffes durch Pilze. *Jb. Wiss. Bot.*, 44, 353–408.

TERVET, I. W. & HOLLIS, J. P. (1948). Bacteria in the storage organs of healthy plants. *Phytopathology*, 38, 960–7.

THOM, C. & HUMFELD, H. (1932). Notes on the association of micro-organisms and roots. *Soil Sci.*, 34, 29–36.
THORNTON, R. H. (1956). *Rhizoctonia* in natural grassland oils. *Nature* (*London*), 177, 230–1.
——, COWIE, J. D. & MCDONALD, D. C. (1956). Mycelial aggregates of sand soil under *Pinus radiata*. *Nature* (*London*), 177, 231–2.
THROWER, S. L. & THROWER, L. B. (1961). Transport of carbon in fungal mycelium. *Nature* (*London*), **190**, 823.
TIMONIN, M. I. (1935). The micro-organisms in profiles of certain virgin soils in Manitoba. *Canad. J. Res.*, 13, 32–46.
—— (1940). The interaction of higher plants and micro-organisms I. *Canad. J. Res.*, 18, 307–17.
—— (1940*a*). The interaction of higher plants and micro-organisms II. *Canad. J. Res.*, 18, 144–56.
—— (1941). The interaction of higher plants and micro-organisms III. *Soil Sci.*, 52, 385–413.
—— (1948). Microflora of the rhizosphere in relation to the manganese deficiency disease of oats. *Proc. Soil Sci. Soc.Amer.*, 11, 284–92.
TOLLE, R. (1958). Untersuchungen über die Pseudomycorrhiza von Gramineen. *Archiv für Mikrobiol.*, 30, 285–303.
TRANQUILLINI, W. (1959). Die Stoffproduktion der Zirbe (*Pinus cembra* L.) an der Waldgrenze während eines Jahres. *Planta*, 54, 107–51.
—— (1964). Photosynthesis and dry matter production of trees at high altitudes. In *The Formation of Wood in Forest Trees*, Ed. M. H. Zimmermann. Maria Moors Cabot Foundation for Botanical Research, New York, pp. 505–18.
TRAPPE, J. M. (1962). Fungus associates of ectotrophic mycorrhiza. *Bot. Rev.*, **28**, 538–606.
—(1964). Mycorrhizal hosts and distribution of *Cenococcum graniforme*. *Lloydia*, 27, 100–6.
TRESCHOW, C. (1944). The nutrition of the cultivated mushroom. *Dansk Bot. Ark.*, 11(6), 1–180.
TRIBUNSKAYA, A. J. (1955). Investigation of the microflora of the rhizosphere of pine seedlings. *Mikrobiologiya* (*Moscow*), 2, 188–92.
TRUSCOTT, J. H. L. (1934). Fungous roots of the strawberry. *Canad. J. Res.*, 11, 1–17.
ULRICH, J. M. (1960). Auxin production by mycorrhizal fungi. *Physiol. Plantarum*, 13, 429–43.
VAN DER PIJL, L. *See* PIJL, L. VAN DER
VERMEULEN, P. (1946). *Studies on Dactylorchis*. Utrecht, 180 pp.
VON HAUSEN, S. S. *See* HAUSEN, S. S. VON
VON MALMBORG, S. *See* MALMBORG, S. VON
WAID, J. S. (1957). Distribution of fungi within decomposing tissues of rye-grass roots. *Trans. Brit. Mycol. Soc.*, 40, 391–406.
—— (1962). Influence of oxygen upon growth and behaviour of fungi from decomposing rye-grass roots. *Trans. Brit. Mycol. Soc.*, 45, 479–87.
WAISTIE, R. L. (1965). The occurrence of an *Endogone* type endotrophic mycorrhiza in *Hevea brasiliensis*. *Trans. Brit. Mycol. Soc.*, 48, 167–78.
WAKSMAN, S. A. & WOODRUFF, H. B. (1940). Survival of bacteria added to soil and the resultant modification of soil population. *Soil Sci.*, 50, 421–7.

WALSH, J. H. & HARLEY, J. L. (1962). Sugar absorption by *Chaetomium globosum*. *New Phytol.*, 61, 299–313.

WARCUP, J. H. (1950). The soil-plate method for isolation of fungi from soil. *Nature* (*London*), 166, 117.

—— (1951). The ecology of soil fungi. *Trans. Brit. Mycol. Soc.*, 36, 376–99.

—— (1951*a*). Studies in the growth of Basidiomycetes in soil. *Ann. Bot.* (*London*), 15, New Series, 305–17.

—— (1957). Studies on the occurrence and activity of fungi in a wheat-field. *Trans. Brit. Mycol. Soc.*, 40, 237–62.

—— (1959). Studies on Basidiomycetes in soil. *Trans. Brit. Mycol. Soc.*, 42, 45–52.

—— & TALBOT, P. H. B. (1962). Ecology and identity of mycelia isolated from soil. *Trans. Brit. Mycol. Soc.*, 45, 495–518.

—— & —— (1963). Ecology and identity of mycelia isolated from soil II. *Trans. Brit. Mycol. Soc.*, 46, 465–72.

—— & —— (1965). Ecology and identity of mycelia isolated from soil III. *Trans. Brit. Mycol. Soc.*, 49, 427–435.

—— & —— (1966). Perfect states of some Rhizoctonias. *Trans. Brit. Mycol. Soc.*, 49, 427–35.

—— & —— (1967). Perfect stages of Rhizoctonia associated with orchids. *New Phytol.* 66, 631–41.

WARREN WILSON, J. *See* WILSON, J. WARREN

WATT, A. S. (1947). Pattern and process of vegetation. *J. Ecol.*, 35, 1–12.

WEISS, F. E. (1904). Mycorrhizas from the lower Coal Measures. *Ann. Bot.* (*London*), 18, 255–65.

WELLS, T. C. E. (1967). Changes in a population of *Spiranthes spiralis* (L.) Chevall. at Knocking Hoe National Nature Reserve, Bedfordshire, 1962–1965. *J. Ecol.*, 55, 83–99.

WENGER, K. F. (1955). Light and mycorrhiza development. *Ecology*, 36, 518–20.

WERLICH, I. & LYR, H. (1957). Uber die Mykorrhiza-ausbildung von Kiefer (*Pinus sylvestris* L.) und Buche (*Fagus sylvatica* L.) auf verschiedenen Standorten. *Arch. Forstwesen.*, 6, 1–64.

WEST, P. M. (1939). Excretion of thiamin and biotin by the roots of higher plants. *Nature* (*London*), 144, 1050–1.

WHITE, D. P. (1941). Prairie soil as a medium for tree growth. *Ecology*, 22, 398–407.

WHITEHEAD, M. D., THIRUMALACHAR, M. J. & DIXON, J. G. (1948). *Microascus trigonosporus* from cereal and legume seeds. *Phytopathology*, 38, 968–73.

WHITNEY, R. D. (1965). Mycorrhiza infection trials with *Polyporus tomentosus* on spruce and pine. *Forest Sci.*, 11, 265–70.

WIENDLING, R. & FAWCETT, H. S. (1936). Experiments in control of *Rhizoctonia* damping-off of *Citrus* seedlings. *Hilgardia*, 10, 1–16.

WILDE, S. A. (1954). Mycorrhizal fungi: their distribution and effect on tree growth. *Soil Sci.*, 78, 23–31.

WILSON, J. M. (1957). A study of the factors affecting the uptake of potassium by the mycorrhiza of beech. *D.Phil. Thesis*, *Oxford University* (typescript).

WILSON, J. WARREN (1951). Micro-organisms in the rhizosphere of beech. *D.Phil. Thesis*, *Oxford University* (typescript).

WILSON, P. M. (1952). The comparative biochemistry of nitrogen fixation. *Advanc. Enzymol.*, 13, 345–75.
WINTER, A. G. (1951). Untersuchungen über die Verbreitung und Bedeutung der Mykorrhiza bei Kultivierten Gramineen und einigen anderen Landwirtschaftlichen Nutzpflanzen. *Phytopath. Zeit.*, 17, 421–32.
—— & MELOH, K. A. (1958). Untersuchungen über den Einfluss der endotrophen Mykorrhiza auf die Entwicklung von *Zea mais* L. *Naturwiss.*, 45, 319.
—— & PEUSS-SCHÖNBECK, H. (1963). Zur Bedeutung der endotrophen Mykorrhiza fur die Entwicklung von Kulturpflanzen. In *Mykorrhiza*, Ed. W. Rawald & H. Lyr. Internat. Mykorrhizasymposium, Weimar, 1960, G. Fischer, Jena, pp. 367–65.
WITHNER, C. L. (1959). *The Orchids. A Scientific Study*. Ronald Press, New York, 648 pp.
WOLFF, H. (1926). Zur Physiologie der Wurzelpilze von *Neottia nidus-avis*, Rich. und einigen grünen Orchideen. *Jb. Wiss. Bot.*, 66, 1–34.
—— (1932). Zur Assimilation des atmosphärischen Sticksstoffs durch die Wurzelpilze von *Corallorrhiza innata* R.Br. sowie der Epiphyten *Cattleya Bowringiana* Viet und *Laelia anceps* Idl. *Jb. Wiss. Bot.*, 77, 657–84.
WORLEY, J. F. & HACSKAYLO, E. (1959). The effect of available soil moisture on the mycorrhizal association of Virginia pine. *For. Sci.*, 5, 267–8.
YEMM, E. W. (1935). The respiration of barley plants I. Methods of determination of carbohydrates in leaves. *Proc. Roy. Soc. B.*, 117, 483–525.
YOUNG, H. E. (1940). Mycorrhizae and the growth of *Pinus* and *Araucaria*. The influence of different species of mycorrhiza-forming fungi on seedling growth. *J. Aust. Inst. Agric. Sci.*, 6, 21–25.
—— (1947). Carbohydrate absorption by the roots of *Pinus taeda*. *Queensland J. Agric. Sci.*, 4, 1–6.
ZAK, B. (1964). Role of mycorrhizae in root disease. *Annu. Rev. Phytopath.*, 2, 377–92.
—— & BRYAN, W. C. (1963). Isolation of fungal symbionts from pine mycorrhizas. *For. Sci.*, 9, 270–8.
—— & MARX, D. H. (1964). Isolation of mycorrhizal fungi from roots of individual slash pine. *For. Sci.*, 10, 214–27.
ZIEGENSPECK, H. (1922). Lassen sich Beziehungen zwischen den Gerhalte an Basen in der Asche art und die Exkretion gestatten? *Ber. Dtsch. Bot. Ges.*, 40, 78–85.
—— (1936). *Orchidaceae: Lebensgeschichte der Blutenpflanzen Mitteleuropas*, vol. I, pt. 4. Wegen Ulmer, Stuttgart, 840 pp.
ZIEGLER, H. (1956). Untersuchungen über die Leitung und Sekretion der Assimilate. *Planta*, 47, 447–500.
ZIMMERMANN, M. H. (1961). Movement of organic substances in trees. *Science*, 133, 73.

APPENDIX

1. *Exudation from Roots of* Pinus—Ch. II, p. 25; Ch. IV, p. 84.

Unpublished work by G. D. Bowen has shown that in aseptic conditions phosphate-deficient plants of *Pinus radiata* exude, in a period of 2–4 weeks, twice the amount of amino-acids plus amides exuded by normal plants. Bowen suggests that the effects of phosphate deficiency upon mycorrhizal intensity may find an explanation in some such manner.

2. *Long and Short Roots and Mycorrhiza Formation*—Ch. III, pp. 47–49; Ch. IV, pp. 82 *et seq*.

An important paper by H. Wilcox on the development of roots and root systems of *Pinus resinosa* (XIV IUFRO Congress Report, 24, 29–39, 1967) has great relevance to the understanding of mycorrhizal development and the possible morphogenic effects of fungal products on the host plants. Wilcox points out that only those lateral roots which are at least half the diameter of their mother roots become long roots. This agrees with Warren Wilson's view (p. 49) that only the roots of largest diameter grow fast and retain functional apical meristems for a long time. Wilcox noted that those emerging laterals which continue growth increase in diameter; 'Roots which come to be recognized as long roots are those which have maintained cell division long enough to elongate appreciably and to augment the size of their meristems'. Since it is not the absolute size but the relative sizes of lateral and mother root which count, Wilcox suggests that mechanisms operate which may be competitive, interacting, or correlative. He suggests that supplementary factors necessary for the maintenance of growth by relatively small laterals are supplied by the mycorrhizal fungi during mycorrhiza formation. He therefore concludes that, though it is possible to doubt the contention that mycorrhizal fungi cause inhibition of root meristems, there is no doubt that they stimulate the growth of infected axes.

This work clearly reinforces the plea made on p. 85 for a study of processes initiating the development and senescence of short roots in the mycorrhizal and uninfected state. Some further consideration is given to these points by J. L. Harley & D. H. Lewis (*Adv. in Microbial Physiol.*, III, 53–81, 1969) in a paper written before Wilcox's valuable contribution became available.

Wilcox's paper also includes observations on the initiation of infection, especially of seasonal infection (*see also Am. J. Bot.* 55, 247 and 688, 1968).

V. Slankis (in XIV IUFRO Congress Report, 24, 84–99, 1967) uses observations on renewed growth of mycorrhizas to reinforce his contention that fungal auxins cause the differences in morphology and histology between the host tissues of uninfected roots and mycorrhizas. He suggests that two modes of behaviour confirm his views:

(i) continued applications of auxins are necessary to ensure that excised roots in culture maintain a similar histology to the host tissues of mycorrhizas, and

(ii) applications of high nitrogen concentrations (such as would prevent initial mycorrhiza formation) to mycorrhizal seedlings cause their mycorrhizal systems to renew growth and become long roots.

Slankis points out that high nitrogen supplies upset auxin production by fungi.

3. *Classification of Types of Ectotrophic Mycorrhizas*—Ch. III, p. 44; Ch. VII, pp. 159 *et seq.*

The greatly increased interest in the classification of ectotrophic mycorrhizas arises from a necessity to describe quantitatively the ecological variability of the mycorrhizal condition, to connect more definitely recognizable types with different species of fungal symbiont, and to put the study of physiological variation on a firmer basis. G. A. Chilvers (*Aust. J. Bot.*, 16, 49–70, 1968) stresses the value of the structure of rhizomorphs and of sheath structure, as seen in whole cleared mounts of mycorrhizas, in classification.

Several papers given at the XIV IUFRO Congress in 1967 were also concerned with classification. B. Boullard (24, 4–16) pointed out that variation in hyphal structure and colour might occur in a single kind, and that difficulties arose from the occurrence of such changes with age. R. Pachlewski (24, 12–28) made the point that synthetic mycorrhizas in culture may resemble natural kinds sufficiently closely to allow the determination of the fungal symbionts of the natural ones. J. M. Trappe (24, 46–59) discussed the whole problem of classification, emphasizing that clamps, septa, sheath structure, fluorescence, and chemical reaction, may all be of value in relating types of mycorrhizas to their causative fungi (*see also* Trappe, *For. Sci.*, 13, 121, 1967).

G. C. Marks (*Aust. Forestry*, 29, 238–251, 1965) and G. C. Marks, N. Ditchburne & R. C. Foster (XIV IUFRO Congress Report, 24, 67–83, 1967), have classified the types and distributions of mycorrhizas present in Australian plantations of *Pinus radiata*. In the latter paper, a method of sampling fixed volumes of soil by washing and wet-sieving was used to estimate the quantities of different kinds of mycorrhizas in chosen soils and their horizons, and also to determine the proportions that were living or dead. For instance, in some plantations there were 580 kg. per acre of mycorrhizas in the top six inches, of which about half were alive.

4. *The Root-cap and Tannin Layer of Ectotrophic Mycorrhizas*—Ch. III, p. 37; Ch. IV, p. 65; Ch. V, p. 97.

The low-power electron microscopic photographs of G. A. Chilvers (*New Phytol.*, 67, 663–5, 1968) confirm the presence and decomposition of root-cap tissue, and the persistence of epidermis, in mycorrhizas of *Eucalyptus bicostatus*, as described by F. A. L. Clowes in *Fagus sylvatica*. The root-cap gives rise to the tannin layer and to pockets of phenolic material in the inner sheath. Tannin-like inclusions were also observed in the epidermal cells in the region of the Hartig net.

R. C. Foster & G. C. Marks (*Aust. J. Biol. Sci.*, 19, 1027–38, 1966, and **20**, 915–26, 1967) have also used electron microscopy to study mycorrhizas of *Pinus radiata*. They point out that the fungal hyphae appear to be affected in activity, or inhibited, by the tannin inclusions of the tannin layer and by pockets of tannin in the sheath.

W. E. Hillis, N. Ishikura, R. C. Foster & G. C. Marks (*Phytochemistry*, 7, 409–10, 1968, and in other unpublished work) have shown that the range of kinds of polyphenols that are extractable from *Pseudotsuga menziesii* and *Pinus radiata* differ. The former plant having a wider range, is, they believe, more discriminating in its ability to develop mycorrhizas with the fungi available to exotics in Australia.

5. *Translocation of Photosynthetic Products*—Ch. V, p. 96–103.

G. R. Lister, V. Slankis, G. Krotkov & C. D. Nelson (*Ann. Bot.*, **32**, 33–43, 1968) confirm other work which indicates that more photosynthetically fixed ^{14}C is translocated to infected than to uninfected root systems under equivalent conditions. The process may be operated by the sink mechanism suggested in the text. D. C. Smith, L. Muscatine & D. H. Lewis (*Biol. Rev.*, in press) review analogous processes in pathogenic, mycorrhizal, and lichen systems, and in animal associations with algae. J. L. Harley & D. H. Lewis (*Adv. in Microbial Physiol.*, Vol. III, 53–81, 1969) have, however, pointed out that hormonal mechanisms may also be operative in the process. L. B. Thrower (*Phytopath. Z.*, **52**, 319–34, 1965) and J. L. Harley (*Trans. Brit. Mycol. Soc.*, **51**, 1–11, 1968) pointed out that, on the analogy of observations on photosynthetic rates during the development of roots and other organs, it is worth investigating whether there is an increased rate of photosynthesis in symbiotic systems. This appears the more worth-while since the tables of Lister *et al.* (op. cit.) seem to show such a possibility; for the highest rates of photosynthesis, translocation and mycorrhizal development are correlated. In the papers of Harley & Lewis and of Smith *et al.* quoted above, this subject receives some further consideration.

It should be noted that the estimates of ^{14}C sugars translocated to the fungal layer by Lewis & Harley (*see* p. 96 in the text) are underestimates, as 20 per cent of the glycogen in the mycorrhizas is found in the core in the Hartig net. Reference should also be made to Foster & Marks (*Aust. J. Biol. Sci.*, **19**, 1027–38, 1966) on this point.

6. *Salt Absorption*—Ch. VI.

P. 117:—The problem of external digestion of inositol phosphates by ectotrophic pine mycorrhizas is being examined by G. D. Bowen and his colleagues at the Waite Institute, South Australia. The activity of mycorrhizas of beech in this regard is being studied by E. M. Bartlett and D. H. Lewis at Sheffield University, England.

P. 123:—V. K. Mejstrik, working in T. M. Morrison's laboratory in New Zealand, reported at the meeting of ANZAAS (January 1968) that he had recognized thirteen sub-sub-types of ectotrophic mycorrhizas of *Pinus radiata* and eighteen of *Nothofagus solandri* var. *cliffortioides*. All the types tested had enhanced phosphate uptake compared with non-mycorrhizal roots, but different sub-sub-types had different optimum temperatures for absorption. V. Mejstrik & U. Benecke, in a paper submitted to *New Phytologist*, have recognized several types of ectotrophic mycorrhizas on *Alnus viridis* in New Zealand. On a wet-weight basis, mycorrhizal roots had much greater uptakes of phosphate than either uninfected roots or nodules.

G. Ritter & H. Lyr (in *Plant Microbe Relationships*, Ed. J. Macrura & V. Vancura, pp. 277–82, 1965) report considerable differences in translocation of ^{32}P to the hosts in different symbiotic combinations with mycorrhizal fungi.

Pp. 134–6:—R. C. Foster & G. C. Marks (*Aust. J. Biol. Sci.*, **20**, 915–26, 1967) showed, using electron microscopic preparations, that bacteria of various kinds were closely associated with the surfaces and interhyphal spaces of mycorrhizas of *Pinus radiata*. It is possible that the nitrogen fixation reported for mycorrhizas could arise through the activity of such closely-attached organisms.

G. Bond's article (*Ann. Rev. Pl. Physiol.*, 18, pp. 107–26, 1967) should also be consulted for a description of nitrogen fixation.

P. 136:—J. M. Trappe (XIV IUFRO Congress Report, **24**, 46–59, 1967) indicates that, of eight mycorrhizal fungi tested, only *Rhizopogon subradiatus* and *Suillus caerulescens* showed the presence of nitrate reductase.

7. *Mycorrhiza and Disease*—Ch. VII, p. 158.

D. H. Marx amplified the results given in Marx & Davey (*Nature*, *London*, **216**, 1139, 1967) in a paper to the XIV IUFRO Congress, 1967 (Report, **24**, 172–81). It was shown that a number of mycorrhizal fungi produced antibiotic compounds which were active against a variety of pathogens. Mycorrhizal fungi differed in the kind of pathogen that they inhibited and also in the antibiotic produced. Tests indicated that both mycelia and mycorrhizas were actively antagonistic to *Phytophthora cinnamoni*.

Much further work is in progress by D. H. Marx and his colleagues at Athens, Georgia. In contrast V. Šašek, in the same volume (pp. 182–9), concluded that his results on the same topic did not suggest any fundamental significance in the protective influence of mycorrhizal fungi against disease. He seemed to regard the observed antibiotic effects as of secondary importance in the selective advantage of the mycorrhizal habit.

8. *Orchid Mycorrhizas*—Ch. XI.

P. 216:—The work of R. Ernst is published in *Bull. Amer. Orchid Soc.*, Dec. 1967, p. 1067.

P. 221 :—G. Harvais & G. Hadley (*New Phytol.*, **66**, 217–30, 1967) investigated further the conditions, both for asymbiotic and symbiotic growth, of seedlings of *Orchis purpurella* in culture, using a drip-feeding technique. They concluded that a careful balance of nutrients is important in the maintenance of a stable relationship between the partners. A steady supply of carbohydrates at low concentrations (such as might originate by digestion of cellulose and translocation of the products by the symbiotic fungus in nature) seems to be required, together with organic substances, stimulating growth, which may be produced by the endophytic fungi.

P. 226 (*also* Ch. IV, p. 88, and Ch. IX, pp. 188–89):—Ella, O. Campbell in a series of papers (*Trans Roy. Soc. N.Z.*, **1**, 289–96, 1962, **2**, 73–82, 1963, **2**, 237 46, 1964, and **3**, 209–19, 1968) concerning chlorophyll-free angiosperms,

Gastrodia cunninghamii, *G. minor*, *G. sesamoides*, and *Thismia rodwayi*, has shown each to be epiparasitic upon a shrub or tree. These fully described cases make it seem probable that so-called saprophytic angiosperms are likely to be epiparasitic, and make the views of Ruinen (1953) concerning the epiphytic orchids appear even more plausible.

9. *Vesicular-arbuscular Mycorrhiza*—Ch. XIII.

Pp. 242 and 267:—A very valuable review by T. H. Nicolson is published in *Sci. Prog. Oxf.*, **55**, 561–81, 1967. The classification of *Endogone* is also considered in T. H. Nicolson & J. W. Gerdemann in *Mycologia*, **60**, 313–25, 1968, and by Barbara Mosse & G. D. Bowen in *Trans. Brit. Mycol. Soc.*, **51**, 469–83, 1968.

P. 249:—A. A. Cridland (*Mycologia*, 54, 230–4, 1962) has also expressed doubts about some of the fossils which have been believed to be mycorrhizal, especially as the fungi appear to be in a variety of organs of the fossil hosts.

P. 248:—B. Fassi (*Allionia*, 11, 7–15, 1965) reports ectotrophic mycorrhiza in *Pinus strobus* due to *Endogone lactiflua*.

P. 267:—An account of the distribution of different kinds of *Endogone* spores in 250 Australian and New Zealand soils is given by Barbara Mosse & G. D. Bowen in *Trans. Brit. Mycol. Soc.*, **51**, 485–92, 1968. It raises important questions about the factors which affect the qualitative and quantitative variation in the *Endogone* spore content of soils.

P. 268:—C. L. Murdoch, J. A. Jackobs & J. W. Gerdemann (*Plant and Soil*, **27**, 329–34, 1967) is the paper cited in the text.

10. *Nitrogen Fixation by Vesicular-arbuscular Mycorrhiza*—Ch. XIII, p. 255, and Ch. XIV, p. 281.

The question of the fixation of nitrogen in the root-region of the Podocarps and of *Agathis* is not yet finally settled. It has much in common with the problem in *Pinus* (*see* p. 134). Leaving aside early unsatisfactory experiments, and arguments on the analogy with Leguminosae, recent experiments show at most only a minor rate of fixation. F. J. Bergersen & A. B. Costin (*Aust. J. Biol. Sci.*, 46, 378, 1964) observed restricted fixation of ^{15}N by nodules of *Podocarpus lawrencii.* They thought that its rate might suffice to support the plant in its pioneering role, although T. M. Morrison & D. A. English (1967) explain this role as due to phosphate absorption by its mycorrhizas rather than to nitrogen fixation.

J. H. Becking (*Plant and Soil*, **23**, 213–26, 1965, and XIV IUFRO Congress Report, 24, 158–71, 1967) observed a small fixation of ^{15}N in the nodules of *Podocarpus rospigliosii* but not in the roots. Moreover there was insignificant transport of fixed ^{15}N to the roots from the nodules. This and other species of *Podocarpus* show nitrogen deficiency symptoms in water cultures which lack a source of combined nitrogen, although the plants may persist in such cultures for a considerable time. G. Bond (*Annu. Rev. Pl. Physiol.*, 18, 107–26, 1964) has discussed this question. He himself had obtained (*Ann. Bot.*, **21**, 513–21,

1957) negative results using ^{15}N to test nitrogen fixation in seven species of *Podocarpus*. He concluded that the few experiments which have been reported in the literature as showing fixation using ^{15}N are not yet sufficient to prove nitrogen fixation by Podocarps.

The fact that *Podocarpus* nodules are not dependent for their formation on infection by any organism, fungus or bacterium, has been confirmed by A. G. Khan (*Nature, London*), 215, 1170, 1967). This emphasizes their great difference from those of Leguminosae, *Alnus* etc.

11. *Phosphate Absorption*—Ch. XIV, p. 282.

An increasing volume of evidence shows that vesicular-arbuscular mycorrhiza is widespread and physiologically significant especially in phosphate nutrition. G. T. S. Baylis, in a paper to ANZAAS (1968), showed that such mycorrhizas are almost universal in those New Zealand plants which do not have other types of mycorrhiza: 'No seedling grows in unfertilized forest soil until an endophyte is established in its roots.' He showed that soluble phosphate was needed to secure growth of uninfected sedlings.

J. G. Obbink & J. V. Possingham (CSIRO, Division of Horticultural Research, Report 1965–67, pp. 35–37) showed that the Grape-vine cultivar *Raisin de Dame* only grew well in soil with quarter-strength Hoagland solution if it was mycorrhizal, and that mycorrhizal infection stimulated growth in phosphate-deficient soil provided it was not also sulphur-deficient.

G. D. Bowen & B. Mosse (CSIRO, Division of Soils, Report 1965–66, and *Trans. Brit. Mycol. Soc.*, in press) observed that mycorrhizal Subterranean Clover and onion roots absorbed phosphate from $KH_2{}^{32}PO_4$ solution twice as fast as did uninfected roots. The primary site of absorption was into the fungal structures.

C. L. Murdoch, J. A. Jackobs & J. W. Gerdemann (*Plant and Soil*, 27, 329–34, 1967) showed that mycorrhizal grew much better than uninfected maize in soils containing insoluble phosphate, but not in those with readily available phosphate.

SUBJECT INDEX

AUTHOR INDEX